ABERDEEN UNIVERSITY STUDIES
NUMBER 144

SCIENCE AND THE RENAISSANCE

VOLUME II

An annotated Bibliography of the Sixteenth-Century Books relating to the Sciences in the Library of the University of Aberdeen

SCIENCE and the RENAISSANCE

An annotated Bibliography of the ∽
Sixteenth-Century Books relating to the
Sciences in the Library of the ∽ ∽
University of Aberdeen

W. P. D. WIGHTMAN

VOLUME II

Published for the University of Aberdeen

OLIVER AND BOYD

EDINBURGH AND LONDON

HAFNER PUBLISHING COMPANY

NEW YORK

FIRST PUBLISHED . . . 1962

© 1962, W. P. D. WIGHTMAN

PRINTED IN GREAT BRITAIN AT THE UNIVERSITY PRESS, ABERDEEN
FOR OLIVER AND BOYD LTD., EDINBURGH

ILLUSTRATIONS

GUIDE TO THE USE OF
THE BIBLIOGRAPHY

THE aim has been to include all books of a scientific character (Dewey .5 and .6) published in the sixteenth century. To these have been added a number of seventeenth-century works (mainly alchemy) where the acceptance of the year 1600 would have imposed an arbitrary break, unwarranted by the development of the subject at that time. On the other hand mere editions or translations without commentary have generally been omitted. While every effort has been made to include every book in the University Library which fulfils these conditions, owing to the dispersal of such books through many special collections the absence of a particular title must not be taken to imply the absence of the book.

The books are catalogued under authors arranged in alphabetical order, the names of authors being, with a few exceptions, those of the British Museum Catalogue. Thus 'Clusius, Carolus' appears as 'L'Ecluse, Charles de' and 'Avicenna' as 'Husain ibn 'abd Allah', synonymous forms being given in parentheses, except in the case of obvious Italian-Latin equivalents. Adequate cross-references are provided. In the case of a few classical authors it is often very difficult to decide whether a work should pass as a 'text' of the original with editorial annotations or as a 'commentary'. To avoid a source of complication unnecessary for a general list of this kind it was decided to list all works under the name appearing on the title page. Names of 'editors' of texts will appear in the index only; but under, e.g. 'Hippokrates', 'Galen', etc., will be given the serial numbers of commentaries listed under another author's name. The fine Paracelsian collection has been treated as a separate section, with a brief introductory guide to this very complex literature: this appears under 'Bombast, Philip' as in BM catalogue.

Each work wherever found, if published under a separate imprint, has been given a serial number. Editions of each work are grouped together in chronological order and distinguished by lower-case letters. The date of the *editio princeps* of every work has been added wherever possible, the authority being (unless stated to the contrary) the earliest cited in the BM Catalogue. This is of course a far from infallible indication, and steps have been taken to check it wherever there was any reason for doubting it. Other bibliographical works have been freely consulted, as shown in the reference list. In the case of works of whose *editiones principes* the Library possesses no copy the date is placed in parentheses.

Provenance has been indicated wherever an identifiable autograph, probably dating from the sixteenth century, has been found. In cases of 'local' interest, later provenance has also been indicated. Stereotyped inscriptions on the volumes of certain large collections have been abbreviated. Where there is evidence of more than one owner the names are quoted in what is believed to be chronological order, numbered as follows: e.g. '(i) "du Puys 1585" ', '(ii) "John Gregorie 1765" '.

Notices of works are generally restricted to details of bibliographical interest; but in those cases where no reference in the *Introduction* has been called for, some indication of the contents may be given. Also, since the Bibliography has been prepared in the hope that it may be of service in the building up of a background to the science of the Renaissance the term 'bibliographical' has been interpreted in a sense wide enough to permit of reference to points of interest in dedication, 'climate of opinion', personal relationships among scholars, etc. For this purpose the sub-titles have often been quoted at length in order that the original 'flavour' might be retained.

ABBREVIATIONS
General

auto	= autograph of owner
Biog. Univ.	= *Biographie Universelle*
col.	= colophon
dedn.	= dedication; ded. = Dedicated to
Ed. p.	= *editio princeps*; ed. = edited by; edn. = edition
fo.	= folio [always in the sense of 'leaf', hence the number is followed by *r* (*recto*) or *v* (*verso*)]

K.C.   = King's College, Aberdeen
Lectori—has been retained as more economical than 'To the Reader'
Mar.   = Marischal College, Aberdeen
n.p.   = no place cited
n.d.   = no date cited
prov.   = provenance
pagn.   = pagination (always in sense of numbered 'faces' to a folio)
secn. heading. = section-heading
sep.   = separate
t.p.   = title-page
trn.   = translation; tr. = translated by
[e.g.] (XVI-2) = latter half of sixteenth century
sig.   = signature (of printed folio)

References to Special Collections under Provenance.
LIDDEL   = Duncan Liddel.
GREGORY   = Gregory Collection (where no auto. of John Gregorie is present)
READ   = Library of Dr. Alexander Read (or Reid)
BUTE   = Donation from the Luton Hoo Library of the Earl of Bute
FORBES   = Library of Sir John Forbes
FORDYCE   = Library of Sir William Fordyce
HENDERSON = Library of Dr. Alexander Henderson of Caskieben
MELVIN   = Library of Dr. Melvin of Aberdeen
JOHNSTON   = William Johnston
CARGILL   = James Cargill
ARBUTHNOT = Alexander Arbuthnot
For an explanation of these names—see chapter 'The Origin of the Collections'.

Bibliographical References
BM   = British Museum Catalogue of Printed Books.
Bibl. Nat. = Bibliothéque Nationale
Duv.   = D. I. Duveen, *Bibliotheca Alchemica*, London, 1949

Copinger	= Supplement to Hain, *Repertorium*, London, 1895-1902
Ferg.	= J. Ferguson, *Bibliotheca Chemica*, Vols. I and II, Glasgow, 1906
GW	= *Gesamtkatalog der Wiegendrucke*, Leipzig, 1925-1932
Johnson	= F. R. Johnson, *Astronomical Thought in Renaissance England*, Baltimore, 1937
Hain	= *Repertorium Bibliographicum*, Stuttgart, 1826
Smith	= D. E. Smith, *Rara Arithmetica*, London, 1908
STC.	= A. W. Pollard and G. R. Redgrave, *Short Title Catalogue of Books printed in England*, London, 1950
Th. Hist.	= Lynn Thorndike, *History of Magic and Experimental Science*, Vols. I–VI, N.Y., 1923-41
ThK	= Lynn Thorndike and Pearl Kibre, *Catalogue of Medieval Incipits*, Cambridge, Mass., 1937
Allan	= D. C. Allan, *The Star-Crossed Renaissance*, Durham, N. Carolina, 1941.
Arber	= A. Arber. *Herbals* 2nd Edn., Cambridge, 1953
Castiglioni	= A. Castiglioni, *History of Medicine* 2nd Edn., New York, 1947

Bibliographical Works relating to single authors have been cited in full with the works of the author referred to.

Imprints

Printer, place (in modern equivalent), date, in that order. No attempt is made to reproduce the imprint *verbatim*, thus printers' names, etc., are abbreviated where no confusion seemed likely. The Latin form is commonly used except where (e.g. Etienne, Froben) the vernacular has become almost a 'household word'. To avoid confusion between *impressit* and *impensis* and their variants, such imprints have generally been rendered by, e.g. 'By Gul. Antonius for Petrus Fischer'. 'Haeredes' has been translated to 'heirs of', and 'Vidua' to 'widow of'.

The title of each work is given as on the title-page of the Library copy, abbreviated where necessary, but not otherwise altered, except as regards the spelling out of printers' abbreviations, and alteration of the punctuation, where this has been necessary in the interest of clarity. Capitals have often been replaced by lower case except for certain initials, to assist in identifying the subject of the work, and where a second author's work is referred to in the title.

Details of publication are taken from the title-page, colophons not being cited except to supplement the title-page, or where there is a discrepancy between them.

THE ORIGIN OF THE COLLECTIONS

IN the vestibule of the old parish church of St. Nicholas, Aberdeen, there is a memorial brass depicting Dr. Duncan Liddel, surrounded by his books and instruments. Graduate of King's College, Aberdeen; Master of Philosophy of the University of Rostock; Doctor of Medicine of the University of Helmstadt; he might still pass unnoticed in these pages. For he made no original discovery, and his name appears in but a few highly-specialised historical works. But it so happened that love of his native town brought him back to Aberdeen, probably in 1606, and there he died in 1613; but not before he had 'mortified all and haill the toune and lands of Pitmedden, with the salmon fishing theirof adjacent thereto on the water of Don'. This was to provide bursaries for six poor scholars in either of the universities then existing in Aberdeen. But in the following year in a further mortification he gave preference to the 'colledge of New Aberdeen', and it was in this college that he provided an endowment for 'ane learned professor of mathematickes weill versed in Euclide Ptolemye Copernik Archimede aliisque mathematicis gif possible he can be hade within this country'. We begin to see why Duncan Liddel is the key figure among those who brought the scientific renaissance to our University. 'Ptolemye and Copernik' were both to be studied; but of far more consequence for our particular study is the fact that he left his magnificent library—the library of a mathematician, astronomer, physician, and (perhaps) alchemist; in short the library of a typical renaissance polymath—to the 'Colledge of New Aberdeen'. We shall return to this later; meanwhile we must retrace our steps and learn something about the two universities.

The Bull of Erection of the University and King's College, as it came a little later to be called, was signed by Alexander VI (Rodrigo Borgia!) on the petition of James IV on 10 February 1494. The document, with the lead *bulla* intact, is in the muniments; its most significant feature, from the point of view of the historian of science, is the naming of Medicine as well as Theology, Civil and Canon Law, and the Liberal Arts, as a lawful faculty.[1] Not that there is any evidence of the prosecution of medical research during the ensuing century; but it betokened an unusual trust in the power of the infant university to cater for the higher studies; moreover some of the early medical works in the Library are also traceable to the sixteenth-century mediciners, the third of whom, Gilbert Skene, wrote the first serious medical work (*Ane Breve Descriptoun of the Peste*, Edinburgh, 1568) to be printed in Scotland. The Collegiate Chapel and halls were, and the former still is, in Old Aberdeen, near the mouth of the Don and a short distance from the ancient cathedral of St. Machar, whose Bishop, William Elphinstone, had been the moving spirit, and was nominated as the first Chancellor.

For reasons sufficiently interesting in themselves and bearing strongly on the political, ecclesiastical and cultural state of contemporary Scotland[2] but of no special relevance to this study, a Charter of Erection for a 'Gymnasium' at (New) Aberdeen was signed on 2 April 1593 by George Keith, fifth Earl Marischal of Scotland. This may seem a foundation sufficiently late in the sixteenth century to be hardly worthy of consideration in a study of the Renaissance; but it must be remembered that the Renaissance in Scotland, like the Scottish spring, came very late, and was telescoped to such an extent as to have left no mark on the countryside comparable, for instance, to the Tudor luxuriance in cottage, barn and palace. Also that it was the products of the Continental scientific renaissance like Duncan Liddel who shaped the early course of studies of the newly-founded Marischal College and University of Aberdeen, as it came to be called until the final complete unification of the two universities in 1860. In this latest form the ancient colleges have no separate existence apart from their buildings.

The need for an awareness of the separate existence of Marischal College (as I shall henceforth call it) is further borne in on us when we come to examine the library imprints on the extant

[1] Statements to the effect that Aberdeen was the first university in Great Britain to establish a Medical Faculty are based on a misconception. See W. P. D. Wightman, 'The Growth of Medical Education in Aberdeen', *Zodiac-Journal of Aberdeen University Medical Faculty*, 4 (1957), 66.

[2] Cf. G. D. Henderson, *The Foundation of Marischal College*, Aberdeen University Studies, No. 123, Aberdeen, 1947.

collections and the documentary histories of the libraries themselves. While no one would wish to belittle the generosity and wide learning of the first Principal of King's College, Hector Boece, whose autograph appears, for instance, on the Ulm edition of Ptolemy's *Geography* (No. 543), it is noteworthy that after the first flush of benefactions to the library of King's College there is a long relatively barren period until 1640. At Marischal College, on the other hand, the magnificent nucleus of Duncan Liddel's library was quickly added to by Thomas Read, Latin Secretary to James VI and I.

It will be convenient at this stage to set out a systematic list of the principal sources of books of the Renaissance period, which may suitably include rather more than passing reference to those early donors who were handing on to future scholars no idly collected volumes, but the very stuff out of which the texture of their own scholarship had been woven.

Apart from isolated volumes the first source in chronological order as well as in importance is the library of Duncan Liddel. But although this donation was to take effect on the donor's death (1613) there is no documentary evidence earlier than 1669-73 of the presence of his books in the library of Marischal College. This is all the more surprising since an interesting MS (M.70) is extant, dated 1624 and comprising a list of the books donated by Thomas Read already alluded to. The catalogue, which contains a few mathematical, botanical, and medical works, was written out by John Garden on a number of sheets, one side of each of which had been used as a record of customs dues in the port of Aberdeen—evidently a paper shortage is no new thing! The make-up of the manuscript indicated that it was not a catalogue of the whole library, so the absence of any mention of Liddel's books must not be taken as evidence that they were not already there at that date.

The earliest extant catalogue of the Library of Marischal College is that of Thomas Gray, the fourth Librarian, who held office from 1669 to 1673. In this catalogue the books are listed in groups according to provenance, each group being broken down into subjects and sizes. The first group is that 'a Thoma Rhaedo R.M. a secretis D.D. Liddelio et Aliis legat . . . nec non de Bibliotheca ecclesiastica translat . . .'. Thomas Gray set out with the laudable intention of indicating (by 'T.R.', 'D.L.', etc.)

the provenance of each book; but very few are so marked after the first six pages (mostly 'T.R.' and theological; but a few marked 'D.L.' indicate the catholic taste, or at least intentions, of the latter). Also the catalogue contains very few indications of date or imprint—a sad contrast, for a 'Librarian', to the admirably detailed earlier inventory of Thomas Read's books. The first group is followed by several others of no inconsiderable size, of which the legacies of William J(h)o(h)nston (1641), Robert Dun (1657), Alexander Read (1633), and William Jamesone (1632) are the most important. Although these gifts fall much later than the end of our period they all contained numerous works which form the foundation of the present collection: many will be found in the Bibliography and may be identified. Many, alas, have disappeared: thus William Johnston's copy of *N. Copernicus de Revolutionibus* was already marked as 'not found' in about 1715, and Kepler's *Epitome Astronomiae* as soon as 1659!

In MS 71 there is bound a second catalogue unsigned and undated but believed to be in the hand of Robert Paterson, fifth Librarian from 1673 to 1717, and Principal of the College from 1679. It embodies the substance of the earlier catalogue, and, in addition, lists of the books bequeathed to the Library by John Dunlop (1714) and John Moir (1713). The former bequest is almost entirely theological, but the latter contains a considerable number of scientific works first published in the sixteenth century, though these lists suffer from the same faults as the older one, namely in the absence of any bibliographical details.

Before we review the remaining bequests to Marischal College it will be proper to take note of a number of bequests to King's College during the earlier part of the seventeenth century. The earliest extant catalogue of the Library of King's College dates only from 1722; and of the 2857 titles listed 239 are *Libri Medici* and 333 *Libri Philosophici*, which presumably include works on mathematics and natural science. From scattered records in the muniments it is known however that numerous comparatively small collections were presented to the College during the seventeenth century. Of these by far the most important for our period are those of Dr. Alexander Read, who in 1640 bequeathed 'to the Colledge of Auld Aberdiene all my medicinall bookes', and of Sir Francis Gordon, who in 1643 gave 'at his return to Scotland after threttie years peregrination 42 fair volumes, most pairt physicall [i.e. medical]'. The names of many of the volumes in these two collections will be found in the Bibliography. In 1696 Matthew Mackaile, a chemist of the Paracelsian persuasion, who wrote numerous works on mineral waters and their analysis, gave nineteen volumes on having conferred on him the degree of Doctor of Medicine: this was apparently a recognised way of compounding for graduation fees. A few sixteenth-century works of this collection are still extant bearing his signature. Lastly we may note that a not inconsiderable number of books bear the bold autograph of Alexander Arbuthnot, first Principal of King's College under the Reformed Church (1569-83).

Although some of the richest sources of books of the renaissance period still remain to be detailed they were collected by men whose lives were passed at a time remote from that period; we shall therefore first look a little more closely into the lives of the early donors. Of whom pride of place must be given to Duncan Liddel.

A Sketch of the Life of Dr. Duncan Liddel of Aberdeen was written in 1790 by John Stuart, Professor of Greek in Marischal College. It is an exemplary piece of scholarship, in which every statement is documented; and in the few cases where the author permits himself a conjecture or judgment he offers it with moderation and due caution. It seems unlikely that much could now be done to fill the gaps in the narrative, since these relate mainly to the last years of Liddel's life when he had ceased to hold any public office, from the records of which information could now be gleaned. On the other hand an examination of every one of the books now extant and believed for one reason and another to have belonged to Liddel has provided some additional information in regard to his movements, and more particularly has produced evidence that many books not autographed by him did actually belong to him. This evidence, steadily accumulating during the latter part of my searches of the Library, has made it appear worthwhile to publish a separate memoir, in which the whole extant *corpus* of his library might be reconstituted and reviewed in such a way as to provide insight into his reactions to contemporary modes of thought. No such reconstitution exists, so far as I am aware, for a scholar of the late sixteenth century [1] comparable to those of the corresponding fifteenth-century bibliophiles,

[1] The recent publication of the Catalogue of the library of Lord Lumley, indeed, fills an important part of the gap, but not the whole of it.

Giovanni Pico della Mirandola, Nicholas Pol, and Hieronymus Münzer (Vol. I, p. 74). With this end in view I shall do no more here than mention that Liddel spent the greater part of his life in Central Europe, ranging from Frankfurt-on-Oder to Rostock and finally Helmstadt, where he taught mathematics and medicine. He returned finally to Scotland in 1606 and it is of the years from then to his death in 1613 that, strange to say, we know least. With regard to the provenance of these books, this will be marked in the present work as 'Liddel' in every case where his autograph appears, and as 'Liddel(?)' in every other case where in my view there is strong evidence for supposing the book to have belonged to him: the details of the evidence, being of less general interest, will be reserved for a future memoir.

Almost of the same generation as Duncan Liddel was Alexander Read, brother of Thomas, Latin Secretary to James VI and I. After graduating at King's College he travelled for a short time on the continent, staying probably at Wittenberg, since he speaks with special reverence of Daniel Sennert, who was at that time the leading medical teacher at that university. Thereafter Read practised as physician and surgeon in Wales and England, where he played no inconsiderable part in the dissemination of the 'new' anatomy—particularly that of Caspar Bauhin—by means of an abbreviated version of Helkiah Crooke's famous compendium *Microcosmographia* (London, 1615). Himself a member of the College of Physicians he lectured for many years to the Barber Surgeons. His appearance in these pages is due to the splendid library of medical (including alchemical) works which he bequeathed to King's College. Perusal of the Bibliography will reveal that it is to him that the University owes such rarities as Libavius, *De Alchimia* (No. 396) and the folio *editio princeps* of Tagliacozzi's famous work; also, what does not appear here, a copy of the *princeps* of William Harvey's immortal work on the circulation of the blood, diligently marked and glossed in Read's own neat hand. To Marischal College he left his books on philosophy and divinity.

Among the remaining early bequests probably only two sets may well have been collected, at least in part, on the continent at about the close of the sixteenth century: these are the works autographed by James Cargill and William J(h)o(h)nston respectively. James Cargill was the brother of Thomas, who was

Rector of the Aberdeen Grammar School and one of the finest classical scholars of his day. James made his mark as a botanist—perhaps the first Scotsman of whom any such record exists: Bauhin acknowledged his help in more than one place, e.g. No. 65. William Johnston, who also had a distinguished brother, Arthur, the Latin poet—was recalled from the continent in 1626 to be the first occupant of the Chair of Mathematics, founded under the will of Duncan Liddel at Marischal College. The Library is very fortunate in possessing a record (MS M.181) of his lectures taken down in 1633 by one of his students. A few years after Johnston had entered upon his duties in the Liddel Chair of Mathematics there was delivered into his hands by George Jamesone, the Scottish portrait painter, a collection of 'mathematicall instrumentis' in pursuance of the testamentary instructions of his brother, William. Unfortunately, of this interesting collection (of which an inventory is preserved) not a trace remains; for some of the contemporary English books on navigation, however, the Library is indebted to the same William Jamesone.

This completes the tale of bequests concerning which there is now only documentary evidence, since the books themselves, such as remain, are dispersed on the shelves. Some of the most important works which fall to be described in the Bibliography were however bequeathed to the Library at a much later date, so that, although in some cases dispersed, they are easily identified by the Librarian's imprint. Such, for instance, are the magnificent collections of Sir William Fordyce, M.D. F.R.S. (1790); the Earl of Bute (who presented 1300 volumes from his Library at Luton Hoo at about the same time); Dr. Melvin (1856—perhaps the finest private library of the Greek and Latin classics in Scotland); Dr. Alexander Henderson of Caskieben (1857); and Sir John Forbes (1859—mainly relatively modern medical works, but containing among a few other sixteenth century works a finer copy of the second edition of the works of Ambroise Paré than is to be seen in the Bibliothéque Nationale). Since the time of the union of the two Colleges the only donation to the Library to furnish more than one or two items in this bibliography is the unique Gregory Collection housed at King's College (as are now both the Melvin and Henderson Collections) since Sir Ian Forbes-Leith, Bart. of Fyvie generously presented it to the University in 1937. It consists of about two thousand printed books and a small number of manuscripts, most of which belonged to one or other of the 'Academic Gregories'. In few of the books is there any indication of ownership other than those bearing the autograph 'John Gregorie', a physician well known in London, Aberdeen, and Edinburgh during the middle years of the eighteenth century. Since the first of these Gregories, James, inventor of the Gregorian reflector, was an early contemporary of Newton their affairs lie well outside our period.

Lastly must be mentioned the Phillips Library of Pharmacology, which is the only one of the collections mentioned here of which a separate printed catalogue has been published. When examined it was found to contain several sixteenth-century works which were not part of the original gift—indeed not a few of Duncan Liddel's books had been added to it. Since the presence in a book of the bookplate specially printed for this collection clearly gives no indication of provenance no mention is made of the fact in the Bibliography.

The sources which have just been passed in review constitute by far the greater number of those listed in the Bibliography; but Librarians have from time to time relaxed their customary struggle to keep up with their contemporary publications and have purchased old books of outstanding interest. In the last few years several notable works have been added.

ANNOTATED BIBLIOGRAPHY

1. ABD AL-MALIK IBN ZUHR (*syn.* Abynzohar, Avenzohar). *Abhomeron Abynzohar colliget Averroys.* (Col. Gregorius de Gregoriis.)

(t.p. Venice 1514)

[*Ed. p.* of Abynzohar's work Venice 1490/1; of Surianus's edn., Venice 1496—GW.]

Incipit (Prohemium): 'Dixit servus regis . . .', ThK 216.

The 'title' (sig. a i *r*) is misleading. The volume contains three works, viz. the *Theyzir* and *Antidotarium* of 'Abhomeron Abynzohar' and the *Colliget* (*Kitab al-Kulliyat*) of Averroes 'correctus atque emendatus per excellentem artium & medicine doctorem magistrum Hieronymum Surianum filium domini magistri Iacobi Suriani de Ariminio physici' (*explicit* of whole volume).

The three 'Books' of the *Theyzir* (*Kitab al-Tayzir* . . .) end sig. 40 *r* (sig. 38 wrongly headed 'secundus') and the *Antidotarium* follows with cont. pagn.

The *Colliget* starts on sig. 45 *r* with the usual *incipit* 'Quando ventilata fuit super me . . .' (ThK quote a MS of 13th cent.).

See No. 470 for another edn. of the *Colliget*.

2. ACHILLINI, ALESSANDRO. *See* No. 165.

3. ACOSTA, CRISTOBAL (*syn.* Costa, Christopher à). . . . *Aromatum & Medicamentorum in Orientali India nascentium liber. Plurimum lucis adferens iis quae a Doctore Garcia de Orta in hoc genere scripta sunt. Caroli Clusii opera ex Hispanico sermone Latinus factus in epitomen contractus & quibusdam notis illustratus.*

(*a*) Off. Chris. Plantin, Antwerp 1582.

Ed. p. of trn. [Spanish original, Burgos 1578.]

Prov. LIDDEL

In his dedn. to William, Landgrave of Hesse (see Vol. I, p. 120), de l'Ecluse states that he came across the work only the previous year during his travels in England where he 'devoured it with avidity'. It had been published in Spanish two years previously.

In the preface of the original (reprinted) Acosta states that he wrote it after meeting Garcia in the East Indies; the latter, however, told him of the difficulty he had experienced in getting his own work accurately printed in Goa (see No. 487). Acosta determined that his improvements on Garcia's work should include 'icones ad vivum'—which Clusius evidently didn't think much of, as in his dedn. (*supra*) he states that he rejected the original pictures except that of the *Caryophollorum arbor*, a dried specimen of which Acosta saw, which had been brought to England by 'Franciskus Dracke nauclerus Anglus' (p. 258). This work has cont. pagn. with de l'Ecluse's edn. of Garcia (No. 488).

(*b*) Same title followed by *Altera editio castigatior & auctior*. Prov. GREGORY, and a second copy.

Off. Plantiniana, Antwerp 1593.

See J. R. Partington, 'Lignum nephriticum', *Ann. Sci.* **11** (1955), 1.

4. ACTUARIUS. *Actuarii Ioannis filii Zachariae Opera.*

B. Turrisanus, Paris 1556
De actionibus et spiritus animalis affectibus eiusque nutritione lib. ii. Ed. *p.*
De urinis lib. VII. [*Ed. p.* Venice 1519 (Latin).]
Methodi medendi lib. VI. [*Ed. p.* Venice 1554 (Latin).]
Prov. LIDDEL

Two volumes in one, sep. pagn., but with only one col. and no separate t.p. Bound in a fine stamped pigskin on wooden boards by Nicholas Goellel of Wittenberg. The t.p. bears the anchor-and-dolphin device of the House of Aldus. Bernard Turrisan was the grandson of André Turrisan, who had bought the type of Nicholas Jenson and who who was associated with Aldus Manutius the elder, who married his daughter. Bernard probably had no press of his own (the col. states 'excudebat Guil. Morelius'), but sold 'in Aldina bibliotheca' all the Italian books for which there was a market in Paris—including some, according to Brantôme, unbecoming the character of a respectable bookshop!

The text was edited by Henricus Mathisius, who translated the *Methodus medendi* from the Greek. The *De actionibus* is in the several times published translation by Julius Alexandrinus of Trent. *De Urinis* (alternatively referred to as *De differentiis urinarum*) is based on the translation of Ambrosio Leone, revised from the Greek codices by the famous French medical humanist, Jacques Goupyl. There were many editions, comprising different groups of works, of Actuarius in (XVI-2). 'Actuarius' was a generic title for the archiaters at Constantinople, and the date of this one—'John, son of Zachary'—is not known.

See *Biog. Univ*

2

5. ACTUARIUS. *De medicamentorum compositione, Ruellio interprete.*
C. Neobarius, Paris 1539
Prov. (i) 'Mgri Iohanne Bruce'. End papers and early pages covered with MS in several 16th-17th cent. hands.
(ii) Alexander Arbuthnot.
Ed. p.
Donated to K.C. by James Morison.
In dedn. to Francis I, Dionysius Corronius says that he has recently revised the translation which Ruel had undertaken shortly before his death.

6. AETIOS OF AMIDA. *Librorum medicinalium tomus primus, primi scilicet libri octo nunc primum . . . editi.*
Heirs of Aldus Manutius & Andrea Asulanus.
Venice 1534
Ed. p. of the Greek text of the first eight books. In his preface, *medicinae studiosis*, the publisher (Paulus Manutius; his father, Aldus, and Andreas Torrisanus being both dead at this time) promised that the remaining eight books would appear after the customary care had been taken to ensure the highest standard of accuracy; the Greek text of these books has in fact never been printed.
Aetios was a Byzantine of the age of Justinian and was held in high esteem as late as Boerhaave (xviii-1); he is now regarded rather as a judicious compiler.

7. AETIOS OF AMIDA. *Contractae ex veteribus medicinae tetrabibli.*
Seb. de Honoratis, Lyon 1560
The four volumes were published separately; only II, III, and IV are now to be found in the Library, all bearing the autograph of Duncan Liddel. See also No. 486.

8. AGRICOLA, GEORGIUS. [*De Re Metallica.*]
(*a*) [t.p. wanting].
(Col. H. Frobenius and N. Episcopius, Basel 1556)
This is the *Ed. p.*; the earlier date (1530), occasionally seen in the literature, is due to the sub-title of the earlier *Bermannus, sive de re metallica* (see No. 9 below—a slight work, cast in dialogue form, bearing on the general character of mining and minerals). *De re metallica* was the product of twenty-five years' study, observation, and reflection. Its dedication to Agricola's patrons, the Elector Maurice of Saxony and his brother August who succeeded him, bears the date 1550. There was some delay in the printing of the work (not surprising in view of the great number of blocks to be prepared for the splendid illustrations) and the author died shortly before its publication.

3

The title of the *Ed. p.* is the same as that of No. 8(b), except for the addition in the latter of 'omnibus nunc iterum . . .'.

Prov. BUTE

(b) *De Re Metallica libri XII quibus officia instrumenta machinae ac omnia denique ad metallicam spectantia . . . describuntur—adiunctis Latinis, Germanisque appellationibus. . . . Eiusdem de Animantibus subterraneis liber ab autore recognitus . . . omnibus nunc iterum (ad archetypum . . . castigatis).*

Froben, Basel 1561

[The *Ed. p.* of *de Animantibus* was Basel 1549.]

(See H. C. Hoover and L. H. Hoover, *Georgius Agricola—De Re Metallica*, translation and commentary, Lond. 1912.)

9. AGRICOLA, G. *De ortu & causis subterraneorum Lib. V. De natura eorum quae effluunt ex terra Lib. iiii. De natura fossilium Lib. x. De veteribus & novis metallis Lib. ii. Bermannus, sive de re metallica dialogus. Interpretatio Germanica vocum rei metallicae.*

Froben, Basel 1546

Prov. (i) 'Io. Bellenden'; (ii) 'Io. Gregory 1767'.

Ed. p. of all the works listed (except *Bermannus*, already printed two or three times since its *Ed. p.* 1530). There are separate dedications, all dated 1545-6 'Kemnicii' (Chemnitz), which is thus probably the oldest mining centre with a continuous activity. The dedication of *Bermannus* (to Heinrich à Conritz) is by Petrus Plateanus, a contemporary writer on mining, who refers to Agricola as 'medicum, virum supra reliquam eruditionem utraque lingua doctissimum'. The name, Bermannus, used for one of the participants in the discussion was probably adapted by Agricola from that of his mining friend, Lorenz Berman.

Erasmus commends the work to Henry's sons, not forgetting to mention the debt owed to Froben, who was of course his personal friend as well as publisher. The testimony of these two men brings out clearly the peculiar virtues of Agricola—virtues which gave him a commanding place in the history of scientific progress in the 16th cent. (See Vol. I, p. 176.)

10. AGRICOLA, IOANNES. *Scholia copiosa in Therapeuticam Methodum, id est, absolutissimam Claudii Galeni Pergameni . . . curandi artem. . . .*

Only edn. BM.   (Col. Phil. Ulhard, Augsburg 1534)

Prov. FORDYCE

The *epistola* of Erasmus is dated Freiburg-in-Breisgau 1533.

A brief sketch of the life of Galen by Ioannes Monardes is included.

4

11. AGRIPPA, HENRICUS CORNELIUS. (*a*) *Splendidae Nobilitatis viri et armatae militiae equitis Equitis aurati ac utriusque Iuris Doctoris Sacrae Caesareae a consiliis & archivis Inditiarii Henrici Cornelii Agrippae ab Nettesheym De Incertitudine & Vanitate Scientiarum & Artium atque excellentia Verbi Dei Declamatio.*

Io. Grapheus, Antwerp 1530

The t.p. bears nothing except this modest commendation of the author and the motto *Nihil scire foelicissima vita.*

Ed. p. Bibl. Nat. gives one of the same date but *sans location.*

The work had a great vogue, being translated into several languages, including English, before the end of the 16th cent. The Melvin Collection contains one printed at London, 1684.

(*b*) *De incertitudine & vanitate scientiarum declamatio invectiva, denuo ab autore recognita.* . . .

No place or printer. 1537.

12. AGRIPPA, H. C. *De Occulta Philosophia libri tres.*

(Col. 1533)

Ed. p. (BM and Bibl. Nat., both of which give Cologne as the place of printing.)

Dedn. to his protector Hermann, Archbishop of Cologne, from Malines 1531.

This work provides an example of what must be the rare procedure of publishing a book whose contents the author had already retracted in a previous publication. The reason for this is stated by Agrippa in his preface—that several corrupt copies were going the rounds, so he thought it safer to publish the juvenile work, little revised by his own hand, rather than to allow an unchallenged circulation of those corrupted by the hands of others. This accounts for the fact that the commendation of the work by Trithemius (No. 688) is dated 1510, at which time evidently Agrippa must have sent him a draft of the work. Although categorically retracted by the author in No. 11 above, the views expressed therein had considerable influence—the BM copy bears the autographs of the amanuenses of Archbishop Cranmer and Lord Lumley. (See No. 572 and 708.)

Prov. JOHNSTON, and a second copy.

13. AGRIPPA, H. C. *Opera, in duos tomos digesta & nunc denuo sublatis omnibus mendis . . . recusa.*

Apud Beringos fratres, Lyon n.d.

One of the numerous undated collected works.

AHMAD IBN MUHAMMAD (= Alfraganus). See Nos. 157, 190, 720.

14. ALBERTI, LEON BAPTISTA. *De Re Aedificatoria.*

Nic. Laurentius, Florence 1485

Ed. p.

Incipit: 'Multas et varias artes que ad vitam. . . .'

15. [ALBERTUS MAGNUS] [t.p. wanting]. Table of contents: 'In hac tabula. . . .'

Paul v. Butzbach, Mantua 1479

Incipit: 'Liber Alberti magni animalium primus qui est de communi diversitate . . .'.

Col. Paulus Iohannes de Butsebbach, Mantua 1479

GW 588.

[*Ed. p.* Rome 1478.]

16. [ALBERTUS M.—supposititious work.] *Albertus. Magnus. De virtutibus quarundam herbarum. De virtutibus herbarum septem planetarum proportionantium. De virtutibus quorundam lapidum. De virtutibus quorundam animalium. De scitu dominii planetarum in qualibet hora. De mirabilibus mundi. Item parvum regimen sanitatis valde utile. Item questiones naturales philosophorum humanae naturae . . . Noviter revisus et castigatus.* [Col. Io. Theobaldus, Antwerp 1519].

[*Ed. p.* GW lists 58 incunable editions including trans. into Italian, French and Catalan.)

Prov. Purchased 1954.

Incipit. 'Sicut vult philosophus in pluribus locis . . .'.

The well known *Liber Aggregationis* or *Liber Secretorum,* for centuries passing under the name of Albertus Magnus, since the Prologue contains the phrase: '. . . immo et egomet, Albertus . . .'. It extols the virtue of 'scientia magicalis' directed to proper ends, since by its means it is possible to avoid evil and follow good. The list of headings indicates the kind of 'knowledge' still being actively prosecuted in (xv-2) and later. The questions on (e.g.) birds, and fishes' eggs (sig. f i) show a naturalistic approach; a recipe for 'Greek fire' (Vol. I, p. 178) occurs on sig d ii *v.*

ALBERTUS M *See* also Nos. 413, 603.

17. [ALBERTUS M.] *Das Buch der Haimligkeyten Magni Alberti von Artzney und tugenden der Kreütter Edelgestayn und von etlichen wol-bekannten Thieren.*

1535

Not in BM, but one with a similar title and same contents was published by I. Knoblauch at Strassburg in 1516. The title of this contains a reference to a plague-tract; No. 17 contains such a tract at the end.

After a *Vorred* supposed to be by Albert ('Da auch ich Albertus . . .') there follow a few fos. of a *Frauenbuch*, thereafter the parts named in the title. (See No. 579).

18. ALBERTUS, SALOMON. *Historia plerarunque partium humani corporis, membratim scripta, et in usum tyronum retractatius edita.*

Heirs of Io. Crato, Wittenberg 1585

Ed. p.

Prov. LIDDEL (?)

A characteristic product of the renaissance in which frequent references are made to Aristotle, Galen, Ezekiel, Job, and the Psalms, but also to the quite recent observations of Falloppio and Colombo. An anatomical text owing much (especially the excellent illustrations) to Vesalius (who, however, is not mentioned), it seeks to reveal the fabric of the human body as a hymn to its Creator 'for the use of young students who ought to be initiated into these sacred matters in order that they may gain for themselves knowledge of philosophy and of the whole of nature'.

ALBERT of SAXONY. See No. 405.

ALBUCASIS. See KHALAF IBN ABBAS.

19. ALDROVANDI, ULYSSE. *Ornithologiae tomus tertius ac postremus.*

Io. Bapt. Bellagamba, Bologna 1603

Ed. p.

Prov. 'Coll. Reg. ex dono James Fraser.' Escutcheon, crest and motto of Sir Christopher Hatton gilt embossed on cover.

While the greater part of the output of Aldrovandi was published in the early years of the 17th cent., this is the third volume of a work which began to appear in 1599. The three volumes comprise a systematic 'history' of birds, arranged according to a relatively 'natural' classification; thus the present volume commences with *Liber decimus nonus, qui est de Avibus palmipedibus*, the first *ordo* dealt with being *Cygnus*. The treatment of each *ordo* is very elaborate, being arranged under stereotyped headings, e.g. *Ordinis ratio, Synonyma, Forma, Descriptio, Auguria, Aequivoca* (i.e. legends and poetic fables), *Usus pennarum, Nidus*, and many others. A similar, but not identical, arrangement is followed for the other *ordines*. The appearance and anatomy are illustrated by splendid in-folio figures in which some attempt is made to illustrate the ecology of the bird by the inclusion of suitable rocks, plants, fish, etc.

Chapters xx-xxii are concerned with the *Anser bassanus* 'vulgo dicta a Solendguse'; the *Gustarda avis Scotica*; and the *Branta seu Bernicla*. The last named includes the celebrated 'Barnacle Goose' (Vol. I, p. 181), of whose (fabulous) generation there is a highly unconvincing picture on p. 174.

20. ALECHAMPS, JACQUES D'. *Historia generalis plantarum in libros XVIII per certas classes artificiose digesta.*

G. Rovillius, Lyon 1587

Ed. p.

Prov. 'Bibliothecae Abredonensi hunc librum donavit Thomas Cargillus civis Abredonensis 15 Jan. 1625'.

The long sub-title indicates the special character of the book, which the printer also refers to in his dedication to Charles Emanuel, Duke of Savoy, son-in-law of Philip II of Spain. He claims that he found the author at work on it twenty years previously, which is not surprising when one discovers that d'Alechamps wrote a commentary on Dioscorides, another on Pliny, and composed a work on surgery. (See Vol. I, p. 197.)

21. ALESSIO (*syn.* Alexis), of Piedmont. *Les Secrets du Seigneur, Alexis Piemontois et d'autres auteurs bien experimentés et approvvés* [sic] . . .

C. Plantin, Antwerp 1561

Ed. p. In his *Au Lecteur* the printer says that this is the fourth and still further revised and rearranged edn. (pp. 201-46 incl. torn out).

Lib. 3 (as in Wecker's edn., No. 723) starts with preservation of orange or lemon peel in a syrup of either sugar or honey—reminiscent of a method widely recommended in 1939, in which, however, the peel was finally dried off for preservation.

The identity of the author of this extremely popular work has never been established. (Ferg. I, 22.)

22. ALEXANDER TRALLIANUS. *Practica Alexandri yatros greci cum expositione glose interlinearis Iacobi de Partibus et Ianuensis in margine posite.*

(Col. Franc. Fradin, Lyons 1504.)

This is probably the first work of Alexander to be printed, but Copinger (No. 378) mentions an undated edition with identical title except '*possite*' for '*posite*'.

Incipit: 'Contingit hec duplex passio cadentibus . . .' agrees with the 9th cent. MS of ThK, 119.

The work begins as usual with cures for baldness, one of which is translated as 'echini chersei cinis, cum pice liquida admixt(us) stercori capri, capillos generat', a marginal note on which states that

8

'echinus est piscis quidem qui cum navi adheserit ipsam immobilem reddit'. The powers of this creature were debated at least from the time of Plutarch (see E. W. Gudger, *Isis*, **13**, 340 & Aberdeen Bestiary). The alleged cure for baldness may be due to a textural error, as a later version gives 'erinacci terrestris cinis pice liquida exceptus . . .' (No. 23, p. 17).

For Simon Januensis (i.e. of Genoa) see Vol. I, p. 240.

Jacobus de Partibus was Jacques Despars (xv-1), commentator on several Arabic medical works.

23. ALEXANDER TRALLIANUS. (*a*) *Alexandri Tralliani Medici libri duodecim Ioanne Guinterio Andernaco interprete & emendatore. Nunc demum . . . Ioannis Molinaei . . . annotationibus illustrati . . .*
Ant. de Harsy, Lyon 1575

[*Ed. p.* Basel 1556.]
Prov. LIDDEL (?)
(*b*) . . . lib. XII.
Rob. Estienne, Paris 1548

Ed. p. Bibl. Nat.

This contains the Gk. text with Latin annotations by Iacobus Goupylus.

24. ALEXANDRINUS, JULIUS. *Salubrium sive de sanitate tuenda libri triginta tres.*
Gerv. Calenius & heirs of Quentelius, Cologne 1575

Ed. p.
Prov. There is a fine autograph on t.p. of Thomas Read, Latin Secretary to James I and VI. Scored and glossed.

25. ALEXANDRINUS, J. *Ad Rembertum Dodonaeum epistola apologetica.*
Heirs of Andr. Wechel, Frankfurt 1584

Ed. p.

A long letter alleging that Dodoens has introduced confusion in regard to the two beans '*Faba*' and '*Phaselus*' by reversing the classical usage.

ALFRAGANUS (= al-Fagarni) See Ahmad ibn Muhammad.

ALHAZEN (= al-Haitham). See Hasan ibn Hasan.

26. ALLAXINUS, IACOBUS. *Medicae aliquot disceptationes quibus recentiorum & Arabum permulti errores ad veterum disciplinam expenduntur.*
Off. Chr. Wechel, Paris 1535

Only edn.

27. ALPHONSINE TABLES. *Divi Alphonsi Romanorum et Hispaniarum Regis astronomicae tabulae in propriam integritatem restitutae . . . Qua in re Paschasius Hamellius . . . operam suam praestitit.*

Chr. Wechel, Paris 1545

Ed. p.

Prov. Liddel. 'I.F.R.D.' stamped on binding.

Dedn. by Lucius Gauricus, who also made additions and corrections to the text.

(For origin of the Tables, see Vol. I, p. 115.)

28. ALTOMARI, DONATUS, A. ab. (*a*) *Omnia opera, quae hucusque in lucem prodierunt nunc primum in unum collecta & ab eodem auctore . . . recognita & aucta. Et denuo a Ioanne Altimaro auctoris filio medico. . . expurgata.*

J. Anielus de Maria, Venice 1574

Prov. GORDON. (Also a duplicate copy.)

(*b*) (An earlier edn. with similar title but not revised by the author's son.)

Gul. Rovillius, Lyon 1565

Prov. 'Io Gregory 1769.'

29. ALTOMARI, D. A. ab. *De alteratione, concoctione, digestione, praeparatione, ac purgatione, ex Hippocratis & Galeni sententia, methodus.*

[*Ed. p.* Venice 1547.]

Gul. Rovillius, Lyon 1548

Prov. 'Liber est Thomas Moor.'

'This is Thomas Moor his book 1600 [?].'

Also a duplicate copy.

30. ALTOMARI, D. A. ab. *De Medendis Febribus.*

M. de Maria Salernitanus, Venice 1569

Not BM or Bibl. Nat. Privilege of Philip II given at Naples 1561.

Prov. LIDDEL (?)

31. ALTOMARI, D. A. ab. . . . *de medendis humani corporis malis ars medica. A Iacobo Rubeo Furciensi sub Ioannis ab Altomari censura . . . recognita, simul cum locis & Galeni numeris integris denuo in margine ad studiosorum emolumentum repositis. Quarta editio.*

Ioan. Ant. de Maria, Venice 1570

Earliest edn. in BM is A. Vincentius, Lyon 1559, but the dedn. of No. 31 is signed by Donatus ab Altomari 'Neap. 1560'.

The epistle of J. Rubeus is addressed to Ioannes, 'suo eximio praeceptori'.

Prov. LIDDEL.

32. AMATUS (LUSITANUS) (pseudonym of João Rodriguez de Castello Branco). (*a*) *Curationum medicinalium centuria prima multiplici variaque cognitione referta. Praefixa est auctoris commentatio in qua docetur quomodo se medicus habere debeat in introitu ad aegrotantem simulque de crisi & de diebus decretoriis* . . .

Petrus Galterus, Paris 1552

Prov. GREGORY.

Ed. p.? (No earlier in BM.)

   (*b*) *Curationum medicinalium centuriae duae prima et secunda.* . . . *Omnia nunc primum opera & lectione cuiusdam doctissimi medici Galli* . . . *repurgata.*

Sebast. Nivellus, Paris 1554

[*Ed. p.* Venice, 1552.]

   (*c*) . . . *Centuriae II priores.* . .

Gul. Rovillius, Lyon 1567

Prov. BUTE

   (*d*) . . . *Centuriae quatuor quarum duae priores ab auctore sunt recognitae duae posteriores nunc primum editae.*

Froben, Basel 1556

Prov. BUTE

   (*e*) . . . *Centuriae duae Tertia et Quarta.*

Gul. Rovillius, Lyon 1565

'Centuria tertia nunc primum . . . edita . . .' (p. 8). Dedn. Ancona 1554.

Dedn. of *Quarta* by Ambrosius Nicandes, Ancona 1553.

Prov. BUTE

   (*f*) . . . *Centuriae duae, quinta et sexta* . . . *Omnium nunc primum* . . . *aedita.*

Gul. Rovillius, Lyon 1564

Dedn. 'D. Iosepho Nassinio, Hebraeo, viro non minus illustri quam sapienti . . .. Thessalonicae anno a creatione mundi 5320.'

Prov. LIDDEL.

   (*g*) Same as (*f*), but last six words omitted.

Gul. Rovillius, Lyon 1576

Prov. BUTE

   (*h*) . . . *Centuria septima. Thessalonicae curationes habitas continens varia multiplique doctrina referta.*

Gul. Rovillius, Lyon 1570

Prov. BUTE

Dedn. 'Thessalonicae anno a mundo creato 5321, secundum vero computum Romanum 1561.'

   (*i*) Another copy of (*h*) bound with (*f*).

Owing to the mixture of *Eds. p.* of one *Centuria* with revised edns. of others it has been considered advisable to list them as various edns. of one work, viz. *Curationum Medicinalium.* . . .

33. APIANUS, PETRUS. *Cosmographia . . . per Gemmam Frisium apud Lovanienses . . . iam demum ab omnibus vindicata mendis ac . . . aucta. Additis eiusdem argumenti libellis ipsius Gemmae Frisii.*

Greg. Bontius, Antwerp 1550

Prov. LIDDEL

The *Ed. p.* of the original work was Landshut 1524 (dedn., also BM); of the work as revised by Gemma Frisius, Antwerp 1529, between which and No. 33 were numerous edns. The book, with its many clear diagrams and volvelles might be described as the 'Bible' of the renaissance world-view (Vol. I, p. 114).

> Contents: *Liber cosmographicus de principiis astrologiae & cosmographiae; Descriptio quattuor partium terrae; Cui adiecta est descriptio regionis Peru nuper inventae; De horarum noctis observatione.*

The above are by Apianus; the *adiecti libelli* of G.F. are: *De locorum describendorum ratione, deque distantiis eorum inveniendis; De usu annuli astronomiae in multis locis aucti.*

The two last works have separate dedns. (Antwerp 1533 and 1554 respectively) but there is no record in BM of separate publication apart from the work of Apianus; but see No. 747.

34. APIANUS, P. (1) *Instrumentum primi mobilis . . . nunc primum et inventum et editum . . .*
(2) *Gebri Filii Affla Hispalensis . . . li. IX de Astronomia, ante aliquot secula Arabice scripti & per Giriardum Cremonensem latinitate donati, nunc vero omnium primum editi.*

Io. Petreius, Nürnberg 1534

Ed. p. Incipit of '(2)' 'Scientia species habet . . .'. ThK 649, 14th cent. MS.

35. APIANUS, P. [*Instrumentum sinium*] *seu primi mobilis nu*[*per*] [*a*] *Petro Apiano inventum nunc autem . . . recognitum & locupletatum . . . Adiectus est & Quadrans Universalis . . . cuius usu quicquid ex tabulis sinium computando elici poterat* [t.p. damaged].

Io. Petreius, Nürnberg 1541

Ed. p.

36. APOLLONIOS OF PERGA. . . . *Conicorum li. iv, una cum Pappi Alexandrini lemmatibus et commentariis Eutochii Ascalonitae. Sereni Antinsensis Philosophi li. duo nunc primum . . . editi. Quae*

*omnia nuper Federicus Commandinus Urbinas mendis quam plurimis
expurgata e Graeco convertit & commentariis illustravit.*

Alex Benatius, Bologna 1566

The work of Serenus—*De Sectione Cylindri* and *De Sectione Coni*—has
separate t.p. and pagn. but same date.

This was the first, and remained the definitive, edn. until Halley's
text of Books I-VII and reconstruction of the lost Book VIII (Oxford 1710),
though an earlier Latin trans. of I-IV was published at Venice 1537.
(Heath, *Hist. Gk. Maths.* **2,** 127.) Books V-VII have been known to the
West only through Arabic trans.

37. ARCAEUS, FRANCISCUS. *De recta curandorum vulnerum ratione . . . libri ii. . . . Eiusdem de febrium curandarum ratione.*

Christ. Plantin, Antwerp 1574

Ed. p. (Engl. trans. 1588).

Prov. LIDDEL.

Cont. pagn. On p. following 286 is a description and illustration of an
orthopaedic device.

38. ARCHIMEDES. *Opera, quae quidem extant, omnia multis iam seculis
desiderata, atque a quam paucissimis hactenus visa, nuncque primum
& Graece & Latine . . . edita . . . Adiecti quoque sunt Eutocii
Ascalonitae in eosdem Archimedis libros commentaria item Graece &
Latine nunquam antea excusa.*

Io. Hervagius, Basel 1544

The editor was Thomas Geschauff 'cognomento Venatorius', who
(in his dedn.) gives an account of earlier work on the MSS of Archimedes. The actual bibliography is highly complex: the following is a
summary of the account given by E. J. Dijksterhuis, Copenhagen 1956
(based on Heiberg's introduction to his definitive edition, Leipzig
1910-15). The prototype for the greater part of all subsequent editions
was a codex prepared for the Emperor Leon VI (IX-2). This finally
disappeared in (XVI-2), but had meanwhile been several times copied.
The chief derivative was the Latin trans. by William of Moerbeke
(1269), still extant: since this contains the work on *Floating Bodies* (not
in the prototype), it is inferred that a second source was available to
William—probably lost in (XIV-1). The Greek text of No. 38 is based
on a copy of the Leon codex made by Cardinal Bessarion; the Latin
version is that made in 1450 by Jacopo of Cremona, at the instigation
of Nicholas V, corrected by Regiomontanus by reference to Bessarion's
Greek copy. No. 38 is thus the *Ed. p.* of both Greek text and Latin
trans. It contains all the works of Archimedes then extant in Greek.
The work on *Floating Bodies* was then known only in William's trans.,
and the *Method* not at all. A Greek codex containing both these works

was discovered by Heiberg in 1906. There was a separate series of Arabic trans., some works from which were translated (e.g. by Plato of Tivoli) earlier than William's trans. from the Greek. Sarton, *The Appreciation of Ancient and Medieval Science during the Renaissance*, pp. 140-1, seems to have failed to recognise the dual source of William's trans.

39. ARCHIMEDES. *De iis quae vehuntur in aqua li. ii. A Federico Commandino Urbinate in pristinum nitorem restituti et commentariis illustrati.*

Alex. Benacius, Bologna 1565

The first critical printed trans. into Latin, based on the Latin trans. of William of Moerbeke: an earlier version by N. Tartaglia being a mere copy. In his dedn. to Cardinal Ranuccio Farnese, Commandinus refers to the fact that no Greek codex had come to light.

A second copy of this work is bound with Commandinus's own work *Liber de Centrogravitatis Solidorum* of same publisher and date but sep. t.p. and pagn.

40. ARCULANUS, IOHANNES. *De Febribus . . . in Avic. quarti canonis Fen. primam . . . expositio.*

Juntae, Venice 1560

There is a double text—the older, followed by Arculanus; and one amended by Benedictus Rinius and Andreas Alpagus. (Some running headlines 'Herculanus'.)

Ed. p.

Prov. GORDON.

41. ARDOINO, SANTE [Ardoynis, Santes de]. . . . *de Venenis . . . nunc castigatissime editum . . . Adjunximus commentarium . . . Ferdinandi Ponzetti Cardinalis.*

[Col: A. Petrus & P. Perna, Basel 1562

[*Ed. p.* Venice 1492, GW.]

Prov. Bute and some 16th cent.—? de la Ramée.

The Commentary has sep. t.p. and dedn. to Augustino Nypho (Nifo) but cont. pagn.

In his Preface Theodor Zwinger (Vol. I, p. 235) gives it as his opinion that the veterinary art is no part of medicine 'cicuta(enim) sturnis, elleborus coturnicibus, scamonea ficedulis, nutrimentum est, homini venenum' (Sig. †5). In the text the properties of magnets are considered relevant to the nature of poisons (e.g. Pietro d'Abano's book on Poisons): this is reminiscent of the persistent doctrine of sympathy and antipathy (Vol. I, p. 280), which, however, is not explicitly mentioned.

14

42. ARGELLATA, PETRUS DE. *Cirurgia magistri Petri de largelata.*
(No printer's name) Venice 1499
Ed. p. Venice 1480—GW, 2321.
Prov. HENDERSON.
Incipit: 'Rogaverunt me socii mei . . .'.
GW. 2324.

43. ARGENTERIUS, IOANNES. . . . *de Consultationibus Medicis.*
Gul. Cavellat, Paris 1552
[Earliest BM, Paris 1557.]
The dedication is by A., but in a foreword *Medicinae studiosis*
Laurentius Gryllius throws a revealing light on the growing dissatis-
faction with the new 'sectarianism' in medicine—'Several spend their
whole time & energy in discussing and collating the teaching of the
ancients, whence, greatly to the detriment of our art, sects arise too
much addicted to the opinions of authorities, & squabbling among
themselves over the dogma to be upheld rather than the truth to be
established.' He holds up Argenterius's work as a corrective.

44. ARGENTERIUS, IO. *De Morbis libri XIIII.*
Laur. Torrentinus, Florence 1556
Prov. Gordon 2nd edn. [*ed. p.* Florence 1536, Bibl. Nat.]
Dedn. 'Cosmo Medici Florentiae Duci Secundo' from 'tua prae-
clarissima Pisana Academia'. The long *Ad lectorem* provides evidence
for the judgment that Argenterius was 'en toute occasion adversaire
déclarée des doctrines de Galien' (*Biog. Univ.*)

45. ARNALDUS DE VILLANOVA. (*a*) *Hec sunt opera Arnaldi de*
Villanova nuperrime recognita ac emendata diligentique opere im-
pressa que in hoc volumine continentur.
Franciscus Fradin, Lyon 1509
The dedn. by Thomas Murchius contains biographical notes on
Arnaldus and is dated 1504. The discrepancy in the notes is accounted
for by the fact that No. 45 is a reimpression of the edn. printed in
1504 for Balthazar de Gabiano by Fradin.
Prov. GORDON.
(*b*) *Arnaldi de villa nova medici acutissimi opera nuperrime revisa:*
una cum ipsius vita recenter hic apposita. Additus est etiam tractatus
de philosophorum lapide intitulatus.
Guil. Hyon, Lyon 1520
Prov. 'Io. Melchior Volandi Io. Francisco ? amico suo.'
'Epilentia' in (*a*) is corrected to 'epilepsia' in (*b*). The contents
are too numerous to list but include the famous *Regimen Salernitatis* in
Arnald's version with extensive commentary. The (rubricated Gothic)

t.p. of (*b*) is a good example of the renaissance tendency to sacrifice clarity to geometrical elegance.

(*c*) . . . . *Opera quae edita sunt hactenus omnia in tomos distincta sex ad diversorum & vetustissimorum codicum collationem . . . recognita multisque locis restituta. . . . Per M. Gualtherum H. Ryff Argentinensem medicum. Tomus primus breviarium practice de medendis singularum humani corporis partium . . . continet.*

Balthas. Pistor, Strassburg 1541

Prov. FORBES.

(*d*) *Arnaldi Villanovani . . . opera omnia cum Nicolai Taurelli . . . in quosdam libros annotationibus.*

C. Waldkirch for P. Perna, Basel 1585

The printer's foreword contains some observations on the relation of Arabic to Greco-Roman medicine. He claims that the Venice and Lyon edns. were corrupt, but that this one is sound. A few short tracts on alchemy and astrology have been added.

46. ARNALDUS DE VILLANOVA. *Praxis Medicinalis.*

Tardif, Lyon 1586

Hac ultima editione . . . repurgata. Cui accesserunt sub finem tractatus eiusdem aliquot.

Ed. *p*. in this form.

[A *Breviarium Practicae* was published Milan 1483 (GW 2526). The 'tracts' refer to alchemy and astrology. See also No. 313.

47. *Articella nuperrime impressa cum quam plurimis tractatibus pristine impressioni superadditis. . . . Petri Pomarii Valentini Hispani ad lectores Hexastychon.*

Iacobus Myt for Const. Fradin, Lyon 1519

[*Ed. p*. Padua 1476—GW; earliest BM, Venice 1491.]

Prov. LIDDEL; also a second copy.

Contents: *Hysagoge Ioanitii.* Phylaretus, *de Pulsibus.* Theophilus, *de Urinis.*

Hippocrates—sixteen works or commentaries.

Io. Damascenus, *Aphorismi.*

Celsus, *Flosculi medicinales.* Arnaldus de Villanova, *Parabole.*

Galen, *Tegni li. iii* in an 'old' translation and in a later by Laurentius Laurentianus.

Avicenna—five works.

'Almansor', *Textus novi ad Almansorem qui est de egritudinibus a capite usque ad pedes.*

Jac. de Partibus, *Summula super antidotario Mesue.*

Descriptio ponderum medicinalium mensurarum et dosium breviario Aiseir [for Theyseir].

16

Since this is the earliest printed 'medical encyclopaedia' its contents have been given in full, except for the detailed titles of Hippokrates, Galen, and Avicenna. See fo. cccc *v* for figure designed as a 'map' for 'flobothomia', which we are told (fo. cccciii) is 'not only useful but necessary' for health and long life. 'Ioanitius' was the famous Arabic translator and scholar Hunain ibn Ishaq. (see Vol. I, p. 211.)

See also No. 313

48. AUBERTUS, IACOBUS. . . . *de metallorum ortu & causis contra chemistas brevis & dilucida explicatio.*

Io. Berion, Lyon 1575

Only edn.

One of the major documents in the 'war' between the followers of Paracelsus and the 'orthodox', mainly Galenists. This tract was promptly 'exploded' by Joseph du Chesne (No. 212). An early alumnus of King's College, Aberdeen, joined in (No. 466). (See Vol. i, p. 177.)

Ferg. I, 54.

49. AUGENIUS, HORATIUS. *Epistolarum & Consultationum Medicinalium libri XXIIII in duos tomos distributi . . . Quibus accessere eiusdem authoris de Hominis partu li. ii nunc primum in Germania . . . repurgati . . . et emissi.*

Heirs of Andr. Wechel, Frankfurt 1598

The 'heirs' of Andreas Wechel were Claudius Marnius and Ioannes Aubrius: their names will not be included hereafter.

[BM fo. edn. 1597, but indication that it is not *Ed. p.*]

Prov. LIDDEL. Bound with No. 500. Also a second copy.

Cont. pagn. for both vols; half-title and sep. dedn. for Vol. II. Sep. pagn. and t.p. for *de Hominis partu*, which is here entitled *Quod homini certum non sit nascendi tempus, libri duo. Adiecimus embryon putrefactum urbis Senonensis cum . . . exercitatione de huius indurationis causis naturalibus.*

50. AUGENIUS, H. . . . *De ratione curandi per sanguinis missionem libri xvii in duos tomos divisi. In quibus extirpatis erroneis opinionibus passim hodie apud novatores medicos vigentibus, omnia ad hoc argumentum pertinentia, secundum Galeni doctrinam explanantur. Hac editione quinta ab . . . erroribus . . . expurgati & nunc primum in Germania correctiore typo decorati.*

Heirs of Andr. Wechel, Frankfurt 1598

Dedn. Padua 1597; based on a similar work, *libri tres*, Venice 1570. Vol. I (text) completed Turin 1584, Vol. II (disputations and responses) Padua 1593 with sep. half-title and new dedn. Cont. pagn.

Prov. LIDDEL.

There is a fig. (p. 285) showing various types of venesection.

51. (Caelius) *Aurelianus Siccensis. . . . de Acutis Morbis lib. iii, de Diuturnis lib. v. Ad fidem exemplaris manu scripti castigati & annotationibus illustrati.*

Gul. Rovillius, Lyon 1567

Ed. p.

52. AURIFABER, ANDREA. *Succini Historia—nunc primum studio et opera Laurentii Scholzii Med. Vratisl. . . . edita.*

Andr. Wechel, Frankfurt [*ca.* 1594]

This brief work is bound with several other works.

It is not recorded in BM; and Bibl. Nat. records only Hanover, heirs of J. Aubrius, 1614. The dedn. of No. 52 by Scholz to Carolus Clusius (see No. 3) and introduction are dated Vratislava 1593.

Prov. LIDDEL.

An early report on the fossils preserved in amber, illustrated by cuts and poem *De Rana et Lacerta Succino Prussiaco insitis*. 'Vratislavia' was the Latin form of 'Breslau'.

AVENZOHAR. See ABD-EL-MALIK.

AVICENNA See HUSAIN IBN ABD-ALLAH.

AVERROES See MUHAMMAD IBN AHMAD.

53. BACCIUS, ANDREAS. *De Thermis . . . li. viii . . . .*
> (*a*) *In quo agitur de universa aquarum natura, deque earum differentiis omnibus, ac mistionibus cum terris cum ignibus cum metallis. De terrestris ignis natura nova tractatio. De fontibus, fluminibus, lacubus. De balneis totius orbis, et de methodo medendi per balneas. Deque laevationum, simulatque exercitationum institutis in admirandis thermis Romanorum . . . .*

V. Valgrisius, Venice 1571

Ed. p.

(*b*) Similar title, and *Demum ab ipso auctore recognitum.*

Same imprint 1588

New privileges (Papal and French, both of 1587).

Prov. GORDON.

Dedn. to his employer Sixtus V. This is the first original encyclopaedic work (see No. 749) on mineral springs which gave rise to a flood of books during (xvi–2) and later. It is 'classical' in tone (see dedn. and p. 436, plan of baths of Diocletian) but is marked by a modern experimental approach.

54. BACCIUS, A. *De gemmis et lapidibus pretiosis, eorumque viribus et usu tractatus, Italica lingua conscriptus. Nunc vero non solum in Latinum*

18

sermonem conversus, verum et iam utilissimis annotationibus &
observationibus auctior redditus a Wolfgango Gabelchovero . . . Cui
accessit disputatio de generatione auri in locis subterraneis illiusque
temperamento.

Matthias Becker, Frankfurt 1603

Not in Ferg. or Duv. BM records this edn. and also *Le XII pietre*
pretiose . . . Rome 1587, which is presumably the original Italian edn.
on which No. 54 is based. Dedn. by Gabelchoverus 1603. Commentary
on text in italic.

Prov. LIDDEL.

55. BACON, ROGER. . . . *de arte chymiae scripta. Cui accesserunt opuscula*
alia eiusdem authoris.

Io. Saurius, Frankfurt, 1603

Excerpts from *Opus Tertium* and from *Liber sextus scientiarum* by way
of intro.

Duv. quotes this edn. omitting first 4 lines of title *Sanioris Medici-nae*
Magistri/D. Rogeri/Baconis Angli. Ferg. quotes full title and list of tracts.

56. BAEZA, LODOICUS. *Numerandi doctrina praeclara methodo exposita*
in qua . . . exponuntur aperte ea, quae ex universa Arithmetica sunt
ad usum potiora.

(Col: Benedictus Prevotius for) Gul. Cavellat, Paris 1556
[*Ed. p.* 1555—Smith p. 269, who states that this '2nd edn.' differs
only in respect of the date on t.p.]

Prov. Purchased 1956. Much glossed in neat 16th cent. hand.

Although setting out the various rules, including the *regula aurea*,
and applying them to various quantities such as money and angular
measure, it is coloured by the classical concern for numerology and
proportion.

57. BAIRO, PIETRO. *De medendis humani corporis malis enchiridion quod*
vulgo 'veni mecum' vocant: cui adiunximus eiusdem authoris tractatum
de peste.

Petrus Perna, Basel 1578

[*Ed. p.* Lyon 1565.]

The dedn., by the well known humanist Theodor Zwinger, states
that Petrus Bayrus was physician to Charles II, Duke of Savoy, and
'while condemning the incantations (*naenias*) and figments of the
Methodists sought to unite the theory of the Dogmatists with the
experimental approach of the Empiricists . . . he regarded Medicine as
a unity'—dated Basel, 1560.

The *De Peste* has section-heading, p. 516 'Anno 1507 Taurini edita'.
This had in fact been published separately at Torino.

58. BARBARUS, HERMOLAUS. *In C. Plinii Naturalis Historiae libros castigationes.*

J. Oporinus, Basel 1534

[*Ed. p.* Rome, 1492.—GW 3340.]

Prov. MELVIN.

The dedication by Barbarus to Alexander (Borgia) VI is dated 1492. The author provides a line-by-line criticism of the text in the light of writers subsequent to Pliny (e.g. Ptolemy) as well as of those preceding him. He claims (*Biog. Univ.*) to have made 6,000 corrections—most of which are merely textual, including the spelling of proper names, but there are many corrections of matters of fact. The spirit is 'literary' (i.e. the weighing of 'authorities') rather than 'scientific'. In the scores of examples I examined I found no attempt to 'weigh the authorities' by reference to *observations*. See Vol. I, p. 188.

59. BARBARUS, H. . . . *in Dioscoridem Corollariorum libri quinque.*

Ioan. Soter, Cologne 1530

[*Ed. p.* n.p. 1516]

The editor, B. Egnatius, draws the reader's attention to the Barbaro family. 'Hermolaus senior, Franciscus avus, & ipse denique Hermolaus noster.' There is no text of Dioscorides, only a running commentary, chiefly in relation to the confusion caused by departing from the usage of the ancients in the nomenclature of plants.

60. BARNAUD, NICHOLAS. *Quadriga aurifera.*

Chris. Raphelengius, Academiae Lugduno-Bat. Typographus,
Leiden 1599

Ed.p. (Not BM) Ferg. I, 73. Duv. p. 44.

After the sack of Antwerp in 1576, Christopher Plantin transferred some of his work to Amsterdam and Leiden, to the latter of which he fled to escape his creditors in 1583. There he bought up the premises in which Louis Elzevir was working in association with the University (founded 1575). When Plantin returned to Antwerp in 1585 he left Elzevir in charge of the original shop and his own son-in-law, Christopher Raphelengius, in charge of the second press—both were thus in a sense 'university presses' (Putnam, Vol. II p. 280f.).

A collection of tracts, viz. *Prima Rota, Tractatus de Philosophia metallorum a . . . viro anonymo conscriptus. Secunda rota, Liber duodecim portarum* (by George Ripley, No. 577). *Tertia rota, Liber de mercurio et lapide* (Ripley) . *Quarta rota, Elixir solis Theophrasti Paracelsi tractatus* (anon.). (Sep. dedns.). There follows with sep. t.p. and pagn. *Triga chemica: de lapide philosophica, tractatus tres, editore & commentatore Nicolao Bernaudo.* (Ref. to Lambspringk as author.)

20

61. BARROUGH, PHILIP. *The Methode of Phisicke, conteyning the causes signes, and cures of inward diseases in man's body from the head to the foote* . . . .

> Thos. Vatrouiller, London 1583

Ed. p. STC 1508. Two copies.

62. BATMAN, STEPHEN. *Batman uppon Bartholome his booke de proprietatibus rerum. Newly corrected, enlarged and amended: with such additions as are requisite, unto every severall booke: Taken foorth of the most approved authors, the like heretofore not translated into English* . . . .

> Thomas East, London 1582

Ed. p. STC No. 1538.

This is so vaguely related to the text of Bartholomaeus Anglicus, *De Proprietatibus Rerum*, as to constitute a new work. The vogue of Bartholomew's book (there is a MS at Oxford of 1296) may be judged by the fact that the Library possesses three incunable edns., one without any colophon, the others by Io. Koelhoff at Cologne 1481 and 1483. The *Ed. p.* was probably Basel, about 1470. An English translation was made by John of Trevisa in 1398, printed by Caxton's successor, Wynkyn de Worde, about 1495. Batman's note to the reader and colophon give the authorities for the 'additions' and also the history of the work itself. How far this edn. was 'corrected and amended' may be judged by the statement that 'printing began in England in 1471' in the reign of Henry VI. In 1471 Henry had already been dethroned ten years (unless the few months in 1471 is to be counted as his 'reign'), and in any case Caxton did not set up his press before 1476. Of the familiarity of the author with the printed books of the renaissance we may judge when we find in the list of 'authorities' that 'Ypocras' (a common form of Hippokrates) was simply 'a philosopher'.

63. BAUHINUS, CASPAR. . . . *de Corporis Humani Fabrica li. iiii. Methodo anatomica in praelectionibus pub. proposita: ad And. Vesalii tabulas instituta: sectionibus publicis & privatis comprobata. Multis denique, novis inventis & opinionibus aucta.*

> Seb. Henricpetrus, Basel (Col. 1590)

(Not BM—'ed. altera' only.) Not Bibl. Nat.
Prov. LIDDEL.

The Preface—'ad Prolem Apollineam' (Ovid's phrase for Aesculapius)—provides a picture of the 'new' teaching of anatomy, popularised by Vesalius and the other anatomists (e.g. Realdo Colombo) mentioned by Bauhin (Vol. I p. 236).

21

64. BAUHINUS, C. [t.p. wanting—title from Bibl. Nat.] *Anatomes C . . . B . . . liber secundus . . . .*

(Col. Seb. Henricpetrus, Basel 1592)

This 'second part' of Bauhin's anatomy explains the *partes similiares* (see Vol. I, p. 217) in contrast with the *externa fabrica*, which as he says in his preface has been dealt with in the first book. He claims that apart from some reference to Falloppio and Sylvius, this aspect has not been developed before. The running headlines throughout are *de partibus similiaribus*.

65. BAUHINUS, C. φυτοπιναξ *seu enumeratio plantarum ab herbariis nostro seculo descriptarum, cum earum differentiis, cui plurimarum hactenus ab iisdem non descriptarum succinctae descriptiones & demonstrationes accessere. Additis aliquot hactenus non sculptarum plantarum vivis iconibus.*

Seb. Henricpetrus t.p.

Basel, 1596 on col. preceding *icones*

Ed. p.

Bauhin dedicated the book to Paschalis Gallus (No. 289) and others who sent him numerous plants from Germany, Italy, and France. There is in addition (B 2 *r*) a long list of those who have sent plants or seeds: this includes 'Iacobus Cargillus Scotus, M. Med. Cand.' The names are as usual followed by their home towns or countries but in one or two cases by the abbreviation 'Cand.', for 'Candidatus'—it is likely that Cargill was a candidate for a medical degree at Basel at this time. On B 3 *r* and *v* there is a long encomium addressed to Bauhin by Cargill, which seems to have been overlooked by the author of *Musae Latinae Aberdonenses* (Aberdeen University Studies No. 43).

The *Ad Lectorem Botanicum* is worth close study (see Vol. 1, p. 198).

66. BAUHINUS, IOHANNES. *De plantis a divis sanctisve nomen habentibus. Caput ex magno volumine de consensu & dissensu authorum circa stirpes desumptum . . . Additae sunt Conradi Gesneri . . . epistolae hactenus non editae a Casparo Bauhino . . .*

Conrad Waldkirch, Basel 1591

Ed. p.

Prov. BUTE.

The *Epistolae* has sep. h.t.

Though the book is written by Jean Bauhin and the letters mainly addressed to him, the work was edited by his brother Caspar, who dedicated it to the son of Ioachim Camerarius (No. 755). The letters, dating from June 1560 (Zürich) to August 1565 (Zürich), are full of contemporary gossip about books, plants, and personalities, such as one would expect from Conrad Gesner (No. 298).

22

67. BAUHINUS, IO. *Historia novi et admirabilis fontis balneique Bollensis in Ducatu Wirtembergico ad acidulas Goepingenses. Mandato . . . Frid. Ducis Wirtemberg . . . a Ioanne Bauhino Ill. eius Cels. Medico conscripta. Adiciuntur plurimae figurae novae variorum fossilium, stirpium, & insectorum quae in & circa hunc fontem reperiuntur.*

No printer's name (J. Foilletus—BM) Montbeliard 1598. *Ed. p.*

Prov. BUTE. Previously 'Ex. lib. Ro. Gray Collegii Med. Lond. socii'.

The dedn. and *Mandatum* cast a light on the happy relationship between noble employer and physician, so often obscured (when it existed) in the 16th C. by sycophantic hyperbole. Dedn. is by Bauhin to John Frederic and Louis Frederic, sons of Duke Frederic, who for his part included 'Mandatum . . . doctiss. atque dilecto et fideli Ioanni Bauhino Medico nostro ordin.' requiring his further assistance for additional proof of the spring's powers. The Duke's consideration extended to others standing in special need of it, as the following quotation from the *Leges Balnei* testifies: 'Whosoever (and how often so ever he) shamelessly comports himself by word of mouth towards honest matrons and young women shall be fined a florin.'

The text is followed by topographic sketches of the countryside taken from different aspects in the neighbourhood of the 'spa'. The additions indicated in the title are wanting in No. 67. The copy examined at BM contains a *Liber quartus* running to over 200 pages and containing not only figs. and legends of 'fossils' but an extensive 'flora' and 'fauna' of the neighbourhood.

BEAUSARDUS, PETRUS. See No. 747

68. BEDA ANGLUS. *De natura rerum et temporum ratione libri duo nunc recens inventi & . . . editi.*

Henricus Petrus, Basel 1529

Ed. p.

Prov. (i) 'Liber M. W. Hay Scott Aberdonen.', (ii) 'Alexander Fraser 1594 amanuensis', (iii) 'Wm. Andersone 1598'.

Sig. a 3 *r* List of Bede's works not yet printed. The make-up of the book is: fos. 1*r*-6*v*, *De Nat. Rerum* argument; fos. 7*r*-11*v*, *De Rat. Temporum* argument; fos. 12*r*-48*r*, *De Nat. Rerum;* Praef. and chaps. 1-64; fos. 48*r*-74*r*, chaps. 65-9 and tables. The 'chapters' of the argument do not correspond with those of the text.

69. BEGUINUS, IOHANNES. *Tyrocinium Chymicum e naturae fonte et manuali experientia depromptum . . . Studio & opera Christophori Gluckradts.* 6th edn.

23

[*Ed. p.* in this form Paris 1621.]   Aug. Boreck (no pl.) 1625
Prov. READ
Ferg. I, 93-4 states that this was the manual issued to Beguin's
students. See Patterson, *Ann. Sci. 2*, 69, Vol. I, p. 180, and No. 467.

70. BELON, PIERRE. *De neglecta stirpium cultura atque earum cognitione
libellus. Edocens qua ratione silvestres arbores cicurari & mitescere
queant. Carolus Clusius Atrebas e Gallico Latinum faciebat.*
                                        Off. Christ. Plantin, Antwerp 1589
Ed. p. of trans. Orig. Paris 1558. Bound with No. 71.
(For Pierre Belon, see Vol. I, p. 196.)
This is one of the earliest 'modern' works on forestry. On p. 85 there
is a reference to Cosimo de' Medici's garden.

71. BELON, P. *Plurimarum singularium & memorabilium rerum in Graecia,
Asia, Aegypto, Iudaea, Arabia aliisque exteris provinciis ab ipso
conspectarum, observationes tribus libris expressae. Carolus Clusius
Atrebas e Gallicis Latinas faciebat.*
                Off. 'Chris. Plantini, Architypographi Regii', Antwerp 1589
Ed. p. of trans. [*Ed. p.* of orig., Paris 1553.]
The 'privilege' of Charles V is dated Brussels 1555; the *Approbatio*
refers to 'ex Gallico verso in Latinum olim cum privilegio Caesareo
impresso' and is dated Antwerp 1589. At this time Plantin was 'typo-
grapher Royal' to Philip II—a somewhat mixed blessing, as Philip let
him down badly over the costly polyglot Bible which Pius V frowned
on. Clusius's (No. 3) dedn. to Wilhelm Friedrich, Landgrave of
Hesse (see Vol. I, p. 120 and No. 95), gives bibliographical details of
Belon's works and is dated Nürnberg 1586. Belon's dedn. (presumably
of the original French work) to François, Cardinal de Tournon,
speaks of 'tuis Turnonensibus Collegiis aliisque quae exstrui iussisti'
and is dated 'tuis aedibus Divi Germani ad prata Parisiensi suburbano
1553'—this is, the great abbey of St. Germain-des-Prés In his preface
Belon states that he was travelling in search of the material during
the years 1546-9.
Numerous cuts of plants and animals (e.g. p. 493, Armadillo).
The list (p. 7) of local names of plants is useful.

BENEDICTUS, ALEXANDER. See Nos. 205, 748.

BENIVIENI, ANTONIO DE, See Nos. 205, 635.

BEROSUS, See No. 752.

72. BERTOTIUS, ALFONSUS. *Methodus generalis & compendiaria ex Hippocratis Galeni & Avicennae placitis deprompta, ac in ordinem redacta* . . . .

Gabrielis Coterius, Lyon 1558

[*Ed. p.* Venice 1556. The BM copy of 1558 lacks all after p. 238.] Chaps. 2 and 3 comprise an interesting review of Galen's enumeration *a priori* of 'possible' diseases, with Avicenna's and Fernel's objections.

73. BESSONUS, IACOBUS. . . . *de absoluta ratione extrahenda olea & aquas e medicamentis simplicibus: accepta olim a quodam Empirico postea vero ab eodem Bessono locupletata & rationibus experimentisque confirmata, liber.*

Andr. Gesner, Zürich 1559

Ed. p.

Bound with Conrad Gesner's *Euonymus,* who in the *Lectori* pays a tribute to Valerius Cordus (Vol. I, p. 193) and reminds other nations how much, in these matters, they owe to the Germans. He refers to Besson as 'viro in opere geometrico & machinarum praesertim variarum vitae humanae utilium inventione incomparabili'. Good cuts of furnaces and other apparatus; also a list of herbs.

74. BIONDO, MICHEL ANGELO. *Della Domatione del Poledro . . . opera necessaria ad ogni imperatore de gli eserciti . . . da incerto philosopho antichamente scritta . . . Novamente percio venuta nelle mani del Biondi, da lui tradutta in lingua materna . . .*

Biondo, Venice 1549

Ed. p.

Prov. BUTE.

75. BIRINGUCCIO, VANOCCIO. *La Pyrotechnie ou Art du Feu.*

Claude Frémy, Paris 1556

[*Ed. p.* Venice 1540.] Duv., who refers to an earlier French edn. 1554.

No. 75 contains a letter from the printer Frémy justifying the French trans. and dated 1555.

This work is held by some scholars to be more original than Agricola's *De Re Metallica,* but the ground covered is in any case different.

76. BLAGRAVE, JOHN. *TheMathematical Jewel,* showing the making and . . . use of a singular instrument so called in that it performeth . . . whatever is to be done, either by Quadrant, Ship, Circle, Cylinder Ring, Dyall, Horoscope, Astrolabe, Sphere, Globe, or any such like heretofore devised . . . .

Walter Venge, London 1585

Ed. p. S.T.C. 3119.

The above list is an almost complete roll of surveying and navigational instruments used in 16th cent. (except for the cross-staff—see No. 197). The book includes end-papers illustrating the improved form of plane astrolabe devised by the author with instructions for making it. Johnson (p. 313) regards it as the best known of its kind during the subsequent century.

The t.p. includes a cut of the armillary sphere, which with all the other illustrations were made from blocks cut by the author.

77. BLAGRAVE, J. *The Art of Dyalling* in two parts.

S. Waterson, London 1609

Ed. p. S.T.C. 3116. Somewhat more advanced than Thos. Fale's work (No. 242).

78. BLUNDEVILLE, THOMAS. *The Theoriques of the Seven Planets . . . Whereunto is added a briefe extract by him made of Maginus his theoriques for the better understanding of the Prutenicall Tables. . . . There is also hereto added, the making description, and use of two most ingenious . . . instruments for sea-men . . . .*

Adam Islip, London 1602

Only edn. S.T.C. 3160.

In the long title and 'To the Reader' Blundeville shows that he is intending a theoretical treatise for practical men—a project formed as the result of the favourable reception of his *Exercises*. He has collected his 'Theoriques' from Ptolemy, Peurbach, Reinhold, Copernicus, and Mästlin (the teacher of Kepler), but he restricts himself to the ordering of the phenomena. He speaks of being greatly indebted to 'my good friend M. Doctor Browne, one of the ordinarie Physicians to her Maiestie'. Only one (Lancelot Browne) of Blundeville's contemporaries is mentioned in the *D.N.B.* as being a physician to Elizabeth and James I, and there is no mention of his having visited Norwich, where Blundeville had with him 'many learned conferences'. The *Instruments for Seamen* has separate t.p. but cont. pagn. They consist of a dip-circle (note that 'dip' is here called 'declination') and an instrument intended for *calculating* latitude from *observed* dip. Instead of the instrument recourse may be had to a set of tables 'annexed to the former treatise by Edward Wright . . . which M. Henry Brigs (No. 474) (professor of Geometrie in Gresham Colledge at London) calculated . . . . (For Wright, see Vol. I, p. 145.) The work is profusely illustrated with clear diagrams.

79. BLUNDEVILLE, T. *The fower chiefyst offices belonging to horse-manshippe, that is to say, the office of the breeder, of the rider, of the keper, and of the ferrer. . . .* [The whole contents of the volume appear on the main t.p. (undated); separate (dated) t.p.s precede the parts.]

Wylliam Seres (Col.) London 1565-6

This is the earliest edn. recorded by BM new catalogue, also S.T.C. 3152, but the sub-title of the second part—*The Arte of Ryding*—reads 'Newly corrected and amended of many faultes escaped in the first pryntynge as well touchynge the matter, as the Bittes: Whereof many were evyll drawen and as evyl cut: but now made perfect through the dilygence of the first author Tho. Blundevill of Newton Flotman in Norff.' This may refer to an adaptation of T. Grisone by T. Blundeville in 1560. S.T.C. 3158.

Prov. BUTE. Numerous cuts of bits, etc.

The section on breeding is of special interest as a mark of the opinions then held on generation.

80. BOCANGELINO, NICOLAO. *Libro de las Enfermedades malignas y pestilentes, causas, pronosticos, curacion, y preservacion.*

Luis Sanchez, Madrid 1600

Only edn. BM

On p. 69 (Cap. X) ('Dondi se ponen las causas que han concurrido para esta constitucion epidemica') 'Ay otra diferencia, y es, quando viene comunicada la peste por seminarios contagiosos'. Reference is made to Luiz Mercado, *Liber de Peste tract.* 1, fo. 24, but not to Fracastoro. [See Vol. I, p. 275]

BOCK, HIERONYMUS. See No. 603

81. BODERIUS, THOMAS. *De ratione & usu dierum criticorum, opus recens natum in quo mens tum ipsius Ptolemai tum aliorum astrologorum hac in parte dilucidatur . . . . Cui accessit Hermes Trismegistus de decubitu infirmorum nunquam antea . . . editus.*

Andr. Wechel, Paris 1555

[*Ed. p.* Paris 1555 Off. A. Parvi.]
Prov. JOHNSTON.

Dedn. to Oronce Fine (No. 262). In his preface the author claims the support of Galen (cited) and Hippokrates for importance of critical days in relation to the motion of the moon.

82. BODINUS, IOANNES. . . . *De Magorum Daemonomania seu detestando lamiarum ac magorum cum Satana commercio libri iv. Recens recogniti & multis in locis a mendis repurgati. Accessit eiusdem opinionum Ioan. Vieri confutatio.*

Nic. Bassaeus, Frankfurt 1590

Ed. p.

Dedn. 'Christophoro Thullenai—primo Praesidi Parisiensis Parlamenti & consiliario in privato consilio regis'. Laoduno 1579. (It will be recalled that the power of the *Parlement* at Paris had become very marked by this time.) In his preface, Bodin, the leading humanist of France, gives a long account of the contemporary circumstances prompting him to compose this work together with a general account of relevant principles. Of special interest is the quotation of a *Determinatio* of the Faculty of Theology of Paris for 1398 condemning 'magic arts' and intercourse with demons and continuing 'neque tamen intentio nostra est in aliquo derogare quibuscunque licitis & veris traditionibus scientiis & artibus,—which perhaps provides a justification for the inclusion of a work on commerce with demons in a bibliography of science! [For Jean Wier, see No. 729.]

BOISSIÈRE, CLAUDE DE. See No. 296

83. BOKELIUS, IOHANNES. *Pestordnung in der Stadt Hamburg* . . . .
Iac. Lucius, Hamburg 1597
Prov. Apparently a presentation copy to Liddel.
An example of a 16th cent. Medical Officer of Health in action (*cf.* Vol. I, p. 269).

84. BOMBAST, PHILIP (*syn.* Theophrastus Paracelsus)
The special problems in cataloguing the books under the name of Paracelsus call for separate treatment. In the first place the numbers involved are in quite a different class from those of any other author. Karl Sudhoff, whose *Versuch einer Kritik der Echtheit der Paracelsicher Schriften* (Berlin 1897, etc.) is unlikely ever to need any major revision, lists 518 printed books alone. This would not in itself demand any special treatment, but for the fact that only about twenty of these works were printed in the lifetime of the author. This of course has precedents in the case of the ancient and medieval writers, but is perhaps unique among authors born after the beginning of printing. The case is further complicated by the fact that almost all the works were written in German and circulated in MS among a devoted following who copied them and almost certainly added to them. When the fame of their author passed beyond the region of his mother tongue (especially in France, according to one of his publishers), and to meet the demand for his works it was thought desirable to prepare Latin translations, it was already a matter of great difficulty to decide which were the authentic works of the master. Three men in particular devoted themselves to the task of translation and editing (Adam von Bodenstein, Gerhard Dorn, and Michael Toxites), and one publisher in particular (Petrus Perna of Basel) was for a time pre-eminent in this

field.   Unfortunately, it seems likely that Dorn at least was not content to let the Master speak for himself, but fabricated a number of MSS in his name. By the beginning of the 17th cent. two scholars of high repute had set about the task of providing definitive editions, Zacharias Palthenius in Latin, Johann Huser in German. The Library is fortunate in possessing both these works, complete sets of both of which are not otherwise recorded by Sudhoff for the British Isles, though the Bodleian Library contains a set of the earlier edition of Huser, upon which the more complete edition of 1603 is based. Sudhoff notes in this connection that the latter does not *supersede* the *Ed. p.* since the language of the former has been over-'modernised' at the risk of change of meaning.

The titles of the separate volumes provide a further problem, in that many of them are collections of tracts, some of which may have been printed earlier in a different collection. Moreover, to certain of these collections, which circulated as such in MS during the author's lifetime, catch-names were affixed by him which were retained when regular publication was undertaken.

Since the exact date and circumstances of particular *Editiones principes* will usually be of interest only to the specialist, who will in any case prefer to refer to Sudhoff direct, the custom adopted elsewhere in this Bibliography of being at some pains to set down the *Ed. p.* has been abandoned in the case of some of the tracts of Paracelsus. A reference to Sudhoff has however been given in every case to enable those interested to proceed at once to the most authoritative source.

Finally, the works, instead of being arranged in chronological order, have been grouped together according to the *bound volumes* in which they occur. This has been done partly on account of the difficulty of deciding what constituted an *Ed. p.* in certain cases, partly on the ground that since, in these cases, all the works are believed to be bound together in a number of contemporary volumes (though the original covers have in a few cases been replaced by brown paper backs) many of which show almost indisputable evidence of having been the property of Duncan Liddel, it was thought that such an arrangement in the catalogue would provide an additional interest without consequent loss for other purposes. The separate '*tomi*' of the Huser and Palthenius editions appear under the same serial *numbers*.

As a general guide to those unfamiliar with the literature of Paracelsus a list of his better known works and of the collections thereof is here appended.

Grosse Wundartzney—('Chirurgia Magna') published separately.

Kleine Wundartzney—('Chirurgia Minor'), also known as 'Bertheonea', published both separately and in collections.

Archidoxa—usually comprising ten sections, including the especially important *De Vita Longa*, which was frequently published separately or in other collections.

Paramirum—on the nature of diseases and the physical world.

Paragraphi—so called, according to the editor, Toxites, because, as his original lectures on Medicine at Basel, they were delivered 'per paragraphos'.

Paragranum—containing a polemical statement of his 'philosophy of medicine', written after the Basel episode, to which he refers—see e.g. Huser (1589), **2,** 6.

The above titles were used by Paracelsus himself. The title *Paradoxa* was used by Palthenius as a short title for his Latin collection. There are two quite distinct versions of the *Paramirum*, commonly distinguished as *Volumen* and *Opus*, the latter being the more mature and comprehensive.

There were two other collections with short titles:—*Compendium* a typical selection from his 'best' books including *De Vita Longa*, made by his early French supporter, Jacques Gohory (No. 13(c)); and a more scholarly *Operum Latine Redditorum Tomus*, I-II by Petrus Perna, which never got beyond the second volume (No. 3).

It has been necessary to quote titles at rather greater length owing to the similarity of many of them. Where no confusion is likely to ensue, I have abbreviated 'Aureolus Philippus Theophrastus Paracelsus'—any or all—to 'A—P—T—P—'.

COLLECTED WORKS

(1a) *Ander Theil der Bücher und Schrifften des edlen hochgelehrten und bewehrten Philosophi unnd Medici Philippi Theophrasti Bombast von Hohenheim Paracelsi genannt. Jetzt auffs new auss den originalien und Theophrasti eigner Handschrifft soviel derselben zubekommen gewesen auffs trewlichst und fleissigst an Tag geben. Durch Ioannem Huserum Brisgoium Churfürstlichen Cölnischen Rhat unnd medicum.*

Dieser Theil begreifft fürnemlich die Schrifften inn denen die Fundamenta angezeigt werden auff welchen Kunst der rechten Artzney stehe und ausz was Büchern dieselbe gelehrnet werde . . . .

C. Waldkirch, Basel 1589

Ed. p.

Prov. Purchased 1955.

Sudhoff, No. 217. This is the only Part of the *Ed. p.* in the possession of the Library. The text follows closely that of the corresponding Part of the 2nd edn. (1.b), but contains (p. 141) a note on the position of alchemy in the scheme of medicine, which is missing from the later edition. The latter, however, contains a number of pages 'Ad Tomum II referenda'. Sig. A ii apparently wanting.

(1b) *(Erster) Theil der Bücher und Schrifften dess edlen hochgelehrten unnd bewehrten Philosophi unnd Medici Philippi Theophrasti Bombast von Hohenheim Paracelsi genannt. Jetzt auffs new auss den originalien und Theophrasti eygener Handschrifft so viel derselbigen zubekommen gewesen auffs trewlichst und fleissigst an Tag geben, durch Ioannem Huserum Brisgoium Churfürstlichen Cölnischen Raht und Medicum.*

(The above is common to all the 'Theile', each of which has its own sub-title as follows):

(Erster) In diesem Theil werden begriffen deren Bücher etliche welche von Ursprung Ursach und Heylung der Kranckheiten handeln in genere.
(Ander) . . . *Dieser Theil begreifft fürnehmlich die Schrifften in denen die Fundamenta angezeigt werden auff welchen die Kunst der rechten Artzney stehe und auss was Büchern die selbige gelernet werde.*
(Dritter). . . . *In diesem Theil werden begriffen deren Bücher etliche welche von Ursprung Ursach und Heylung der Kranckheiten handeln in Specie.*
(Vierdter) . . . *In diesem Theil werden gleichfals wie im Dritten solche Bücher begriffen welche von Ursprung Ursach und Heilung der Kranckheiten in Specie handeln.*
(Fünfter) . . . *In diesem Theil und dessen Appendice werden die Bücher ad Medicinam Physicam gehörig begrieffen.*
(Sechster) . . . *In diesem Theil werden begriffen solche Bücher in welchen von Spagyrischen Bereytungen natürlicher Dingen die Artzney betreffend gehandelt wird.*
(Siebenter) . . . *In diesem Theil sind verfasst die Bücher in welchem fürnehmlich die Krafft tugenden und eigenschafften natürlicher Dingen auch derselben bereitungen betreffend die Artzney beschrieben werden: neben eingemischten Sachen zur Alchimey dienstlich.*
(Achter) . . . *In diesem Tomo (welche der Erste unter den philosophischen) werden solche Bücher begriffen darinnen fürnehmlich die Philosophia de Generationibus & Fructibus quatuor Elementorum beschrieben wirdt.*
(Neundter) . . . *Dieser Tomus (welcher der ander unter den philosophischen) begreifft solche Bücher (darinnen allerley natürlicher unnd übernatürlicher Heimligkeiten Ursprung Ursach Wesen und Eigenschafft gründtlich und warhafftig beschrieben werden).*
(Zehender) . . . *Dieser Theil (welcher der Dritte unter den philosophischen Schrifften) begreifft fürnehmlich das treffliche Werck Theophrasti* Philosophia Sagax, *oder* Astronomia Magna *genannt: sampt etlichen andern Opusculis, und einem Appendice.*

Heirs of Io. Wechel, Frankfurt 1603

All 'Theile' have separate pagination and t.p., except that *Ander Theil* has half-title and continuous pagn. with *Erster*. Prov. The whole work is heavily scored and glossed in Liddel's hand up to about p. 105 of *Neundter Theil*.

This collection is taken by Sudhoff (Nos. 254-5) as the most nearly definitive edition of the Paracelsian corpus (but see p. 29). He cites no complete set of this edition in the British Isles, but the Bodleian Library has a complete set of the *Ed. p.*

The British Museum records a copy of *Erster–vierdter Theil* of a variant of this edition which, according to Sudhoff (Nos. 226-9), differs from the usual one of this date and 'was certainly not printed in 1590, perhaps 1599'. Incidentally the BM note, viz: 'no more published in this edition', is misleading in the absence of any record of the 'standard' edition, since it might easily lead one to suppose that there were no further volumes published until 1603 or even later.

> (2) *Nobilis clarissimi ac probatissimi philosophi & medici Dn. Aureoli Philippi Theophrasti Bombast, ab Hohenheim, dicti Paracelsi, Operum Medico-Chimicorum sive Paradoxorum Tomus genuinus (primus).*

Every 'Tomus' has separate sub-title as follows:

> (*Primus*) *Agens de causis, origine ac curatione morborum in genere. Recenter Latine factus, & in usum asseclarum novae & veteris philosophiae foras datus.* [Not repeated after sub-titles of remaining 'Tomi'.]
>
> (*Secundus*) *Tradens fundamenta quibus verae & genuinae medicinae ars superstructa & ex quibus solis illa addisci possit.*
>
> (*Tertius*) *Agens de causis, origine ac curatione morborum in specie.*
>
> (*Quartus*) *Agens itidem ut tertius de causis, origine ac curatione morborum in specie.*
>
> (*Quintus*) *Agens de libris ad medicinam physicam spectantibus.*
>
> (*Sextus*) *E chymicis primus, continens processus & preparationes Spagyrica rerum naturalium ad usum medicinae; multaque alia de tinctura physicorum & coelo philosophorum: de cementis item & gradationibus.*
>
> (*Septimus*) *E chymicis secundus, continens vires efficias & proprietates rerum naturalium & earum quoad medicinam praeparationes: cum multis alchimicam scientiam secretis spectantibus.*
>
> (*Octavus*) *E philosophicis vero primus, continens philosophiam de generationibus & fructibus quatuor elementorum.*
>
> (*Nonus*) *E philosophicis vero secundus, continens arcana naturalia, supernaturalia eorumque caussas, origines, substantias & proprietates.*
>
> (*Decimus*) *E philosophicis vero tertius, continens philosophiam sagacem & astronomiam magnam.*

(*Undecimus*) *E philosophicis vero quartus, continens astronomiam magnam cum artibus incertis, transmutationibus metallorum, magicis adversus morbos figuris, & planetarum sigillis.*

(Bound with *Bertheonea sive Chirurgia Minor*, 1603.)

A Collegio Musarum Paltenianarum, Frankfurt 1603-5 All 'Tomi' have separate t.p. and pagn.

No complete set mentioned by Sudhoff for British Isles, but the Wellcome Historical Medical Library now has a complete set (Sudhoff, Nos. 259-63; 269-74).

(3) *Aurioli Theophrasti Paracelsi Eremitae, Philosophi summi operum Latine redditorum Tomus I and II.* P. Perna, Basel 1575.

Prov. 'Ex dono D   Forbesii fortissimi viri Nationis Scoti illustrissimi D.   Moderatoris anno Domini 1587.'

Also another copy of Vol. II. The volumes are separately bound. This, the *Ed. p.*, was never completed. Sudhoff (Nos. 165-6) summarises from the preliminary matter the nature of this ambitious project, explained by the printer, and introduced by Adam von Bodenstein, the translations used being those (for the most part already published) by Dorn, Toxites, and Georg Forberg. Perna's own foreword to the reader gives a good idea of the difficulty of making a complete collection of works 'quae author toto orbe fere peregrinus & vagus, ut gallina ova, sparsa reliquit'. The works contained in the collection are listed at the beginning of each volume, both of which are separately but continuously paged, except that in Vol. II the interesting *De Morbis fossorum metallicorum*, in which an attempt is made to understand diseases of mineworkers, is separately paged at the end. The publisher intended to bring out a third volume, but there is no evidence that it ever appeared.

SINGLE WORKS

(4) *Ettliche Tractatus des hocherfarnen unnd berumbtesten Philippi Theophrasti Paracelsi der waren Philosophi und Artzney Doctoris*

I, Von Natürlichen dingen. II, Beschreibung etilcher [sic] kreutter. III, Von Metallen. IIII, Von Mineralen. V, Von Edlen Gesteinen. . . .

Heirs of Christian Muller, Strassburg 1570

Ed. Michael Toxites. Sudhoff (No. 120) says that there is on the whole a close correspondence with the MSS of Paracelsus.

(5) . . . *De summis naturae mysteriis commentarii a Gerardo Dorn conversi multoque quam antea fideliter characterismis & marginalibus exornati auctique.* . . .

Off. Perniana Conr. Waldkirch, Basel 1584

D*   33

Sudhoff No. 201 (from which the above is largely copied, the t.p. being badly damaged). A new edn. of that of Basel (Sudhoff No. 125). Sudhoff rejects *de Spiritibus Planetarum* from the authentic works of Paracelsus. The case of *Medicina Coelestis* is doubtful; it was included by Huser, who used the same blocks for the illustrations, but its ' paternity' is by no means established. The tract *De Occulta Philosophia* was first printed with the *Archidoxa*, Basel 1570.

> (6) *Theophrasti Germani Paracelsi medicorum et philosophorum omnium in universum facile principis, De restituta utriusque medicinae vera praxi, liber primus. Gerardo Dorn doctore physico ac interprete Germanico in hunc ordinem recolligente.*
>
> Jac. du Puys, Lyon 1578

Prov. 'Thomas Moresinus D.M.' (see No. 466).

Sudhoff No. 181.

The dedn. to François Valois, Duc d'Alençon, son of Henry II of France, emphasises the necessity for a Latin translation. The work is really a 'Practica' drawn from Paracelsus's writings. It contains a note (p. 206) on Paracelsus's use of 'alcool vini, per hoc intelligens aquam ardentem rectificatam. . . .' (See Vol. I, p. 172.)

Although the whole volume carries a running title 'liber primus' there does not seem to have been any further 'liber'.

> (7) P—A—P—. *utriusque medicinae doctoris celeberrimi, centum quindecim curationes experimentaque, e Germanico idiomate in Latinum versa. Accesserunt quaedam praeclara atque utilissima a B.G. a Portu Aquitano annexa. Item abdita quaedam Isaaci Hollandi de opere vegetabili a Rouillasco Pedemontano.*
>
> Io. Lertout (n.p.) 1582

Sudhoff No. 190. A curious collection edited by Bernard Pénot (see No. 506) and not previously printed. Sudhoff regards the 'cures' as being unconnected with Paracelsus. For Isaac of Holland see No. 357.

> (8) P—A—Th—P—. *Bombast Eremitae, summi inter Germanos Medici & Philosophi Chirurgia Magna in duos tomos digesta. . . . Nunc recens a Iosquino Dalhemio Ostofranco Medico Latinitate donata. . . .*
>
> Petrus Perna, 'Argentorati', [Strassburg] 1573

Two volumes in one, together with *Chyrurgia Minor quam alias Bertheneam intitulavit . . . ex versione Gerardi Dorn.*

All three works have separate t.p.'s and pagn.

Sudhoff (Nos. 146-8) states that the location 'Argentorati' is false. The history of the edition is related by Petrus Perna to Petrus de Grantrye and dated 'Basileae XIII Cal. Sept. Anno 1573'. Sudhoff regards this edition as a rather poor translation of Bodenstein's edition of 1564 (Strassburg) which bore the title *Opus Chyrurgicum*.

The 'Great Surgery', one of the few books written by Paracelsus and printed in his lifetime, was published at Ulm in 1536 under a rambling title containing the words 'grosse wundartzney von allen wunden . . .' dedicated to the Emperor Ferdinand I. This dedication and a second written by Paracelsus are reprinted in No. 9 in German and in Latin translation in No. 8. The Ulm edition was disavowed by Paracelsus in a foreword to the reader printed in the Augsburg edn. of the same year. Sudhoff (No. 15) regards the latter as the only authentic edn.

The *Chirurgia Minor* appeared first in print as *Drei Bücher von Wunden und Schaden* (note that the *Grosse Wundartzney* was also originally in three books) published by the heirs of Christ. Egenolph at Frankfurt in 1563.

> (9) *Chirurgia warheittige Beschreibunge der Wundartzney des hoch-gelehrten und wolerfahrnen Medici T—P— Der erste Theil darin . . . auf vorgehende ausz den originaln geschehene Correction alle und jede Bücher so zuvor unter derer Titul der Grossen Wundartzney ausz ganzen begrieffen werden. . . .*
>
> Conrad Waldkirch, Basel 1586

The *Ander Theil* follows with sep. t.p., pagn., and foreword.

Sudhoff (Nos. 210, 211) states that this edn. is identical with that of 1585 (Nos. 206, 207).

Prov. First 100 pp. scored and glossed in D.L.'s hand.

> (10) *A—T—P—Archidoxorum seu de Secretis Naturae Mysteriis libri decem. Quibus nunc accesserunt libri duo, unus de mercuriis metallorum, alter de quinta essentia. Manualia item duo, quorum prius chymicorum verus thesaurus, posterius praestantium medicorum experientiis refertum est ex ipsius Paracelsi autographo.*
>
> P. Perna, Basel 1582

Prov. HENDERSON.

Sudhoff No. 191. An edition of *Archidoxa* revised from several previous ones. Of the following works, Sudhoff, following Huser, rejects as spurious the first two, and accepts the latter's conclusion that while the last two were certainly found in Paracelsus's hand they may well have been mere compilations. (The volume contains also two works by G. Gratarolus. See No. 312.)

> (11a) *Medicorum et philosophorum summi A—T—P— Eremitae, de Tartaro libri septem. . . . Opera et industria . . . Adami a Bodenstein in lucem propter commune commodum microcosmi primo editi, nunc vero auctiores & castigatiores denuo excusi.*
>
> P. Perna, Basel 1570

[*Ed. p.* from the same House 1563. Sudhoff Nos. 54 and 126].
The dedn. to Cosimo de'Medici lacks the last (dated 1563) fo. (8*r*), the verso of which has the Paracelsus portrait. 'Libri septem' is incorrect: it is given correctly in the *Ed. p.* The dedn. gives a summary account of the 'Tartarus' theory of disease, so important in the Paracelsian system. There is also a preface by Paracelsus himself.

> (11b) *D. Theophrasti Paracelsi Chirurgia Vulnerum, cum recentium tum veterum, occultorum & manifestorum ulcerum, &c cui libri duo, prior de Contracturis: de Apostematibus, Syronibus & Nodis alter, accesserunt, per interna & externa medicamenta curam eorum veram continentes. Ex Gerardi Dorn e Germanico in Latinum versione.*
> P. Perna, Basel [1569]

Sudhoff (No. 112) stated that this is the *Ed. p.* of the translation, but his collation gives 'ocultorum[!]', and he does not mention any edition in which this is corrected.

> (11c) *Libri XIIII Paragraphorum P—T—P— philosophi summi & utriusque medicinae doctoris praestantissimi. Nunc primum a docto Toxite in communem utilitatem restituti. . . .*
> Christianus Mylius, Strassburg 1575 (all from col.)

The t.p. and several sigs. are wanting, which is all the more regrettable, since according to Sudhoff (No. 160) the dedication (to Io. Egolphus, Bishop of Augsburg) contains information as to the conditions under which the lectures were first given in Basel. Paracelsus 'taught partly in Latin, partly in German' . . . 'Paracelsus himself wrote (*fecit*) the *Paragraphi* in Latin; but in a barbarous style, such as even learned men used in that century . . .'. Toxites admits that he had had to 'clean up' the text from the various MSS, intercalating explanations and altering the phrasing at his own discretion. [Latin original quoted at length by Sudhoff.]

Sudhoff points out that the tract had already been published in German ('Dreyzehn Bücher . . .') Basel 1571 (No. 130). For the apparent discrepancy in the number of 'books' see Sudhoff I, 276.

> (11d) *Libellus Theophrasti Paracelsi utriusque medicinae doctoris de Urinarum ac pulsuum iudiciis: tum de Physionomia quantum medico opus est. Accessit de Morborum physionomia fragmentum.*
> Sam. Emmelius, Strassburg 1568

Sudhoff (No. 97) states that in neither work could Toxites (translator) have had access to the original MS. Toxites discusses Paracelsus at length in his dedn.

> (11e) *Medicorum ac philosophorum facile principis T—P— Eremitae libri v de Vita Longa, incognitarum rerum, & hucusque a nemine*

36

tractatarum referitissimi [sic], *una cum commendatoria Valentii de Retiis, & Adami a Bodenstein, dedicatoria epistola.* . . .

P. Perna, Basel 1566

Ed. p. Perna, Basel 1562.

The two dates are those assigned by Sudhoff (Nos. 46 and 503). The supposed *Ed. p.* does not contain the superfluous 'i' in 'refertissimi'; in No. 11e it has been cancelled in ink.

Prov. Scored and glossed in D.L.'s hand and bound with other works (within a modern cover), whose t.p. and last page bear the usual D.L. autographs.

Sudhoff regards 'Valentius de Retiis' as an apocryphal 'Great Alchemist' similar to 'Basil Valentine' and others.

In his *Epistola* Adam Bodenstein gives the classical summary of the 'Spagyrist' (Paracelsian) 'chemistry' which, he says, 'Theophrastus . . . plurimis docet libris' and 'in quaecunque omnia corpora dissolvuntur ultimo, ex iisdem quoque componuntur, & in sulphur, Mercurium & salem omnia corpora ultimo resolvuntur, quod experientia manifestissime demonstrat' ('Into whatsoever things all bodies are ultimately broken up, out of the same are they also composed, and all bodies are ultimately resoluble into sulphur, mercury, and salt, which experience demonstrates most clearly').

(11f) *Liber Paramirum viri Areoli* [sic] *Theophrasti Paracelsi, in quo universalis, theorica physices & chirurgiae origines, & causae morborum traduntur. . . . Accesserunt huic. . . . De modo pharmacandi, De Xeondochio* [sic], *De Thermis. A quodam docto & Theophrasticae medicinae studioso nunc primum e Germanico in Latinum sermonem conversi.* . . .

P. Perna, Basel 1570

Sudhoff (No. 122) believes this to be the first of Georg Forberger's translations.

(11g) *Aurora Thesaurusque Philosophorum, T—P— Germani philosophi, & medici prae cunctis omnibus accuratissimi. Accessit monarchia physica per Gerardum Dorneum, in defensionem Paracelsicorum Principiorum, a suo praeceptore positorum. Praeterea Anatomia viva Paracelsi qua docet autor praeter sectionem corporum & ante mortem patientibus esse succurrendum.*

(Device of) Palma Guarinus, Basel 1577

The *Anatomia* is wanting.

Sudhoff (No. 177) thinks it likely that the two works stated to be by Paracelsus were in fact Dorn's own fabrication.

(12a) *Metamorphosis Doctoris Theophrasti von*[!] *Hohenheim der zerstörten guten Künsten unnd Artzney Restauratoris gewaltigs unnd*

37

nutzlichs schreiben. . . . Durch Doctor Adamen von Bodenstein den ankloppfenden und suchenden filiis Sapientiae zu Nutz mit allem Fleisz publiciert unnd in Truck verfertigt.

[Samuel Apiarius, Basel] 1572

Prov. Scored and glossed in two hands, the less frequent being almost certainly that of Liddel.

Sudhoff (No. 137) says that the name (*Metamorphosis*) was coined by Bodenstein for a collection, some of which are well known as authentic, others, e.g. fo. 2r and v *Theophrasti Paracelsi vom Stein des Weisen*, probably spurious. In his dedn. Adam speaks of the difficulty of reading Paracelsus's works.

> (12b) *Ein schöner Tractat P—T— Paracelsi des berumbtesten und erfarnen Teutschen Philosophi und Medici. Von Eygenschafften eines volkomnen Wundartzets. Aus Theophrasti handgeschrifft erstmahls in Truck geben. . . .*

Christian Muller, Strassburg 1571

Sudhoff No. 133. A collection (tabled t.p. *v*) of surgical tracts.

> (12c) *Drey herrliche Schrifften Herren Doctors Theophrasti von Hohenheim: Das erst vom Geist des Lebens und seiner Krafft. Das ander von Krafft innerlicher geistlicher und leiblicher glider. Das dritt von Krafft eusserlicher Glider und Sterckung der inneren . . . durch Adamen von Bodenstein academischen promovierten der philosophey unnd medicin doctorem.*

Samuel Apiarius, Basel 1572 (col.)

Sudhoff (No. 136) reports that Huser spoke as if he had seen the original MSS. Sudhoff thinks that Bodenstein added the 'academischen promovierten' in order to steal a march over the rival 'editors' of Paracelsus—Dorn and Toxites—who were certainly not.

> (13a) *Archidoxorum Aureoli P—T—P— de Secretis naturae mysteriis libri decem . . . . His accesserunt libri de Tinctura physicorum, de Praeparationibus, de Vexationibus alchimistarum, de Cementis metallorum, de Gradationibus eorundem. Singula per Gerardum Dorn e Germanico sermone Latinitati* [sic] *nuperrime donata.*

P. Perna, Basel 1570

Sudhoff No. 123 (and *passim* for other editions of *Archidoxorum*). This was the *Ed. p.* of this actual edition—there was, however, a spate of *Archidoxa* in this year, but in German, some with Latin section titles.

Prov. LIDDEL(?)

Dorn in his dedn. gives an outline of the 'Spagyrist' philosophy. Since this work was commonly regarded as the key to the Paracelsian

38

therapy (Sudhoff, p. 185, but see Vol. I, p. 249) it may be as well to give the table of contents: *De mysteriis microcosmi, De renovatione & restauratione per primum ens. De separatione elementorum ex metallis, etc. De arcanis, ut, primae materiae etc. De magisteriorum extractionibus ex metallis, etc. De specificis, ut odoriferis, etc. De remediis extrinsecis, ad vulnera, etc. De vita longa.* The order and actual contents vary slightly in the different editions, *De vita longa* having previously been printed separately more than once.

(13b) *A—P—T—P— Philosophorum atque Medicorum excellentissimi de Summis Naturae Mysteriis libri tres . . . Per Gerardum Dorn e Germanico Latine redditi.*

P. Perna, Basel 1570

Sudhoff No. 125.

This work contains the following parts: (1) *De spiritibus planetarum.* (2) *De occulta philosophia.* (3) *De medicina coelesti sive de signis Zodiace & mysteriis eorum.* Of these, Sudhoff regards '(1)' as certainly a forgery, probably by Dorn, and '(2)' and '(3)' as quite possibly so.

(13c) *T—P— philosophiae et medicinae utriusque universae, Compendium ex optimis quibusque eius libris: cum scholiis in libros IIII eiusdem* De Vita Longa, *plenos mysteriorum, parabolarum, aenigmatium. Auctore Leone Suavio, I.G.P. Vita Paracelsi. Catalogus operum & librorum. . . .*

Petrus Fabricius, Frankfurt 1568

Ed. p. Rovillius, Paris, n.d. but (from *Epistola* by 'I.G.P.'), not before 1567; Sudhoff (No. 89) takes it to be the end of that year.

'I.G.P.' (= Jacques Gohory, *Professor*) explains his concealment of his name and use of 'Leo Suavius' in the first of his two *epistolae.*

Prov. No auto. but heavily scored and glossed in D.L.'s hand.

This work contains a great deal of matter relevant to the life of Paracelsus, including a letter from Erasmus (p. 18) and a very clear 'effigies Paracelsi, et apothegma, *Alterius non sit, qui suus esse potest'.*

(14) *Astronomica et Astrologica . . . A—T— von H—P— genandt . . . Opuscula aliquot jetzt erst in Truck geben und . . .verzeichnet.*

Heirs of Arnold Byrckman, Cöln 1567

Ed. p. (BM new cat.)

Purchased 1955.

Sudhoff No. 85. Fine portrait of Paracelsus. Apparently devoid of astronomy in the usual sense, it sets forth the ideas of cosmic influence. Latin captions in the margins.

See also Nos. 209, 600.

39

85. BONATUS, GUIDO . . . . *De Astronomia tractatus X, universum quod ad iudiciariam rationem nativitatum aeris tempestatum attinet, comprehendentes. Adiectus est Cl. Ptolemaei liber fructus, cum commentariis utilissimis Georgii Trapezuntii.*

(No imprint or device exc. Basel 1550)
Only edn. BM.

In the dedn. to Gul. Pagetus (Wm. Paget), Nicolaus Pruknerus describes how when Martin IV was attacking the inhabitants of Forlì they were advised by Bonatus, 'astrologorum suae aetatis ut primario, ita et doctissimo" about the day and hour to give battle to the besieging troops, thus relieving the city. An index to every kind of astrological query is provided. George of Trebizond was the first humanist translator of Ptolemy. The *Commentary* is on the (probably spurious) *Centiloquium* of Ptolemy—an astrological 'quiz'. This work of Bonatus was highly commended by Thomas Digges—see No. 194.

BONUS, PETRUS. See No. 682.

86. BORDINGUS, IACOBUS. *ΦΥΣΙΟΛΟΓΙΑ, ΥΓΙΕΙΝΗ, ΠΑΘΟΛΟΓΙΑ. Prout has medicinae partes in inclytis Academiis Rostochiensi & Haffniensi publice enarravit. Omnia ex manuscripto autoris . . . emendata.*

Steph. Myliander, Rostock 1591
Ed. p. Earliest BM same imprint 1615.
Prov. LIDDEL.

Dedn. by Levinus Battus from Rostock 1591 refers to Bording's call to Denmark by Christian III and contains other information about Rostock.

The *Hygiene* has new t.p., pagn. and registrn., the *Pathology* has half-title only with cont. pagn. and registrn.

Page 6 shows that the *methodus compositiva* and *resolutiva* were still commanding attention.

BOROUGH, WILLIAM. See No. 480.

87. BOTTONI, ALBERTINO. . . . *de Morbis Muliebribus.*

P. Meietus, Padua 1585
Ed. p.
Prov. BUTE.

Running head-lines throughout *Liber primus*, but there is no indication that there is, or was, any *secundus*.

88. BOURNE, WILLIAM. [*The Regiment of the Sea.*]
[t.p. wanting: no Col.]
'To the Reader' is signed 'W.B.' and begins: 'I have thought it good

now in this third impression . . .', which makes it probable that it was issued in 1577 (Johnson, p. 309).

[*Ed. p.* 1574.]

Ten edns. listed in S.T.C. 3422-3431.

Prov. 'William Jamesone'.

89. BOURNE, W. *A booke called the Treasure for Traveilers devided into five bookes or partes, contaynyng very necessary matters for all sortes of travailers, eyther by sea or by lande.*

Thos. Woodcocke, London 1578

Ed. p.

Prov. 'Williame Jamesonn'.

Dedn. to 'Wm. Winter, Knight, Maister of the Queenes Maiesties Ordinaunce by sea, Survaior of her highnesse marine causes &c. . . . He referes to himself as 'a poore Gunner serving under your worth-inesse . . .', and in the *Preface to the Reader* as 'altogether unlearned' but the self-disparagement rings rather false (cf. Vol. I, p. 145).

The parts include use of instruments, latitude and longitude, 'Art statical and hydrostatical', natural causes of sand banks, etc., and 'brief note, taken out of M. Dee's Mathematicall Preface that goeth before Euclides Elements' (No. 235(g)). In the text he refers to Digges's *Pantometria* (No. 194) and his own *The Art of Shooting in Great Ordnance* (*Ed. p.* Lond. 1587). Johnson notes (p. 177) that Woodcocke published in the same year Bourne's *Inventions & Devises*, where arrangements of lenses and mirrors for magnification of distant objects are treated. (See Vol. I, p. 161.)

90. BOVILLUS, CAROLUS (syn. de Bouelles, C.) *Physicorum elementorum . . . libri decem. . . .*

Io. Parvus and Iodocus Badius Ascensius (Col: Paris 1512)

Ed. p.

Prov. Bp. Leslie to Andrew Strachan, who gave it to King's College 1635.

Charles Bouelles was one of the 'Erasmian' humanists, which included Jacques Lefèvre d'Etaples (No. 381) and Josse Clichtove of Nieuport, who worked unsuccessfully for radical reform within the Roman Catholic Church. The printer mainly responsible for publishing their work was Henri Estienne, but Badius lectured on Latin literature and published works for the Faculty of Theology: he was also the first to use Greek letter-type founts. His daughter, Petronilla, an accomplished latinist, married Henri's son, Robert Estienne, in whose cosmopolitan household the scholarly tradition was conveniently maintained by making Latin the medium of domestic discourse.

This happy relationship between printers, Faculty, and individual scholars was brought to an end by the realisation by conservative elements of the power of the press, and in particular by the fanatical bigotry of Noel Becher ('Beda'), Syndic of the Sorbonne: hence the rapid decline of the pre-eminence of Paris printers and the complementary rise in Geneva, Lyon, Leiden, Frankfurt. (Putnam *op. cit.*)

No. 90 is a valuable historical document in that it presents the pietistic natural philosophy derived from the Thomist adaptation of Aristotle. Numerous cuts, including one on the t.p., illustrating work in the office of Badius.

91. BOVILLUS, C. *Géometrie pratique composée par le noble philosophe maistre Charles de Bouelles & novellement par luy reveue, augmentée & grandement enrichie.*

Guil. Cavellat, Paris, 1555

[*Ed. p.* Paris 1547.] but dedn. dated 'Novioduni 1542'.

The later chapters on harmony of bells, disposition of the elements, etc., have no apparent connection with geometry—least of all *practical* geometry. But his normal Aristotelianism seems to be here tinged with Pythagorean number-mysticism.

92. BRADWARDINE, THOMAS. *Geometria speculativa Thome Bravardini . . . recoligens* [sic] *omnes conclusiones geometricas studentibus artium & philosophie Aristoteles valde necessarias simul cum quodam tractatu de quadratura circuli noviter edito.*

Jehan Petit (Col: Paris 1511)

Sig. A ii *r* Heading [*Breve compendium artis geometrie a Thoma bravardini ex libris Euclidis Boecii & Campani peroptime compilatus et dividitur in quattuor tractatus. . . .*]

Incipit: 'Geometria est arithmetice consecutiva . . .'. ThK 277 14th cent.

Explicit: 'Et sic explicit Geometria Thome brevardini cum tractaculo de quadratura circuli bene revisa a Petro Sanchez Cirvelo . . .'.

[*Ed. p.* Paris 1495, GW 5002.] For P. Cirvelo see No. 160.

93. BRAHE, TYCHO. The works of Tycho Brahe were published in such peculiar circumstances as to require a few words of introduction. His first epoch-making work *De nova et nullius aevi memoria prius visa stella iam pridem anno a nato Christo* 1572, *mense Novembri primum conspecta, contemplatio mathematica* was published, together with a number of short tracts, at Copenhagen 1573. The edition was so small that it is now known in only a very few copies and is represented at Aberdeen only in the facsimile published at Copenhagen in 1901. This first work was followed by an ambitious project to refound the science of astronomy on new principles: two volumes only appeared, the second with a title giving no indication that it was a continuation of the first except the words 'Liber Secundus'. To make matters worse the first volume bears throughout the running-title *De nova stella anni* 1572, though these words appear only as an addition on the t.p. Tycho's remaining work was to consist of all his astronomical correspondence, of which only one volume appeared. Lastly, the circumstances of publication were further complicated by the fact that all these works were first set up at the author's private press at Uraniborg on the island of Hven, partly distributed therefrom in private circulation, and later, the printing having been completed at Prague, were regularly published, first at Nürnberg, later at Frankfurt. The details will appear more precise as the various works are passed in review.

> *Tychonis Brahe Dani Astronomiae Instauratae Progymnasmata, quorum haec prima pars de restitutione motuum solis & lunae stellarumque inerrantium. Et praeterea de admiranda nova stella anno* 1572 *exorta luculenter agit. Excudi primum coepta Uranibergi Daniae ast Pragae Bohemiae absoluta tractat.* . . .

Godefridus Tampachius, Frankfurt 1610 [Not in '(*a*)']

(*a*) This copy lacks sigs. x i-ii and (:) i-iv, beginning at B *r Caput Primum*. It lacks also everything after p. 786 (catchword 'Conclu—') which includes Tycho's *Conclusio* and Kepler's *Appendix*. It is therefore reasonable to suppose that it was one of the copies privately circulated from Hven.

Prov. LIDDEL. There is evidence that he visited Hven on June 24, 1587.

(*b*) Title as above. Dreyer states that it differs from the first regular edn. (Prague 1602) only in having new t.p. and a re-setting of the first 16 pp. Dedn. (by Tycho's heirs) to the Emperor Rudolph II dated Prague, Aug. 1602. It appears that Tampachius bought most of the sheets of the 1500 copies which are known to have been ordered.

Prov. ANDREW YOUNGSON, and another copy.

The prelims. include a *Privilegium regis Scotorum* (James VI) dated '*nostra Sancruciana* [Holyroodhouse] 1583' which confirms the statement in the Appendix (written in fact by Kepler, though ostensibly by Tycho's heirs) that the work was completed by 1592. An unusual feature is a *carmen* by James VI in honour of Tycho.

94. (*a*) *T— B— D— de Mundi aetherei recentioribus phaenomenis liber secundus. Qui est de illustri stella caudata ab elapso fere triente Novembris Anni* 1577 *usque in finem Januarii sequentis conspecta.*

43

Uraniburgi cum privilegio. Col: *Uraniburge. In insula Hellesponti Danici Hvenna imprimebat authoris typographus Christophorus Uveida. Anno Domini MDLXXXVIII.*

There is no evidence that this 'edn.' was ever 'published', but it is known that several copies were distributed to learned correspondents. As the t.p. of the *Epistolarum Astronomicarum* . . . (with which it is bound in a contemporary vellum music score) and its own colophon-page bear Liddel's autograph, it is reasonable to suppose that he was one of the recipients. Page 189 shows the famous Tychonic system. (See Vol. I, p. 127.)

(*b*) *T— B— D— de Mundi aetherei recentioribus phaenomenis, liber Secundus. . . . Excudi primum coeptus Uraniburgi Daniae, ast Pragae Bohemiae absolutus.*

G. Tampachius, Frankfurt 1610

There are added to the original 'edn.' a new t.p.; a dedn. to Ioannes Barutius of the Imperial Privy Council, signed by Franciscus Tengnagel (Tycho's son-in-law) 'ex Musaeo Uranico Pragae' 1603; and a *Lectori.* The *Prooemium* was also re-set, the original sheets beginning on sig. A *r*.

Two copies, one of which ('d.d. Andrew Youngson 1646') has the original 'Uraniborg' col. (see No. 94a), the other, 'Pragae Bohemorum absolvebatur typis Schumanianis' . . . 1603.

95. *T— B— D— Epistolarum astronomicarum libri. Quorum primus hic illustriss: et laudatiss: principis Gulielmi Hassiae Landtgravii ac ipsius mathematici literas, unaque responsa ad singulas complectitur.* (Col: *Uraniborgi ex officina typographica authoris—* MDXCVI.)

 (*a*) Prov. LIDDEL.

 (*b*) Second issue, with a new t.p. having a portrait of Tycho *in verso*, by Levinus Hulsius, Nürnberg 1601.

 (*c*) Third issue, with a different t.p., lacking portrait, by Tampachius, Frankfurt 1610.

All have the same col. There were no further volumes. Full of valuable historical material relating to, e.g. John Craig (the 'protector' of Liddel at Frankfurt) who attacked Tycho's theory of comets; 'my singular good friend, John Dee', and Tycho's alchemical views.

96. BRANCALEONI, GIOVANNI FRANCESCO. *De balneorum utilitate . . . dialogus ex Hippocrate, Galeno, caeterisque medicorum peritissimis decerptus & nunc primum aeditus.*

Christ. Wechel, Paris 1536

Ed. p.

PLATE I

URANIBVRGI

Jn Infula Helleſponti Danici Hvenna imprimebat
Autheris Typographus Chriſtophorus Uveida.

ANNO DOMINI.
M. D. LXXXVIII.

Colophon of the privately circulated Uraniborg 'edition' of Tycho Brahe's *De Mundi aetherii recentioribus phaenomenis* (No. 94a). Duncan Liddel's autograph.

PLATE II

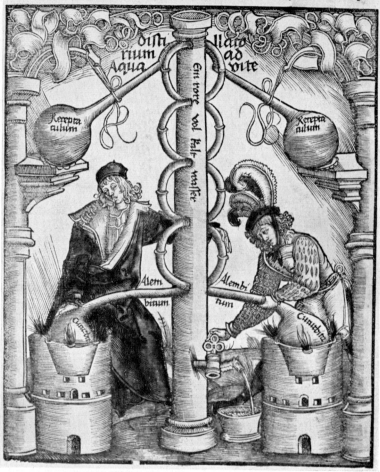

Title-page of Hieronymus Braunschweig's treatise on distillation, showing air- and water-cooled stillhead (No. 101).

97. BRASAVOLA, ANTONIO MUSA. *Antonii Musae Brasavolae in octo libros Aphorismorum Hippocratis . . . & Galeni commentaria & annotationes.*

H. Froben and Episcopius, Basel 1541

Ed. p.

Prov. GORDON.

98. BRASAVOLA, A. M. *Examen omnium loch, id est linctum suffuf, id est pulverum, aquarum, decoctionum, oleorum, quorum apud Ferrarienses pharmacopolas usus est, ubi de Morbo Gallico . . . tractatur.*

Heirs of Lucantonius Junta, Venice 1553

Ed. p.

For Brasavola's part in the reform of pharmacology, see Vol. I, p. 245. The words 'loch' and 'suffuf' are corruptions of those used by the Arab physicians. The name 'Musa' was honorific, having been conferred on Antonio Brasavola by Francis I to mark a comparison with Antonius Musa, the famous physician to the Emperor Augustus. The text contains (fo. 294*v*) the significant statement 'Morbus Gallicus est in declinatione universali . . .'.

99. BRASAVOLA, A. M. *Examen omnium simplicium quorum usus in publicis est officinis. . . . Ab ipso authore recognitum & auctum.*

Anton Vincentius, Lyon 1556

[Ed. p. Lyon 1537]

Prov. BUTE.

The text takes the form of a dialogue between *Senex pharmacopola*, 'Herbarius', and Brasavola starting on a walk 'through these inhospitable and inaccessible alps and through such horrifying declivities hardly giving a passage to the timid hare and goats . . .'.

100. BRASAVOLA, HIERONYMO. *De officiis medicis libellus.*

Benedictus Mamarellus, Ferrara 1590

Earliest in BM is 1599.

Hieronymo was the son of Antonio.

101. BRAUNSCHWEIG, HIERONYMUS. *Liber de arte distillandi de compositis* [followed by an expanded title in German]. *Das Buch der waren Kunst zu distillieren die Composita und Simplicia und das Buch Thesaurus Pauperum . . . von den Buchern & Artznei und durch Experiment von mir Heronimo Brunschwich.*

[Joh. Gruninger?] Strassburg 1512

Ed. p. of the work in this form: the earlier work (1500, from the same House) had the title *Liber de arte distillandi de simplicibus* and being consequently on a much smaller scale is often referred to as *Das kleine Distillierbuch.*

Numerous fine cuts (hand coloured in this copy), many of which are frequently repeated. The t.p. shows distillation of 'aqua vite' with cold-water-cased still-head.

102. BR(A)UNSCHWEIG, H. *Haus Artzneibüchlein.*

Nicholaus Bansee (Col. Frankfurt 1580)

Prov. LIDDEL.

The whole title is far too long to quote—as is often the case in books published in German in the 16th cent.—but it gives an indication of its nature and use both as a 'family doctor' and a 'Tractätlin von allerley Gebranten Wassern . . .' 'an vilen orten gemehrt und gebessert . . .'. This appears to be a late edn. of the (1st ?) edn. of 1529 (BM).

'Gebranten' presumably means *separated* by fire, i.e. distilled. The work is Galenical and astrological in character.

103. BRESSIUS, MAURITIUS. *Metrices astronomicae libri quatuor. Haec maximam partem nova est rerum astronomicarum & geographicarum per plana sphericaque triangula dimensionis ratio, veterique impendio expeditior & compendiosior.*

Aegidius Gorbinus (Col. Excudebat Petrus le Voirries in Mathematicis Typographus Regius.) Paris, 1581.

Two copies.

A practical manual containing tables of sines, etc., and sexagesimal arithmetic. Bressius describes himself as 'Regius et Rameus Mathematicarum Lutetiae Professor'.

104. BRIGHT, TIMOTHY. *Hygieina, id est de sanitate tuenda medicinae pars prima . . . cui accesserunt de studiosorum sanitate li. iii Marsilii Ficini.* [The latter work is wanting in No. 104.]

(*a*) Io. Wechel, Frankfurt 1588.

(*b*) Off. Paltheniana, heirs of Petrus Fischer and Io. Rhodius, Frankfurt 1598.

[*Ed. p.* London 1581.]

Prov. LIDDEL, and a second copy.

Both edns. have same dedn. to William Cecil, Lord Burghley, in which the author refers to the Second Part—*Therapeutica*—in the press. The Note to the Reader refers to the 'fruitful mode of philosophising' of Petrus Ramus (No. 546) and to the latter's tragic death on St. Bartholomew's Eve. Ramus (de la Ramée) was of course noted for his strongly critical attitude to the Aristotelian philosophy, so it is rather surprising to find that Bright's text is based on the time-worn Aristotelian classification of dry—hot—wet—cold (but see No. 105). Bright, physician to St. Bartholomew's hospital, London, is credited by *D.N.B.* with re-inventing shorthand.

105. BRIGHT, T. *Therapeutica hoc est de sanitate restituenda, medicinae pars altera.*

[same imprint & dates as No. 104]

Prov. LIDDEL, bound with No. 104; and a second copy.

The dedn. to Burghley—'praeclaro ordinis, quem vocant, Garterii militi'—is dated 1582. It contains the significant admission: 'In this *Therapeutica*, just as in the *Hygieina*, we are compelled to dissent from authors of high repute.'

106. BRUELE, GUALTHERUS. (*a*) *Praxis Medicinae Theorica et Empirica . . . in qua . . . morborum internorum cognitio, eorundemque curatio traditur. Thesarus Innocentia.*
Earliest BM same imprint, 1585.

Chris. Plantin, Antwerp 1579

Prov. LIDDEL; and a second copy. End-paper inscribed 'In arce Helsingburgensi Anno 1580 ix Julii'. Part of binding paper is from Erastus . . . *de Lamiis* . . . Basel 1579 [?]. Cover embossed 'I.F.R. 1579'.

(*b*) Off. Plant. Christ. Raphelengius, Leiden 1599

Dedn. to Earl of Huntingdon includes a reference to his having 'kept himself many months in the labyrinth' of the works of Paracelsus hoping to gain something from 'writings, by many exalted above the stars'; but in his forthcoming *De Corollariis philosophicis* he intends to show up the 'new hypotheses' of Paracelsus, Lull, and Geber as not worth the trouble of reading. He recalls also 'the memory *ignitarum impressionum* [aurora?] which appeared in the northern heavens about the ninth hour of the night on the first day of St. Michael 1575'.

107. BRUELE, G. *Methodus tum pro cognitione, tum pro curatione Pestis inserviens.*

Jac. Roussin, Lyon 1586

Ed. p.? Not BM.
Prov. BUTE.

In his dedn. he explained how as a result of the 'intestine wars' and disturbed state of his country, Belgium, he had had to content himself with publication of his views on the plague instead of a collection of his works. A sober, but not very original, account with a bias towards the *contagion* theory of origin. The following may be worth noting . . . 'prius tamen scarificationes fient profundiusculae, ut sanguis, venenatus eliciatur utque *seminaria* mali plenius superentur' [*ital.* mine].

47

108. CLAVIUS, CHRISTOPHORUS. *Gnomonices libri octo, in quibus non solum horologiorum solarium sed aliarum quoque rerum quae ex gnomonis umbra cognosci possunt, descriptiones geometrice demonstrantur.*

Franciscus Zanettus, Rome 1581

109. BRUNFELS, OTTO. *Catalogus illustrium medicorum sive de primis medicinae scriptoribus per Ot. Brunfelsium.*

Io. Schott, Strassburg 1530 ('xix December 1529' after 'Errata') Only edn. Bibl. Nat. and BM

Prov. READ—bound with several other works glossed and scored in the red ink and characteristic clear hand.

Index of proper names starting with Apollo and Aesculapius! But fairly well documented.

110. BRUNFELS, O. *Iatrion medicamentorum simplicium continens remedia omnium morborum quae tam hominibus quam pecudibus accidere possunt . . . digestum in libros quatuor.*

No imprint or col. Strassburg 1533 from dedn. Only edn. BM which gives G. Ulricher as printer with no indication of lack of imprint.

In the dedn. to Laurentius Schenckbecher, Provost of St. Thomas's Church, Strassburg, Brunfels refers to 'Herbario meo excolendo' (No. 111).

A collection from various authors of all periods—unusual in combining cures for men and beasts. Bound in three vols.

111. BRUNFELS, O. *Kreüterbuch contrafayt vollkummen nach rechter warer beschreibung der alten leeren und ärtzt. Sampt einer gemeynen inleytung der kreüter urhab erkantnusz brauch lob und herrlicheit. . . .*

Hans Schotten, Strasszburg [*sic*] 1534

This is the second German edn. of the famous *Herbarum Vivae Eicones* (Arber, p. 274, and Vol. I, p. 186).

In this 'pocket edn.' they are much reduced as compared with the lovely copies in the folio *Ed. p.* 1532. (BM copy examined.)

A comparison of the *Christwurz* (sig. e ii *r*) with the photograph of the cut in the *Ed. p.* revealed a detailed correspondence, but a lateral inversion. (See Vol. I, p. 26.)

112. BRUNFELS, O. *ONOMAΣTIKON Medicinae, continens omnia herbarum, fruticum . . . lapidum preciosorum, metallorum, . . . definitionum medicinalium, instrumentorum medicinae, unguentorum, . . . clysterium, . . . morborum pecudum, animantium omnis generis nomina propria . . . anatomiae . . . philosophiae naturalis,*

48

magiae, achimiae [*sic*] & *astrologiae ex optimis* . . . *vetustissimis autoribus, cum Graecis, tum Latinis. Opus recens, nuper multa lectione* OTHONIS *Brunfelsii* . . . *congestum in gratiam eorum qui se priscae* MEDICINAE *dediderunt.*
Io. Schott, Strassburg, 1534
Ed. p.
Prov. GORDON.
The first 'medical dictionary'—an invaluable storehouse of facts—and fancies.
See also, No. 268.
BURIDAN, JEAN. See No. 405.

113. BUTEO, IOANNES (*syn.* Jean Borrel). . . . *Logistica, quae & Arithmetica vulgo dicitur, in libros quinque digesta* . . . *Eiusdem ad locum Vitruvii corruptum restitutio, qui est de proportione lapidum mittendorum ad balistae foramen, libro decimo.*
Gul. Rovillius, Lyon 1560
[*Ed. p.* Lyon 1559.]
Prov. JOHNSTON.
Interesting in relation to the period when, under the stimulus of the 'commercial' works, arithmetic was moving from its classical Euclidean-Boëthean form to become an instrument of calculation. *Liber tertius* deals with algebraic problems by means of a kind of geometrical 'false position'. The section on Vitruvius is a brief commentary on *tormenta* for hurling stones. (Smith, p. 292.)

114. BUTEO, IO. *Opera Geometrica.*
Michaelis Iovius, Lyon 1559
[*Ed. p.* Lyon 1554.]
Prov. 'Io. Gregory 1769' and two earlier owners.
Numerous cuts. Includes some geometrical curiosities, e.g. the dimensions of Noah's Ark and a discussion of squaring the circle.

115. BUTINUS, IOHANNES *Hippocratis Aphorismi digesti* . . . *cum brevi & dilucida expositione singulis Aphorismis ex Galeno supposita. Eiusdem Hippocratis Praenotionum* . . . *libri tres cum explanatione eodem ex fonte hausta.*
(a) Io. Tornaesius & Gul. Gazeius, Lyon 1555. *Ed. p.*
(b) Io. Tornaesius, Lyon 1580
Prov. (a) LIDDEL.

116. CAESALPINUS, ANDREAS. *De Plantis libri xvi.* . . .
Georgius Morescottus, Florence 1583 [Col: 1581 permission]
Ed. p.
Prov. FORDYCE, previously owned by his brother, David, 1749.

In his dedn. to Francisco de'Medici the author refers to his work at the Pisan Academy and to the numerous gardens where foreign plants sent from the most distant parts are kept for the public use. Book I contains the first *biological* introduction to plant study (with the possible exception of Albertus Magnus) since Theophrastos.

117. CAESALPINUS, A   *Quaestionum peripateticarum lib. v.*
   [*Ed. p.* Venice 1571.]
   Daemonum investigatio peripatetica . . ., secunda editio [*Ed. p.*
      Florence 1580].
   Quaestionum medicarum libri ii.
   De medicament. facultatibus lib. ii . . . nunc primum editi.
                                               Juntae, Venice 1593
      Prov. Two copies: 'Jo. Gregory 1772 1··6', and Bute.
   The name of the first work indicates that Caesalpinus was a prominent Aristotelian—in many ways an enlightened one, whose point of view probably influenced such eminent 'reformers' as Harvey and Ray. See Vol. **I**, p. 198.

118. CAESALPINUS, A. *De Metallicis li. iii.*
   (t.p. Recusi curante Conrad Agricola). (Col. . . . ex officina
      Katharinae Viduae Alexandri Theodorici, Nürnberg 1602.)
   [*Ed. p.* Rome 1596.]
   Prov. LIDDEL (?)
   Ph. Scherbius, in his dedn., justifies the new edn. on the grounds that no copies were to be found in Germany and more often than not they were sought in vain in Rome.
   The term 'metals' embraced 'fossils' of all kinds. On p. 27 Caesalpinus states that from a certain clay 'they make shells . . . and the goldsmiths little vessels called crucibles (*crucibula*). . . .'

119. CAIUS, IOHANNES. . . . *Opera aliquot.* . . .
                                     A. M. Bergagno, Louvain, 1556
   Contents:
   1. *De medendi methodo libri duo ab ipso authore recogniti atque aucti.*
      [*Ed. p.* Basel 1544.]
   2. *De Ephemera Britannica. Liber unus iam primum excusus.* [Engl.
      1552.]
   3. *Galeni de libris suis li.i; de ordine librorum suorum li.i; de ratione*
      victus secundum Hippocratem in morbis acutis li.i non ante aeditus.
   4. *De Placitis Hippocratis & Platonis li. primus, eodem Iohanne*
      Caio interprete.

50

Wants pp. 233-92 incl.—the first two tracts on Galen. Cont. pagn. throughout, but half-title for (2) and sep. dedn. Section-headings only for remainder. Frontispiece portrait.
Prov. GREGORY.
(2) is the classical personal observation of the 'English sweats'. See also No. 459 [Montanus *Opuscula Varia*] and Vol. I, p. 222.
CALZOLARI, FRANCISCO. See No. 434.

120. CAMERARIUS, ELIAS. *Observatio et descriptio novi sideris quod in principio Octobris Anno Christi 1572 forma stellae primae magnitudinis apparuit, & per totius anni curriculum fulsit, etiam nunc in tertia magnitudine splendet. In qua observatione demonstratur novum hoc sidus non in elementari sed in aetherea regione existere. . . .*
Frankfurt on Oder 1573
Ed. p.
Prov. LIDDEL.
One of several contemporary accounts of the famous 'supernova' of 1572, first observed by Tycho Brahe (Vol. I, p. 122 and No. 93). An account of the quadrants used is included. Note that Camerarius claims that the star appeared at the beginning of October instead of November.

CAMERARIUS, IOACHIMIUS [The Younger]. See Plague Tracts, No. 755.

121. CAMUTIUS, ANDREAS. *Disputationes quibus Hieronymi Cardani magni nominis viri conclusiones infirmantur, Galenus ab eiusdem iniuria vindicatur, Hippocratis praeterea aliquot loca diligentius multo, quam unquam alias, explicantur.*
Hiero. Bartolus, Pavia 1563
Ed. p.
The text provides a list of Cardano's annotations (signed by him 1561) on Galen's exposition of the *Aphorisms* of Hippokrates. This is followed by a long dialogue between Cardano and Camutius on the same, and a shorter one on the question of the existence of the (Aristotelian) 'sphere of fire'.

122. CAMUTIUS, ANDREAS. . . . *excusio brevis praecipui morbi, nempe cordis palpitationis Maximiliani secundi Caesaris simul ac aliorum aliquot virorum illustrium praeter naturam affectuum.*
Georgius Marescotus, Florence 1578
Only edn.
Prov. BUTE. Front-paper inscribed 'Londini 1655' and in same hand 'e Bibliotheca Io. Riolani'.

51

Consilia concerning the palpitation (and numerous other signs and symptoms) of the Emperor Maximilian II; also correspondence with Julius Alexandrinus (No. 24).

123. CAPIVACCIUS, HIERONYMUS. *Nova Methodus Medendi.*
Io. Feyrabend(t), Frankfurt 1593
Neither BM nor Bibl. Nat. record this edn. Earliest is the folio of 1606.
The text is based on thirty-four public lectures given in the University of Padua.

124. CAPIVACCIUS, H. *Practica Medicina . . . publice olim proposita. Nunc vero recens elimata, librisque septem & capitibus interstincta. Studio et opera Johannis Hartmanni Beyeri. . . .*
Off. Paltheniana, Frankfurt 1594
Ed. p.
Prov. LIDDEL.

125. CAPIVACCIUS, H. *. . . de Urinis tractatus. Nunc primum plane recens ex Bibliotheca D. Laurentii Scholzii Medici Vratisl. in lucem prodiens.*
Bonaventura Faber, Szervestyë 1595
Ed. p.
Henricus Osthausen in a note to the reader explains that Scholz (No. 630) had undertaken the editing of his work at his own and other requests. See also No. 424.

126. CARDANO, HIERONYMO (*syn.* Girolamo) *. . . Libelli duo. Unus de supplemento almanach. Alter, de Restitutione Temporum et Motuum Coelestium. Item Geniturae lxvii insignes casibus & fortuna, cum expositione.*
J. Petreius, Nürnberg 1543
The dedn. suggests that the work had been published before. Cardano himself (*De Vita Propria*) alludes to this fact, but BM cites this earliest edn. (I.A. Castellioneus, Milan 1538) as if it were a separate work.

There is a fine portrait on t.p. Following the *Geniturae* are two short articles, *Cognitio qualitatum aeris* and *Encomium Astrologiae*. The *Geniturae* is of course a *locus classicus* of the 'art' of constructing a 'nativity'. The first two books are 'straight' texts of astronomy, with special reference to chronology.

127. CARDANO, H. *Contradicentium Medicorum liber continens contradictiones centum octo.*

(*a*) Hieronymus Scotus, Venice 1545. Earliest cited in BM is Marburg, 1607. Bibl. Nat. earliest is (b) below.

Prov. LIDDEL.

(*b*) *Contradicentium Medicorum libri duo, quorum primus centum octo, alter vero totidem disputationes continet. Addita praeterea eiusdem autoris de Sarza Parilia, de Cina Radice, eiusque usu. Consilium pro dolore vago, Disputationes etiam quaedam aliae. Accesserunt praeterea Iacobis Peltarii* [sic] *contradictiones ex Lacuna desumptae cum eiusdem axiomatibus.*

Iacobus Macaeus, Paris 1564

Prov. 'Richardus Foster'.

The *alter liber* has sep. t.p., dedn. (to Cosimo de'Medici, Pavia, 1547) and pagn. The *addita* have cont. pagn. with secn. headings. The work of Peletarius (No. 505) has sep. t.p.—*De Conciliatione Locorum Galeni sectiones duae*—with dedn. 1564. The *alter liber* bears the date 1565; Bibl. Nat. dates publication 1564-5.

The list of 'Contradictions' gives a good insight into Cardano's versatile and acute mind.

128. CARDANO, H. *De Subtilitate libri xxi.*

(*a*) J. Petreius, Nürnberg 1550.

Ed. p.

(*b*) Philibertus Rolletius, Lyon 1554 (. . . *ab ipso autore recogniti atque perfecti*).

(*c*) Officina Petrina, Basel (Col. 1560) (*Ab authore illustrati . . . cum additionibus*).

(*d*) Barth. Honoratus, Lyon 1580 (*Ab ipsa authoris recognitione. . .*).

Prov. (*a*) Read.

The t.p. of (*c*) has 'Addita insuper Apologia adversus calumniatorem, qua vis horum librorum aperitur'.

His views on heat have a 'modern' character, especially 'Frigor nihil est actu, sed sola caloris privatio' (p. 41 and p. 93 in *a*).

129. CARDANO, H. *De Rerum Varietate libri xvii.*

(*a*) Henricus Petrus, Basel 1557.

Ed. p.

(*b*) Mattheus Vincentius, Avignon 1558.

(*c*) B. Honoratus, Lyon 1580.

Prov. (*c*) 'Jacobus Cargillus'—End-papers have inscription by 'Thos. Cargill, Aberdonensis 1635'.

The 'variety' includes marvels as well as facts; but the necessity for caution in rejecting these stories is shown by the record (p. 417) of a change of sex, which might have been doubted by some of the more sceptical of his contemporaries, but of which there are now, of course, numerous well authenticated cases. (Cf. No. 522.)

130. CARDANO, H. *Somniorum Synesiorum, omnis generis insomnia explicantes libri iiii. . . . Quibus accedunt. . . .*[see below].
Henricus Petrus (Col. Basel 1562)

Ed. p.

The 'additions' (listed on t.p.) comprise ten shorter works, of which *De Libris Propriis* is of special importance, since it not only lists all (?) his writings but gives information concerning the order and circumstances in which they were written. His rambling style, however, makes the search for any particular item a laborious affair. Pagn. and titling vary in respect of these additions.

131. CARDANO, H. *Tomus I . . . Ars Curandi Parva, quae est absolutiss. medendi methodus, & alia, nunc primum aedita opera in duos tomos divisa. . . .*
Henricpetrus, Basel, n.d.

Ed. p. BM and Bibl. Nat. Col. of Tomus II is dated 1566 (seen at BM).

Prov. LIDDEL.

The *Ars Curandi parva* ends on p. 303 and is followed (cont. pagn.) by *Consilium pro reverendissimo D.D. Ioanne Archiepiscopo Sancti Andreae Ecclesiae Cathedralis Regni Scotiae*, laborante difficultate spirandi. There follows (p. 470) *Ephemeris . . .* written by Cardano to Archbishop Hamilton from Edinburgh, 1552, and (p. 498) a letter of thanks from the archbishop reporting (by 1544) 'vita quodammodo recuperata', due doubtless to the sound sense mixed with much dubious physiology in Cardano's virtual 'treatise' on asthma. The remainder of *Tomus I* contains a further *consilium* on asthma and several on various complaints and concludes with an *apologia* in respect of Cardano's criticism of Galen (see also dedn.). A few pages are wrongly numbered.

Tomus II (seen at BM) proceeds without break of pagination and is likewise made up of numerous separate pieces. Contents and index of both volumes at the beginning of Tomus I.

132. CARDANO, H. *. . . de Methodo Medendi sectiones quatuor. . . .*
G. Rovillius, Paris 1565

Only edn. Bibl. Nat. Not BM.

The dedn. is dated 'MDXLV'—presumably a mistake for 'MDLXV', since on p. 296 there is a ref. to the case of Abp. Hamilton, whom Cardano treated in 1552.

Hippocratic and Galenic in tone, warning (p. 4) against the errors of those who have followed the Arabs. The importance of individual diets is stressed: where there is weakness of the stomach and a person bolts (*devorat*) his food instead of chewing it, he had much better refrain from meat and dry foods and make use of concentrated (*spissae*) meat extract ('meliora sunt contusa ex carnibus quam carnes ipsae'). This treatment (he asserts) certainly greatly benefited the Archbishop of St. Andrews (pp. 295-6).

133. CARDANO, H. . . . *Commentarii in Hippocratis de Aere, Aquis et Locis opus, quo nullum, teste Galeno, omnibusque medicis & philoso-phicis, praestantius extat, in quo, sine dubio, sicut Hippocrates autor, ita Cardanus, commentator seipsum superat. In cviii lectiones divisum, quae medicis, philosophis, geographis, astrologis, aliisque facultatibus plurimum profuturae sunt. Poterunt omnes domi Car-danum incomparabilem professorem audire sine peregrinationis molestia & sumptibus, non minori cum fructu, quam si Bononiae essent. Accedunt praeterea ante quoque non impressa* . . . .[See below].

Off. Henricpetrina, Basel n.d.

BM cites 1570 as *Ed. p.*—the dedn. is undated, but there is a reference in the text to 'this year (1568)'.

This work comprises the text of Hippocrates divided into short sections with long commentaries. Of special interest is that on (Hippo-crates's) statement 'oportet autem & astrorum exortus considerare; praecipue canis deinde Arcturi & Pleiadum occasum'. ('We ought to take into account also the risings of the stars: especially the setting of the Dog, and then of Arcturus & of the Pleiades'.) This is a *locus classicus* for the importance of a knowledge of astronomy to the physician (p. 137).

The 'additions' include some *Consilia*, an *Oratio*, and *De Fulgure* by his son, Gianbattista.

134. CARDANO, H. *Opus novum de Proportionibus Numerorum motuum ponderum, sonorum, aliarumque rerum mensurandarum, non solum geometrico more stabilitum sed etiam variis experimentis & obser-vationibus rerum in natura, solerti [sic] demonstratione illustratum, ad multiplices usus accommodatum, et in V libros digestum. Praeterea Artis Magnae, sive de Regulis Algebraicis, liber unus, abstrusissimus & inexhaustus plane totius arithmeticae thesaurus, ab authore recens multis in locis recognitus et auctus. Item de aliza regula liber, hoc est algebraicae logisticae suae, numeros recondita numerandi subtilitate secundum geometricas quantitates inquirentis, necessaria Coronis, nunc demum . . . edita.*

Col. Henricpetrus, Basel 1570

55

[The *Ed. p.* of the *Ars Magna* was published by Petreius at Nürnberg 1545 (the dedn. to Andrea Osiander in No. 134 is of that year.] The remainder are *Edd. p.*
There is a duplicate of No. 134 bound with No. 135.
Ars Magna has sep. t.p. and pagn.

135. CARDANO, H. *Opus novum . . . de Sanitate Tuenda, ac vita producenda. . . . in quattuor libros digestum. A Rodolphio Sylvestrio . . . eius discipulo recens . . . editum.*

Seb. Henricpetrus, Basel (Col. 1582)

There are two copies of this work, in one of which the colophon is damaged. The date on the other agrees with the first entry BM. The work is largely concerned with diet and regimen, especially in old age. See Vol. I, p. 29 f., and (*a*) Henry Morley *Life of Girolamo Cardano*, 2 vols., London 1854, (*b*) *The Book of My Life*—trans. and ed. Jean Stoner, London 1931, (*c*) Oystein Ore, *Cardano, The Gambling Scholar*, Princeton 1953.

136. CARRICHTER, BARTHOLOMAEUS. *Practica, aus den fürnemesten Secretis, weiland des Edelen unnd hochgelehrten Herren Bartholomei Carrichters . . . Von allerhand Leibskranckheyten. Von Ursprung der offenen Schäden und ihrer Heylung.*

Antonius Bertram, Strassburg 1590

[*Ed. p.* Strassburg 1575]
Prov. LIDDEL.
Ed. M. Toxites from Hagenau 1574.

137. CARRICHTER, B. *Kräutterbuch des ehrnuesten [ehrneusten?] und hochgelehrten Herren D— B— C— darinnen begriffen under welchem zeichen Zodiaci auch in welchem gradu ein jedes Kraut stehe wie sie in Leib unnd zu allen Schäden zu bereyten und zu welcher zeit sie zu colligieren sein.*

Anton. Bertram, Strassburg 1589 (Col. 1588)

[*Ed. p.* Strassburg 1575].
Dedn. by M. Toxites (the editor).
An astrological 'materia medica' arranged according to the grades of the signs. Ref. to Paracelsus on sig. Aiiii.

138. CASMANNUS, OTHO *Astrologia, Chronographia et Astromanteia, seu commentationum disceptationumque physicarum syndromus methodicus et problematicus iii. De stellarum natura, affectione, motibus & effectibus. Ex Dei Verbo, philosophis et astrologis, cum veteribus, tum praecipue recentioribus . . . concinnatus.*

Zach. Palthenius, Frankfurt 1599

Pagn. in *Pars. II* proceeds from '99' to '200' and continues to '244', then reverts to '145'. On p. '238' *seq.* it is proved by 'reasoning & authority' that *astromanteia* is 'erroneous, false, vain & superstitious'— a list of authorities being provided:—Reminiscent of recent views on 'bourgeois-imperialist genetics'.

139. CASTRO, RODERICUS A, *Tractatus brevis de natura et causis Pestis, quae hoc anno MDXCVI Hamburgensem civitatem affligit.*

Jac. Lucius Junior, Hamburg 1596

Prov. 'D. Ioanni Caselio suae aetatis philosophorum facile principi mittit Georg . . . . Traje [ctinus].' Liddel (?).

Ded. to the Senate of Hamburg, it sets forth an account of the plague with a reasoned statement of the precautions necessary for preventing its spread and ridding the city of its present infestation. Hippocratic in tone and reference. (See Vol. I, p. 269.)

140. CELSUS, AURELIUS CORNELIUS (*a*) . . . *Medicinae li.* . . . *viii quam emendatissimi, graecis etiam omnibus dictionibus restitutis.* . . .

Aldus et Andr. Asulanus socer, Venice (Col. 1528)

[*Ed. p.* Florence 1478. GW 6456.] No. 140 is the *Ed. p.* of the Aldine Edn.

Prov. Two copies, one autographed by Sir Wm. Fordyce, the other Bute.

The *Liber de Medicina* of Q. Serenus Sammonicus is included.

(*b*) *de Re Medica libri octo.*

Gul. Rovillius, Lyon 1566

Annotations and corrections of R. Constantinus, physician to Queen Johanna of Navarre, mother of Henry IV. The poem of Sammonicus and the *Poema de Ponderibus & Mensuris* of Rhemnius [sic] are included.

(*c*) *de Re Medica libri octo.*

Off. Plantiniana—F. Raphelengius, Leiden 1592

Commentaries of H. T. Brachelius and B. Ronsseus (No. 583) added, also two letters of 'Celsus'.

Prov. BUTE.

See also No. 406.

141. CHALMETEUS, ANTONIUS. *Enchiridion Chirurgicum, externorum morborum remedia tum universalia tum particularia brevissime complectens. Quibus morbi venerei curandi methodus probatissima accessit.*

Andr. Wechel, Paris 1567

Ed. p. The dedn. is, however, dated 1560.

Prov. (i) 'Loys de la balle' (deleted), (ii) 'Robertus Forbasius Scotus'.

57

This work was the basis of numerous editions and translations departing in varying degrees from the original (see No. 142). John Banister, 'Maister in Chirurgerie', brought out in 1575 *A needeful, new, and necessarie treatise of chyrurgerie . . . drawen foorth of sundrie . . . wryters, but especially of A. Calmeteus Vergesatus etc.* The *Morbus Venereus* was re-published as late as 1728.

142. CHALMETEUS, A. *Enchiridion Chirurgicum praeter tutissimam Chalmetei Practicen brevem externorum affectuum theoriam complectens. Ad haec accessit Luis Venereae cum . . . . descriptione methodica curatio. Adiecimus praeterea ne quid in hoc compendio desideraretur concisos de febribus ex probatis authoribus tractatus quatuor.*

Ioan. Lertout, Lyon 1588

Prov. LIDDEL.

The *Lectori* suggests the note of a publisher who was also versed in medicine—'rogatus ab amicis ut aliquod Enchiridion in peregrinantium chirurgorum gratiam conscriberem . . .' he decided that Chalmeteus's work would meet the requirement.

143. CHAMPIER, SYMPHORIEN. . . . *Liber de quadruplici vita.* . . .

Stephanus Gueynardus & Jac. Huguetanus, Lyon 1507

M. P. Allut *Etude biographique et bibliographique sur Symphorien Champier,* Lyon 1859. No. vii, p. 148 *seq., Ed. p.*

The contents on t.p. are too numerous to quote, but the book consists of two main parts: the *Quadruplex Vita* proper, including a guide from the precepts of theology, philosophy, and medicine; and the *Tropheum Gallorum,* embodying a history of France and especially of the triumph of Louis XII at Genoa. On sig. a vi r is a cut of the martyrdom of St. Symphorien (repeated sig. h *v* etc.). Sig. G ii *v* has the device of Jannot Deschamps.

144. CHAMPIER, S. *Practica nova in medicina. Aggregatoris Lugdunensis domini Simphoriani Champeri de omnibus morborum generibus ex traditionibus Grecorum, Latinorum, Arabum, Penorum, ac recentium autorum, Aurei libri quinque. Item eiusdem aggregatoris liber unus de omnibus febrium generibus.*

(Col. Iohannes Marion, Lyon 1517)

Allut, No. xxiv, p. 198.

Ed. p.

The introductory encomium of the art of medicine is cast in the form of a 'history'.

145. CHAMPIER, S. *Rosa Gallica aggregatoris Lugdunensis domini Symphoriani Champerii omnibus sanitatem affectantibus utilis & necessaria. . . .*

Off. Ascensiana (Col. 1518) Paris
Venundatur ab Iodoco Badio (t.p.)

Prov. Bute.

Allut (No. xii, p. 167 *seq.*) is of the opinion that the *Ed. p.* was from the same House, Paris 1514, and that the edn. of 1512, Nancy, alleged by Brunet is apocryphal, the error being due perhaps to the author's note [fo. cxxxv M (5) *r*] that the *Margarita Pretiosa*, which follows the *Rosa*, was finished at Nancy. Allut, however, had never seen the Badius edn. of 1518, which Brunet reported.

The work is concerned with diet and regimen and based on numerous earlier authorities named on t.p.

146. CHAMPIER, S. *Hortus Gallicus pro Gallis in Gallia scriptus veruntamen non minus Italis, Germanis, & Hispanis quam Gallis necessarius, Symphoriano Campegio Equite aurato ac Lotharingorum Archiatro authore, in quo Gallos in Gallia omnium aegritudinum remedia reperire docet, nec medicaminibus egere peregrinis, quum deus & natura de necessariis unicuique regioni provideat.*

Followed by: *Campus Elysius Galliae amoenitate refertus* and *Periarchon, id est de principiis utriusque philosophiae.*

Melchior et Gaspar Trechsel Fratres, Lyon 1533

(Iodocus Badius learnt his craft with Trechsel, whose daughter he married.)

All have sep. t.p., pagn., panegyrics, and dedns.

Allut, No. xxxix, p. 245 *seq.*

Ed. p. The collation agrees with No. 146, except that only I 8 r is unnumbered.

See also No. 510.

147. CHESNECOPHERUS, NICOLAUS. *Isagoge optica cum disceptatione geometrica de universo geometriae magisterio, hoc est, Geodaesia, rectarum per radium & aliis quaestionibus philosophicis, iuxta auream P. Rami methodum concinnata et in inclyta Marpurgensi Academia pro insignibus magisterii philosophici consequendis publice ad disputandum proposita sub praesidio philosophi acutissimi, Dom. M. Rodolphi Goclenii. . . .*

Ioan Wechel, Frankfurt 1593

Ed. p.

Prov. Liddel.

Dedn. to Charles of Sweden in praise of the science of optics and repeating the story of Archimedes's use of mirrors ('urentia specula ex aere') for setting the Roman ships on fire. The text gives a very good idea of what a young Master of rather exceptional talents would know about optics—there are references to most of the leading Greek and medieval workers.

148. CHEYNEIUS, IACOBUS. . . . *De priore Astronomiae parte seu de Sphaera li. ii.*

Ioannes Bogardus, Douai 1576

Only edn.

James Cheyne was Professor of Mathematics and Philosophy in the newly founded University of Douai—one of the spearheads of the counter-reformation.

A lucid introduction to the pre-Copernican astronomy for young students to whom Ptolemy was, for various reasons, inaccessible. On p. 26 are tabulated his own observations (at Aberdeen) of the risings and settings of prominent stars.

149. CHEYNEIUS, I. *De Geographia li. ii. Accessit Gemmae Phrysii . . . de orbis divisione & insulis rebusque nuper inventis opusculum longe quam antehac castigatius.*

Lodovicus de Winde, Douai 1576

Only edn.

Commonly bound with No. 148. Gemma Frisius's work has separate t.p. and pagn. with a Note to the Reader dated Antwerp 1530.

These works mark a change from the medieval practice (e.g. Sacrobosco *De Sphaera*), in which no distinction was made between cosmography and geography.

150. CHEYNEIUS, I. *De Sphaerae seu globi coelestis fabrica, brevis praeceptio . . . Eiusdem principium aliquot sphaerae usuum.*

Ioannes Bogardus, Douai 1575

Only edn.

Commonly bound with No. 148 & No. 149 this contains instructions for the construction and use of a kind of armillary sphere (see Vol. I, p. 118).

Dedn. to his young pupils Nicolaus and Ioannes à Montmorency.

151. CHEYNEIUS, I. *Scholae duae, una de perfecto philosopho, altera de praedictionibus astrologorum, habitae Duaci anno salutis humanae 1576.*

Ioannes Bogardus, Douai 1577

Only edn.

Bound with No. 152. The second of these public lectures is of special importance in that it may be taken as an authoritative statement of a view of astrology acceptable to the Roman Catholic Church. It bears the *nihil obstat* of the theologian, Arnoldus Massius, in the words 'Hae duae orationes doctae sunt et catholicae, nec quicquam continent quod pias aures offendat'.

152. CHEYNEIUS, I. . . . *Analysis in xiiii libros Arist. de Prima seu Divina Philosophia, eiusdem in eosdem libros scholia.* . . .
(*a*) Io. Bogardus, Douai 1578.
(*b*) . . . *Analysis & scholia in Aristotelis xiv libros de Prima seu Divina Philosophia. Editio nova, emendata & notis aucta opera & studio Io. Rudolphi Lavateri Tiguri.*
Gul. Antonius, Hanover 1607

153. CHEYNEIUS, I. . . . *Succincta in Physiologiam Aristotelicam analysis.*
Aegidius Gorbinus, Paris 1580
Only edn.
Dedn. to 'D. Mariam Scotiae Reginam Gallieque Dotariam' dated from the College of St. Barbe, Paris (whither Cheyne went on leaving Douai) Sept. 1579. Mary Stuart had helped to endow a 'college' founded by Cheyne for training priests for the future re-conversion of Scotland.
For further details see W. P. D. Wightman 'James Cheyne of Arnage', *Aberdeen Univ. Rev.* **35** (1954), 369-83.

154. CHIOCCO, ANDREA. *Quaestionum philosophicarum et medicarum libri tres.*
Hiero. Discipulus, Verona 1593
Ed.p.
The *Lectori* contains an interesting passage bewailing the pitiable condition of earlier centuries when the best men wasted their time on mere dialectics or on the nature of time, motion, substance, etc. The author's own *Quaestiones* include the then vexed questions of *antiperistasis* and metallic remedies.
Bound with this work is a large fragment of L. Mercatus *De essentia . . . febris*, which includes (with cont. pagn.) Matthaeus Curtius (see No. 187), *De medendis febribus*, the last pp. of which are in MS.

155. CHRISTIANUS, ANDREA. *Enchiridion medicum de cognoscendis curandisque tam externis quam internis humani corporis morbis. Ex Victoris Trincavelii praelectionibus de Compositione Medicamentorum atque Morbis Particularibus concinnatum.*

Typis Oporinianis per Balth. Han & Hiero. Gemusaeum, Basel 1583
Ed. p.
Prov. LIDDEL.
Dedn. to the sons of Frederick II of Denmark. For Trincavelli see No. 684.

156. CHRISTMANNUS, IACOBUS. *Observationum Solarium libri tres, in quibus explicatur verus motus solis in zodiaco & universa doctrina triangulorum ad rationes apparentium coelestium accommodatur.*
Lazarus Zetzner, Basel 1601
Ed. p.
Prov. LIDDEL.
Dedn. to Otho Brahe dated Heidelberg (where Christmannus was then Professor of Logic) 1601.

157. CHRISTMANNUS, I. *Muhamedis Alfragani Arabis Chronologica et Astronomica Elementa, e Palatinae bibliothecae veteribus libris versa, expleta & scholiis expolita. Additus est commentarius qui rationem calendarii Romani, Aegyptici, Arabici, Persici, Syriaci, Hebraei explicat.*
Heirs of Andr. Wechel, Frankfurt 1590
Ed. p. BM (under Ahmad ibn Muhammad).
Prov. LIDDEL (?).
Dedn. to Io. Casimir, Elector Palatine, refers to a gift of books by Huldrich Fugger (at huge expense) to Frederick IV, Elector Palatine, and names a number of works in the Palatine Library. Dated Heidelberg (where Christmannus was now Professor of Hebrew) 1590.

The work contains 32 chapters of the *Liber de Aggregationibus scientiae stellarum & principiis coelestium, quem Ametus filius Ameti dictus Alfraganus compilavit* 30 *capitulis* [Note on no. of chs., p. 7] (ThK, 448). The text is followed by the long *Commentarius* of Christmannus, subdivided into sections on the various calendars, and is wound up with a long discussion *De Connexione Annorum.*

158. CHRISTMANNUS, I. *TractatioGeometrica. De QuadraturaCirculi in decem capitula distributa. Adversus errores tam veterum quam recentiorum mechanicorum. . . .*
Off. Paltheniana, P. Kopfius, Frankfurt, 1595
Dedn. dated 1595, also from the Hejra and in the Hebrew style. Contains a re-affirmation of Aristotle's denial of the possibility of 'squaring' the circle, and an examination of Archimedes' mechanical method.
Prov. See. No. 156.

159. CHRISTMANNUS, I. *Epistola chronologica ad clarissimum virum Iustum Lipsium. Qua constans annorum Hebraeorum connexio demonstratur & non leves aliorum errores luculenter refutantur.*

Iosua Harnischius, Heidelberg, 1591

Not BM. Earliest Bibl. Nat. 1593.

Prov. LIDDEL.

160. CIRVELLUS, PETRUS (*syn* Cirvelli, Pietro). *Habes lector Iohannis de Sacro Busto Sphere textum una cum additionibus ... Intersertis praeterea quaestionibus domini Petri de Alliaco. Omnia . . . castigata et rursus coimpressa.*

Iohannes Petit, Paris 1515

The details of imprint are from Col. at end of main text: there is no col. at end of the subsequent *Disputatorius Dyalogus*.

Ed. p. (slightly different title) Paris 1498 (wrongly printed as 1468) (Hain 5363). Neither BM nor Bibl. Nat. records 1515 edn.

Prov. ' . . Steuart & suos'.

The foliation becomes irregular towards the end, but the sigs. correspond to the registration.

Incipit: 'dixit Iohannes tractatum de Sphaera mundi . . .'.

sig. Vv has fine cut of sphere, etc.

For Petrus de Alliaco (Cardinal Pierre d'Ailly), see Vol. I. p. 132.

161. CLAVEUS, GASTON. *Apologia crysopoeiae et argyropoeiae adversus Thomam Erastum. . . . In qua disputatur & docetur an quid, & quomodo sit chrysopoeia & argyropoeiae . . . Nunc primum a Bernardo G. Ponoto [sic] a Portu S. Mariae Aquitano cum annotationibus marginalibus edita.*

Heirs of Eustathius Vignon n p 1598

Not in Young Collection (Ferg.) exc. as secn. of *Theatrum Chemicum,* Strassburg 1659. Duv. (p. 182) lists the edn. of Sutorius, Ursellis 1602, and states that *Ed.p.* was Nevers (Claveus's birthplace) 1590. Not in BM exc. Cologne 1612 under a different title.

In the dedn. to Landgrave Maurice of Hesse Penotus complains that 'our age is full of impostors . . . kings and magnates are deceived ', but in the contents goes on to describe the 'true and right theory (*ratio*) for procreating (*progignendi*) the philosopher's stone!' The whole book illustrates very well the extraordinary confusion of scientific criticism and 'theory' which existed only in the minds of the 'adepts'. (See Vol. I, p. 179.)

CLAVIUS, CHRISTOPHORUS. See Nos. 108, 235(*h*), 620, 678.

162. [CLICHTOVEUS, IODOCUS.] *In hoc libro continetur intro-ductorium astronomicum theorias corporum coelestium duobus libris complectens: adiecto commentario declaratum.*

Henri Estienne, Paris 1517

BM cites this as *Ed. p.* confirmed by dedn. by Clichtoveus to Petrus Gorraeus. There is also an undated dedn. by Jacobus Faber (Jacques Lefèvre of Etaples) which may account for the citation in BM of another earlier *Astronomicon* from the same publisher. No. 162 consists of only 8 fo's and is evidently incomplete. (For Clichtoveus and Le Fèvre see No. 381 and Vol. I, p. 11.)

163. CLIVOLO, BARTHOLOMAEUS a. *De Balneorum Naturalium Viribus li. iv.*

Mathias Bonhomus, Lyon 1552

Sigs. a iii, a iv, wanting.
Ed. p.
Prov. BUTE.

CLUSIUS, CAROLUS. See Nos. 3, 70, 456, 488, and Appendix B.

164. [CLOWES, WILLIAM. *De Morbo Gallico.*]
(Col. Thos. East for Thos. Cadman, London 1585)
t.p. wanting.

[A briefe and necessarie treatise touching the cure of the disease called *Morbus Gallicus* . . . Newly corrected and augmented—from BM Cat.]

[*Ed. p.* John Daye, London 1579 STC 5447 'A Short and Profitable Treatise . . .'.]

Prov. BUTE. Earlier hand, 'Sum liber Anthonii Banister . . .'.
Contents include: *The nature & propertie of quicksilver, by F. Baker, Gent., Maister in Chirurgerie. An epilog, collected and gathered by John Banester, Gent., Maister in Chirurgerie, & practitioner in phisicke . . . John Gerard Chirurgion to his friend Maister William Clowes.*

165. COCLES, BARTHOLOMAEUS. *Chyromantie ac Phisionomie Anastasis cum approbatione magistri Alexandri Achillinis.*
(Col. Hier. de Benedictis, Bologna 1523)
[*Ed. p.* 1503]

The *Respondeo* of Achillini (a famous medical Averroist philosopher of the 15th cent.) is of interest as a specimen of the dialectical method of 'discovery'.

64

166. CODRONCHI, GIOVANNI BAPTISTA. . . . *de Vitiis Vocis li. ii. In quibus non solum vocis definitio traditur & explicatur, sed illius differentiae, instrumenta & causae aperiuntur. Ultimo de vocis conservatione, praeservatione, ac vitiorum eius curatione tractatur . . . Cui accedit Consilium de Raucedine. . . .*

Heirs of Andr. Wechel, Frankfurt 1597

Ed. p.
Prov. LIDDEL.

167. COLUMBA, GERARDUS. *Disputationum medicarum de febris pestilentis cognitione et curatione libri duo. Nunc primum . . . castigati & . . . in Germania editi. Accessit tractatus singularis Nicolai Maccheli . . . de Morbo Gallico eiusque curatione.*

Heirs of Romanus Beatus, Frankfurt 1601

[*Ed. p.* Messina 1596.]
Prov. LIDDEL.

The index of *quaestiones* contains several of interest, e.g. the rival views of Giovanni Francesco Pico della Mirandola, and Roger Bacon on the cause of tides. Page 1 of the text refers to Fernel's recent distinction between pan-, epi-, and endemic diseases.

168. COLUMBUS, MATTHAEUS REALDUS. *De Re Anatomica libri xv.*

(*a*) Andr. Wechel, Paris 1572. Prov. Liddel, and a second copy 'Io. Gregory 1769 £0, s i, d i'.

(*b*) M. Lechler, sumpt. Petri Fischeri, Frankfurt 1593; title continues 'Hisce iam accesserunt Ioannis Posthii . . .Observationes Anatomicae. . . '.

[*Ed. p.* Venice 1559.]

This work contains (in *liber septimus*), the famous passage on the pulmonary circulation (see Vol. I, p. 219). It is evident that Columbus had by no means grasped the matter clearly, but care is necessary in the translation of anatomical terms as used at that time.

COMMANDINUS, FEDERICUS See Nos. 39, 332, 494

169. CONSTANTINUS AFRICANUS. . . . *operum reliqua, hactenus desiderata nunc primum impressa ex venerandae antiquitatis exemplari, quod demum est inventum. In quibus omnes communes loci, qui proprie theorices sunt, ita explicantur & tractantur, ut medicum futurum optime formare & perficere possint. Quaecunque enim Galenus iusto* *fusius habet clara & docta brevitate perstringit & apud Hippo-cratem obscurioribus mirabilem lucem addit disciplinarum omnium, praecipue dialectices, praesidiis instructissimus autor.*

Henricus Petrus, Basel (1539 Col.)

Ed. p.

Dedn. 'Reverendissimo Patri Domino Desyderio abbati Montis Cassinensis totius ordinis Ecclesiastici decori praecipuo, C— A— omnia secunda exoptat'.

Contains also with cont. pagn. Antonius Gaizo, *de Somno ac eius necessitate quidque faciat ad bonam digestionem.*

13th cent. MS at BM has same incipit 'Oportet (eum) qui medicine vult' (ThK 473).

170. COPERNICUS, NICOLAUS. *De lateribus et angulis triangulorum, tum planorum rectilineorum, tum sphaericorum, libellus . . . ad plerasque Ptolemaei demonstrationes intelligendas . . . Additus est Canon semissium subtensarum rectarum linearum in circulo.*

Io. Lufft, Wittenberg 1542

Ed. p.

This work, ed. by Joachimus Rheticus, forms (less the tables at the end) Chs. xiii and xiv of Bk. 1 of the *De Revolutionibus* (No. 171). In the latter a short paragraph . . . 'Haec obiter de triangulis . . .' was added to conclude Bk. 1.

171. COPERNICUS, N. . . . *de Revolutionibus Orbium Coelestium libri vi. Habes in hoc opere iam recens nato & aedito, studiose lector, motus stellarum, tam fixarum quam erraticarum, cum ex veteribus tum etiam ex recentibus observationibus restitutos & novis insuper ac admirabili-bus hypothesibus ex quibus eosdem ad quodvis tempus quam facillime calculare poteris. Igitur eme, lege, fruere.*

Io. Petreius, Nürnberg 1543

Ed. p.

Prov. LIDDEL, and a second copy.

The former has table of *errata* bound at end (*in recto*, and blank *in verso*). The latter has second t.p. with the table of *errata in verso*. The prelims. comprise *Ad lectorem de Hypothesibus huius operis* (the 'apology' attributed to Andreas Osiander by Kepler), the letter of commendation (Rome 1536) from Cardinal Nicholas Schonberg, and the dedn. by Copernicus to Paul III. (Vol. I, p. 112.)

172. COPERNICUS, N. . . . *de Revolutionibus Orbium Coelestium li. vi. In quibus stellarum et fixarum et erraticarum motus, ex veteribus atque recentibus observationibus, restituit hic autor. Praeterea tabulas* *expeditas luculentasque addidit ex quibus eosdem motus ad quodvis tempus mathematum studiosus facillime calculare poterit. Item de Libris Revolutionum Nicolai Copernici Narratio prima per M. Georgium Ioachimum Rheticum ad D. Ioan. Schonerum scripta.*

Off. Henricpetrina, Basel (Col. 1566)

The *Narratio Prima* is prefaced by a letter from Achilles Gassarus to Georgius Vogelinus saying that the *Narratio* had been sent to him by his greatest friend Rheticus (dated 1540). The table of contents contains Reinhold's *Prutenic Tables.*

> Prov. The Liddel copy contains MS (in his hand) of the beginning of the *Commentariolus.* There is a second copy of this work: it differs in being the only work in the volume, whereas Liddel's is bound with the *De Triangulis* of Regiomontanus (No. 557).
> Io. Petreius, Nürnberg 1533.

173. CORDAEUS, MAURITIUS. *Hippocrates Coi, medicorum principis, liber prior de morbis mulierum ad Henricum III Galliarum . . . Mauricio Cordaeo Rheno interprete & explicatore.*

Dionysius Duvallius, Paris 1585

Ed. p.

Prov. BUTE.

Greek and Latin texts with a very full commentary on a work generally regarded as not by Hippocrates himself.

See also No. 738.

174. CORDUS, VALERIUS. *Pharmacorum omnium quae in usu potiss. sunt componendorum ratio. Vulgo vocant Dispensatorium sive Antidotarium. Ex optimis autoribus tum recentibus quam veteribus collectum, ac scholiis utilibus illustratum, in quibus imprimis simplicia diligenter explicantur. . . . Opera et studio Collegii Medici inclytae Reipub. Norimbergensis iam primum multo emendatius ac selectis compositionibus auctius . . . editum.*

(a) Christ. Lochner & Io. Hofmann, Nürnberg 1592.
[*Ed. p.* 1546, Arber.]
Prov. GORDON.

This is the revised edn. (folio), of the original work, which after some years' use in Saxony in MS, had been shown to the authorities in Nürnberg by Cordus when 'seeking Italy for the sake of his studies he had reached our city'. (From the preface to the *Ed. p.* reprinted in No. 174). (See Vol. I, p. 193.) Copies of the following 'unofficial' edns. are also in the Library.

67

(*b*) *Novum V— C— Dispensatorium, hoc est pharmacorum conficien-dorum ratio quam ab ipso authore olim auctam, nunc a quam plurimis mendis expurgatam . . . edimus. Appendix ex scriptis D. Jac. Sylvii ad pharmacopolas antea impressa.*

Off. Valgrisiana, Venice 1563

Not in BM.

(*c*) *V— C— D— sive Ph— conficiendorum ratio a Petro Coudebergo Pharmacopaeo Antwerpiano plus quadringentis erroribus liberata . . . Adiecto V— C— novo libello. Quid porro a Matthia Lobelio praestitum sit.*

Franc. Raphelengius. Ex Off. Plantiniana, Leiden 1590
Also a copy of the edn. of 1608, and of Jo. Maire, Leiden 1627

175. CORDUS, V. *Annotationes in Pedacii Dioscoridis Anazarbei de Medica Materia libros V longe aliae quam ante hoc sunt evulgatae. Eiusdem Val. Cordi Historiae Stirpium lib. iiii posthumi nunc primum . . . editi, adiectis etiam Stirpium iconibus & brevissimis annotatiunculis. Sylva qua rerum fossilium in Germania plurimarum, metallorum, lapidum & stirpium aliquot rariorum notitiam brevis-sime persequitur, nunquam hactenus visa. De artificiosis extractionibus liber. Compositiones medicinales aliquot, non vulgares. His accedunt Stocc-Hornni et Nessi in Bernatium Helvetiorum ditione montium & nascentium in eis stirpium, descriptio Benedicti Aretii, Graecae & Hebraicae linguarum in schola Bernensi professoris. . . . Item Conradi Gesneri de Hortis Germaniae liber recens una cum descriptione Tulipae Turcarum. . . .*

(Col. Iosias Rihelius, Strassburg 1561)

t.p. device and date only.
Ed. p. of the *Dioscorides* 1549.

The 'title' is of the nature of a table of contents and has been repro-duced *in extenso* to give some idea of the extraordinary richness and interest of the work. The whole was edited by Gesner from MSS, partly Cordus's own, partly those of a pupil of the latter. Most of the 'additions' have sep. dedns., the whole being dedicated to the College of Physicians in the University of Wittenberg. There is a list of annota-tions of the text of Dioscorides. The *Historia Stirpium* (generally referred to in the text as *Historia Plantarum*) contains numerous fine cuts. The *Sylva* is a record of Cordus's famous journey through Italy in 1542, from the hardships of which (including malaria) he died at the age of 29. The *De Artificiosis Extractionibus* contains (fo. 226*v*) *De oleo vitriolo faciundo.* Gesner's article (fo. 297*v*) *de Scilla et eius speciebus . . . iudicium ad F. Calceolarium Pharmacopolam Veronensem* is of special interest.

68

176. CORNARIUS, IANUS. *De Peste, libri ii. Pro totius Germaniae, imo omnium hominum salute.*

Io. Hervagius, Basel (Col. 1551)

Only edn. BM.

Prov. BUTE.

The dedn. to Hulderich Mordisius, Chancellor to the Elector Maurice of Saxony, opens with an account of the spread of the plague.

CORNARIUS, I. *See* Hippocrates.

177. CORNARIUS, DIOMEDES. *Consiliorum medicinalium . . . tractatus.*

Michael Lantzenberger, Leipzig 1599

This volume contains also:

Observationum medicinalium partim ab autore D— C— partim ab aliis . . . praemeditationes, Historiae admirandae rarae ab eodem D— C— collectae. Oratio in Funere . . . Wolfgangi Lazii Viennensis . . . (sep. t.p. Michael Zimmermann, Vienna 1565) (Public disputations).

Sig. Nn 4 *v* contains a ref. to the 'boy born with the golden tooth'— a *cause célèbre* in which Duncan Liddel played a decisive part, see No. 348. The whole volume is terminated by the Col. of Lantzenberger.

178. CORTES, MARTIN. *The Arte of Navigation containing a compendious description of the Sphere, with the making of certayne instrumentes and Rules for Navigations, and exemplified by many demonstrations, written by Martin Cortes Spaniarde, Englished out of Spanishe by Richard Eden, & now newly corrected and amended in divers places. Whereunto may be added at the wyll of the byer another very fruitefull and necessary booke of Navigation translated out of Latine by the sayd Eden.*

Widow of Rd. Jugge, London 1584

[*Ed. p.* 1561 STC 5798.]

The dedn. to 'Wylliam Garerd Knyght and Maister Thomas Lodge Aldermen of the Citie of London . . .' calls for 'the mayntenaunce of such Arts and Sciences as the Commonwealth can not well be without'. The original work (Cadiz 1551) was ded. to the Emperor Charles V and refers to his capture of Francis I (at Pavia, where it is said, the latter fought with a lance; his bad faith during his imprisonment nearly led to the unusual spectacle of a duel between a king and the emperor!). Johnson (p. 141) regards this as the best contemporary (non-Copernican) text on navigation. (Vol. I, p. 146.)

COSTA. *See* Acosta.

179. COSTAEUS, IOANNES. *Disquisitionum Physiologicarum . . . in primam primi Canonis Avic. sect. libri sex. . . .*

Io. Rossius, Bologna 1589

Short quotations from text, but mainly commentary. Page 90 *seq.* reviews the Copernican 'revolution' under the traditional medieval *Disputatio an terra moveatur tota* ('Disputation as to whether the earth is moved as a whole').

180. COSTAEUS, IOANNES. *De Igneis Medicinae Praesidiis, libri duo.*

Rubertus Meiettus, Venice 1595

Ed. p.
Prov. LIDDEL (?)

A study of the nature of 'heat' and 'hotness' and its relation to disease and therapy.

181. COTTA, J. *A short discoverie of the unobserved dangers of severall sorts of ignorant and inconsiderate practisers of physicke in England. Profitable not only for the deceived multitude and easie for their meane capacities but raising reformed and more advised thoughts in the best understandings.*

Wm. Jones, London 1612

Ed. p.
Prov. READ.

Of special interest in regard to the vexed question of the relation of astrology to medicine. See esp. pp. 97-103.

The work has the distinction of being bound with Harvey's *De Motu Cordis, Ed. p.* Frankfurt 1628, glossed in Read's neat hand.

182. CRATO À KRAFTHEIM, IOANNES. . . . *Consiliorum & Epistolarum Medicinalium liber. Ex collectaneis clariss. viri Dn. Petri Monavii Vratisl. . . . selectus. Et nunc primum a Laurentio Scholzio Sil. Medico Vratislaviensi . . . editus.*

Heirs of A. Wechel, Frankfurt 1591

Followed, each with similar t.p. and sep. pagn., by *Liber secundus* (1592) and *Liber tertius* (1592). And, with sep. t.p. but pagn. cont. with last, *Joh. Cratonis . . . Μικροτεχνη seu parva ars medicinalis. Nunc primum studio & industria Laurentio Scholzio . . . edita* (1592). Bound separately, with similar t.p., *Liber Quintus* (1593).

Prov. Both vols. 'Carolus Lumisden'. The whole set, together with *Liber Quartus* 1593, is duplicated in three vols. with Liddel's auto. In one of these there follows *De Morbo Gallico Commentarius nunc primum studio & opera Laurentio Scholzio . . . editus* (1594)

See p. 496 (in the *Μικροτεχνη*) of Vol. I (Lumsden) for views on the *cause célèbre* of chemical remedies, with the comment 'sed cum istorum hominum insania non libet rixari'. Breslau was at that time a thriving centre of medical activity.

183. CRATO À KRAFTHEIM, I. *Commentarius de vera praecavendi et curandi febrem pestilentem contagiosam ratione . . . Germanico idiomate conscriptus. Nunc in Latinum . . . conversus, additis plurimis remediorum formulis nunquam antea editis studio et opera Martini Weinrichii Vrat. Addita est eiusdem assertio Latina pro libello Teutonice edito.*
> Heirs of A. Wechel, Frankfurt 1594 Col.

[*Ed. p.* 1585. German edn. not BM.]
Prov. LIDDEL. Cont. pagn. with other works.

184. CROLL, OSWALD. (*a*) *Basilica Chymica, . . . In fine libri additus est eiusdem authoris tractatus novus de signaturis rerum internis.*
> Philippus Albertus, Geneva 1631

[*Ed. p.* Frankfurt 1608 (Ferg. I, 185, who cites this as 2nd Latin edn.]
(*b*) *Basilica Chymica & Praxis Chymiatricae or Royal and Practical Chymistry in Three Treatises . . . A translation . . . augmented . . . by John Hartman.* [The translation is apparently not by Hartman.]
> John Starkey, London 1670, [Ferg. I, 187]

185. CUBE, IOHANN VON [t.p. wanting—Kreuterbuch von Kube.]
> (Col. Christ. Egenolph, Frankfurt 1550)

[*Ed. p.* in this form 1535 BM, but the ultimate basis of the work is the *Gart der Gesuntheit*, P. Schoeffer 1485.]
It comprises sections on distillation, animal creation (incl. human anatomy in a crude form), precious stones, 'forestry', herbs. There are fine cuts (all pirated, according to Arber, p. 70) of distillation apparatus, and cutting and polishing machinery.

For an account of the tradition of the German *Herbarius* see Arber, p. 22 *seq.* and Vol. I, p. 186.

186. CURIO, IOANNES (Editor). *Medicina Salernitana, id est conservandae bonae valetudinis praecepta . . . Cum . . . Arnoldi Villanovani . . . exegesi . . . Nova editio . . . aliquot medicis opusculis . . . auctior.*
> Jac. Stoer n.p. 1599

[*Ed. p.* 1551.]
Prov. FORDYCE.
The 'medica opuscula' are by Fernel, Diokles of Karystos, etc.

187. CURTIUS, MATTHAEUS. . . . *in Mundini Anatomen Commentarius.*

Theobaldus Paganus, Lyon 1551

[*Ed. p.* 1550.]

Dedn. 'Andreae Vuaselio' by the editor, Iohannes Canappaeus. Proeemium, presumably by Curtius to his students, gives an outline of how he proposes to treat the subject. The main body of the work consists of the text of Mondino with commentary (in smaller fount). The *incipit* 'Quia dixit Galienus . . .' agrees with ThK ref. to *Fasciculus Medicinae*.

See also No. 154.

188. DANEAU, LAMBERT (*syn.* Danaeus). *Physice Christiana sive Christiana de rerum creatarum origine et usu disputatio. Tertia editio. Aliquot locis ab ipso autore aucta.* . . .

Eustachius Vignon, Geneva 1588

[Earliest BM, Geneva 1550 'Tertia editio'.]

Preface and dedn. dated 1575.

An attempt to derive the whole of 'physics' from Holy Writ (cf. p. 86 *De figura terrae* with Isaiah xl. 22), but quotes freely from e.g. Galen and Pliny.

189. DARIOT, CLAUDE (tr. Fabian Wither). . . . *Introduction to the Astrologicall Judgement of the Starres . . . lately renued, and in some places augmented and amended by G. C. Gentl. Whereunto is annexed a most necessarie Table for the finding out of the planetarie and unequall houre, under the Latitude* 52 Gr. 30 Ml [English Midlands] *exactly calculated by the said F.W.* [Fabian Wither in dedn.] *Also . . . a briefe treatise of Mathematicall Phisicke, entreating . . . of the natures and qualities of all diseases incident to the humane bodies by the naturall influences of the coelestiall motions. Never before handled in this our native language. Written by the said G.C. practitioner in Phisicke. Astra regunt homines, et regit astra Deus.*

Dariot's *Ad Astrorum Iudicia facilis introductio.* . . . *Quibus accessit fragmentum de morbis & diebus criticis ex astrorum motu cognoscendis,* was printed Lyon 1557, part of it being translated into French in 1588 (Lyon) and Englished (F.W.) London 1583 (BM).

Second t.p. after sig. S.4.

See Allen, p. 170. The work follows Ptolemy in general (but notes one departure), yet takes Regiomontanus as final authority (Vol. I, p. 101).

190. DASYPODIUS, CUNRADUS. [*Hypotyposes Orbium Coelestium quas appellant Theoricas Planetarum, congruentes cum Tabulis Alphonsis et Copernici seu etiam Tabulis Prutenicis.*]

T.p. wanting: title from heading of first page of text. No imprint or col. In dedn. to William, Landgrave of Hesse, Dasypodius refers to Erasmus Reinhold ('whom I had almost called a second Ptolemy') and his commentary on the *Theoricae* of Peurbach (No. 513a). He states that the MS was brought to him by his brother-in-law, and, though some believe it to have been written by Reinhold, this is uncertain and the book ought therefore to be published as by an unknown author, no one wishing to take the credit from him, whoever he may be. (See No. 512.)

No copy in BM (personal communication by the courtesy of the Keeper of Printed Books). No record in Bibl. Nat. or Bodleian. The only other copy I have traced is at Trinity College, Dublin (EE.o.81), listed as 'Argent. 1568'.

The main text is preceded by (1) *Ex Alfrag. de Ortu et Occasu Planetarum & de occultationibus eorum sub radiis solis,* (2) *Ex libro tertio Epitomae Ioannis de Regiomonte in Almagestum Ptolemaei. Dies naturales duplici causa inaequales esse.*

191. DEE, JOHN [t.p. wanting]. [*Ioannis Dee Londinensis de Praestantibus quibusdam naturae virtutibus* προπαιδευμαια ἀφοριστικα] from BM, where there is one complete copy of this *Ed. p.* and another (bound with Leowitz *Brevis et perspicua ratio*) wanting all but the *Epistola Nuncupatoria.* No. 191 (apparently complete) is bound with several works including one by Leowitz.

(Col. Henry Sutton, London 1558)

A rough collation of this copy is as follows: (i)·ộ·i and·ộ·iii (wrongly signed—text continuous) comprise *Epistola Nuncupatoria* addressed to Gerard Mercator and dated London 1558; (ii) a i—f ii, on the *verso* of the last of which the text ends and the Col. is printed. The *Epistola,* bound in at the *end,* contains a list of works 'per medias meas, maximasque difficultates ita a me mihi composita scriptaque extant ut eadem . . . in publicum producere . . . exoptem maxime'. Evidently Dee wrote most of these for his own satisfaction: very few seem to have been published. (For Dee and Mercator, see Vol. I, p. 141.)

STC 6463. (See also No. 235 (g).)

192. DESIDERIUS, GUIDO. *Epitome operis . . . morbis curandis Valesci de Taranta in septem congesta libros.*

Io. Tornaesius & Gul. Gazeius, Lyon 1560

Earliest BM.

Dedn. refers to Valesco's composition of this book 230 years previously and to the necessity for freeing the text from errors and crudities of style—dated 1554. The *Epitome* is followed (sep. h.t. and pagn.) by *Epitome Chirurgiae Valesci de Taranta* by the same editor. Valesco was a Portuguese member of Montpellier in early 15th cent. (Castiglioni, p. 339.)

193. DESSENNIUS, BERNARDUS (*syn.* Cronenburgus, B.). *Medicinae veteris et rationalis adversus oberronis cuiusdam mendacissimi atque impudentissimi Georgii Fedronis ac universae Sectae Paracelsicae imposturas, defensio.... Accessit ... purgantium medicamentorum ... divisio ...*

Io. Gymnicus, Cologne 1573

Ferg. I, 188.
Prov. 'Sum Thomae Langto . . . . 1574'.
Followed (with sep. t.p. and pagn.) by *Purgantium Medicamentorum ... particularis divisio a nemine hactenus tradita. ...*

194. DIGGES, LEONARD. *A Geometrical Practise, named Pantometria, divided into three bookes, Longimetria, Planimetria, and Stereometria containing rules manifolde for mensuration of all lines, superficies and solids: with sundry straunge conclusions both by instrument and without, and also by Perspective glasses, to set forth the true description or exact plot of an whole region ... lately finished by Thomas Digges his sonne, who hath also thereunto adioyned a mathematicall treatise of the five regulare Platonical bodies, and their metamorphosis ... into five other equilater uniforme solids geometricall ... hitherto not mentioned of by any geometricians.*

Henrie Bynneman, London 1571

Ed. p. STC 6858.
Prov. (i) A. Arbuthnot 1579 and (ii) 'This book pertaineth to Master James Downie'.
Dedn. to Sir Nicholas Bacon (father of Francis).
Thomas Digges's *To the Reader* contains an account of wonders performed by his father with 'proportionall glasses', of which there were then still living *oculati testes*; also a reference to other volumes 'compiled in the Englishe tongue, desiring rather with playne and profitable conclusions to store his native language and benefite his countrymen, than by publishing in the Latin rare and curious demonstrations, to purchase fame among straungers'.

A highly important work (see Vol. I, p. 140) for the study of the growth of science outside the universities in England. The use of 'perspective glasses' gives L. Digges a strong claim to be the inventor of the telescope nearly half a century before Lipperhey.

74

195. DIGGES, L. [finished by his son, Thomas]. *An Arithmeticall Militare Treatise, named Stratiotikos: compendiously teaching the science of numbers, as well in fraction as integers, and so much of the rules and aequations algebraicall and arte of numbers cossicall, as are requisite for the profession of a soldiour. Together with the moderne militare discipline, offices, lawes and dueties in every wel governed campe and armie to be observed:* . . . *Whereto he [sc.* Thomas] *hath also adioyned certaine questions of great Or[dnance] resolved in his other treatize of Pyrotechney and gr[eat] artillerie, hereafter to be published.*

H. Bynneman, London 1579
Ed. p. STC 6848.

In his *To the Reader* Thomas elaborates so much on the importance of the 'Auntient Romane Discipline for the Warres' that Fluellen himself might have received this from his inspiration! For 'numbers cossicall' see Vol. I, p. 91.

196. DIGGES, L. (and Digges, T.). *A prognostication everlasting of right good effect,* . . . *augmented by the author* . . . *containing* . . . *rules to iudge the weather by the Sunne, Moone, Starres, Comets, Rainbow, Thunder, Clowdes, with other extraordinary tokens, not omitting the aspects of planets with a briefe judgement for ever, of Plentie, Lacke, Sickness, Dearth, Warres, etc. opening also many naturall causes worthie to be knowne. To these* . . . *are ioyned divers generall pleasant tables.* . . .

Felix Kyngstone, London 1605
[*Ed. p.* 1555. S.T.C. 6860.] No. 196 is of the 11th edn. cited by S.T.C. The first edn. to be corrected by Thomas was the 5th, 1576.

The *Prognostication* is followed [sig. Mr) by Thomas Digges's *Addition,* consisting of *To the Reader,* a chart and text of *A Perfit Description of the Coelestiall Orbes according to the most ancient doctrine of the Pythagoreans: lately revised by Copernicus, and by geometricall demonstrations approved.* (See Vol. I, p. 117.)

197. DIGGES, L. *A Booke Named Tectonikon, briefly shewing the exact measuring & speedie reckning all manner of lands, squares, timber, stone.* . . . *And in the end a little treatise adioyning, opening the composition and appliancie of an instrument, called* [long description here omitted] *the profitable staffe.*

F. Kyngston, London 1626
[*Ed. p.* 1562 STC.] This edn. not STC.
Virtually an illustrated text book on the cross-staff.

75

[198. DIOSCORIDES, PEDACIUS.]

The corpus of those works which at any time have been ascribed to Dioscorides comprises (1) *De Materia Medica libri v;* (2) *Liber de Venenis . . .;* (3) *De iis, quae virus ejaculantur, animalibus libellus . . .;* (4) *De facile parabilibus tam simplicibus quam compositis medicamentis ad Andromachum li. ii.* (Titles as given in Sprengel, C. Pedacius Dioscurides, Leipzig 1829-30—the first definitive modern text). Of these (1) is universally accepted as genuine. The Greek *Ed. p.* (Aldus, Venice 1499) bears the title περι ὕλης ἰατρικης (five books) but contains in addition περι δηλητηρίων φαρμακων (sixth book) and περι ἰοβολων (seventh, eighth, and ninth books) both of which are regarded as spurious (GW 8735, cf. Choulant). (Aldus also included the two similar works of Nikandros of Kolophon—see Nos. 476). Subsequent edns. include varying combinations of (1), (2), (3), and after 1565 (4), with titles which must be interpreted with care. (2) and (3) often appear as one 'book', since they were both addressed to 'Areus'. (*See* J. L. Choulant *Handbuch der Bücherkunde für die ältere Medizin*, Leipzig 2nd enlarged edn. 1841 and M. Wellman in Pauly-Wissowa, **5**, 1131 *seq.*)

199. DIOSCORIDES, P. (*a*) [. . . *de Medica Materia libri sex, interprete Marcello Virgilio Secretario Florentino, cum eiusdem annotationibus, nuperque excusi.*] [Title from copy in Bodleian Library].
>Col. Heirs of Phil. Junta, Florence 1518

(Sigs AA 1-8, a i-a ii, wanting).

[*Ed. p.* of the text by Aldus Manutius, Venice 1499 (Hain 6257)].

No. 199 is *Ed. p.* of Virgilius's version.

Prov. Robert Gray, Hepburn, Ioannis Cameronii Monachi, and others.

The text is printed in a large fount, the translation in smaller, but is often longer than the corresponding part of the text.

>(*b*) *ΔΙΟΣΚΟΡΙΔΗΣ*, etc. The effective title follows *in verso* in Greek and Latin, viz. *Pedacii Dioscoridis de Materia medica libri sex. Eiusdem de venenatis animalibus libri duo, quibus canis rabidi signa, & curatio eorum continetur, quibus venenata animalia morsum defixerint.*
>>Ioan. Bebelius, Basel 1526

Ed. p. in this form, the whole of which except the dedn. is in Greek. *Liber sextus* is the combined (2) and (3).

Prov. CARGILL.

>(*c*) *Dioscoridis libri octo Graece et Latine. Castigationes in eosdem libros.*
>>Petrus Haultinus, Paris 1549

76

Prov. Augustine and Thomas Bernhere (different hands) and a
second copy.

The Printer's *Lectoribus* suggests that Jacques Goupyl was generally
responsible for the edn., but the heading of the Latin version (parallel
columns to Greek) states 'Io. Ruellio Suessionensi interprete'. (2) and
(3) are here distributed in *Libb. VI, VII, VIII.*

> (d) *Pedanii Dioscoridis Anazarbei, de Medicinali materia, libri sex,*
> *Ioanne Ruellio Suessionensi interprete.* . . . *Additae annotationes* . . . *e*
> *selectiori medicorum promptuario.*

Balth. Arnolletus, Lyon 1550

Prov. LIDDEL.

Numerous cuts of plants. *Liber sextus* is the combined (2) and (3).

Dedn. has a ref. to Mattioli's text—this must have been the commentary in Italian, since the first Latin edn. of that work was 1554.

> (e) [t.p. wanting—Title from copy in Library of University of
> Edinburgh]. *P— D— A— Simplicium medicamentorum reique*
> *medicae libri vi. Interprete Marcello Vergilio* . . .*Quibus accessit* . . .
> Basileae 1532.
>
> (Col. And. Cratandrus & Io. Bebelius, Basel 1532)

Not in BM or Bibl. Nat.

Prov. 'Joachinus (?)'

In what remains of the *Lectori* (a 2 r) the publisher (?) gives reasons
for preferring this version of Marcellus Virgilius (No. 199). The *Liber
sextus* comprises (2) and (3).

200. DIOSCORIDES, P. . . . *Ad Andromachum, Hoc est de Curationibus*
Morborum per medicamenta paratu facilia libri ii. Nunc primum &
Graece editi & partim a Ioanne Morbano medico Augustano, partim
vero post huius mortem a Conrado Gesnero in linguam Latinam
conversi: adiectis ab utroque interprete Symphoniis Galeni aliorumque
Graecorum medicorum.

Iosias Rihelius, Strassburg 1565

Ed. p.

This is the 'Ευπορι σ τα (viz. '4'), generally regarded as spurious.
Prelims. comprise a dedn. by Gesner and long letters to him from
Achilles Gasserus and Io. Crato concerning the much admired young
man, Io. Mabanus, and containing much biographical matter and
discussion of the attribution of the work to Dioscorides.

See also Nos. 610, 59, 431.

DIVERSUS, SALIUS. *See* Salius-Diversus.

201. DODOENS, REMBERT (*syn.* Dodonaeus, Rembertus). *Historia*
frumentorum, leguminum, palustrium et aquatilium herbarum ac

eorum quae eo pertinent. . . . Additae sunt imagines vivae exactissimae, iam recens non absque haud vulgari diligentia & fide artificiosissime expressae, quarum pleraeque novae & hactenus non editae.

Off. Chris. Plantin, Antwerp 1569

Ed. p. 1566 (Arber). Dedn. 1565.

Prov. Liddel (Bound with No. 202). Embossed binding '1569'.
In the *Ad Lectores* 1565 he speaks of preparing 'florum ac odoratarum, coronariarum, & umbelliferarum herbarum historiam'. (*See* 202 and 203.)

202. DODOENS, R. *Florum, et Coronariarum odoratarumque nonnullarum herbarum historia.*

Off. Chris. Plantin, Antwerp 1568

Ed. p. (Arber).

Prov. (i). *Liber Ioannis Henrichii Dassiliensis pastoris Hamburgensis ad S. Jacob* (?) (*ii*) Liddel.

A ref. to the growing interest in the division of genera into species (*see* Vol. I, p. 196) is 'De Satyrio Basilico . . . Quod recentior aetas Satyrum basilicon nominat non unius est generis. Est enim maius unum; alterum minus; quoddam folio non maculato; aliud maculato' ([The plant] which has in modern times been called Satyrum basilicon is not of one genus. For there is one greater and a second less; one with an unspotted leaf, another with a spotted).

203. DODOENS, R. *Purgantium aliarumque eo facientium, tum et Radicum, Convolvularum ac deleteriarum herbarum historiae Libri iiii . . . Accessit appendix variarum & quidem rarissimarum nonnullarum stirpium ac florum quorundam peregrinorum . . . icones omnino novas nec antea editas singulorumque breves descriptiones continens, cuius altera parte umbelliferae exhibentur non paucae, eodem auctore.*

Chris. Plantin, Antwerp 1574

Ed. p. (BM and Arber). But dedn. (to Philip II of Spain) and Permission both 1572, the *Praefatio* 1571.

Prov. Liddel.

Appendix has sep. t.p., cont. pagn. The group *Umbelliferae* corresponds roughly to the modern family, but includes two Saxifrages (pp. 494-6).

204. DODOENS, R. . . . *Stirpium Historiae Pemptades sex sive li. xxx.*

Chris. Plantin, Antwerp 1583

Ed. p. (Arber).

Dodoen's *magnum opus* (Arber, p. 82 *seq.*). The prelims. are interesting as giving an insight into the 'co-operative society' of Dodoens, De L'Obel, and De L'Ecluse—in respect of both figures and text. There is an impressive list of gardens from which a collection of plants was made, showing a wide geographical distribution.

78

205. DODOENS, R. *Medicinalium observationum exempla rara. Accessere et alia quaedam.*

Christ. Plantin, Leiden 1585

[*Ed. p.* Cologne 1581 (Bibl. Nat.—BM cites 'Hardervici 1521' without noting that this is an original misprint for '1621'].

In his dedn. and *Lectori* Dodoens urges the importance of post mortem dissection for the shedding of light on diseases and their causes otherwise hidden from observation. Of equal importance are the records made by experienced physicians of rare diseases, especially of such as are not described by the Fathers of medicine.

The text consists of a number of his own 'case histories' followed by some chosen mainly from the *Liber de abditis nonnullis ac mirandis morborum & sanationum causis* written by Antonio Benivieni (Florence 1514), with *scholia* added by Dodoens himself, and a few from Alexander Benedictus (1483), Matthias Cornax and others. On p. 122 *seq.* occurs Benivieni's classic account of the beginning of the Morbus Gallicus. Dodonaeus's annotation refers to the recording of similar symptoms by William of Saliceto, Bernard of Gordon, and Valescus of Taranto 'ex congressu cum sordidis meretricibus aut immundis mulieribus'—all long before the names 'morbus Gallicus' or 'lues Venerea' had been heard of. Of the latter name he says, 'Luem Veneream hunc morbum nunc rectius recentiores appellant citra alicuius gentis invidiam'. (Cf. Vol. I, p. 275.)

On p. 237 is an interesting ref. to incontinence of the bladder due to 'lapsus' (slipped disc?) of the last vertebra whereby the nerve 'qui constringendi vesicam vim habet' was impaired (attrito).

206. DONATUS, MARCELLUS. *De medica historia mirabili libri sex. Opus nunc denuo editum & ab ipsomet auctore emendatum . . . & auctum*

Juntae, Venice 1597

[*Ed. p.* Venice 1588.]

Prov. Two copies. Bute and 'Io. Gregory 1769'.

The dedn. to his master, Vincentius Gonzagha, Prince of Mantua, dated 1586.

In his *Lectori* he refers to errors in the previous (Mantuan) text, and now that (through the kindness of the Prince) he has leisure for writing he has put off the task no longer.

Under *morbi novi* (p. 227 *seq.*) he refers to ancient authors' accounts; also to the *morbus Gallicus;* English sweats; Hector Boece on plagues in Scotland, whence he quotes the account of Vespasian's expedition to the Orkneys, where at 'Pomona (hodie Kyrkualium vocant)' the inhabitants were so long lived and healthy that they had little use for the services of a physician!

207. DONDIS, IACOBUS DE. *Promptuarium Medicinae, in quo non solum facultates simplicium & compositorum medicamentorum declarantur, verum etiam quae quibus morbis medicamenta sint accommodata ex veteribus medicis . . . monstratur.*

Juntae, Venice 1576

Ed. p.
Prov. GORDON.

DONZELLINI, GIROLAMO. See Nos. 306, 755.

208. DORN, GERARD. *Lapis Metaphysicus, aut Philosophicus, qui universalis medicina vera fuit patrum antiquorum, ad omnes indifferenter morbos: etiam eos quos incurabiles vocarunt illi qui curare non potuerunt. Et ad metallorum tollendam lepram, fabricandos lapides preciosos etc. . . .* [No name or place] 1570.

BM queries Frankfurt of this date. Ferg. I, 220 does not mention this work. Duv. p. 176 records a single vol. of 3 pts, of which this is the third. (See Th. *Hist.* **5**, 630-5.)

The *Elenchus* of topics is a valuable guide to the questions which interested contemporary alchemists. In the *Praefatio* Dorn joins issue with the 'moderns' who are denying the existence of 'our' *quinta essentia*. Sig. 4 *v* provides a good example of debased scholastic verbiage which clogged the wheels of progress even to the end of the 16th cent.

209. DORN, G. *Dictionarium Theophrasti Paracelsi continens obscurorum vocabulorum, quibus in suis scriptis passim utitur, definitiones a G— D— collectum & plus dimidio auctum.*

[Device of] Christoff Rab. t.p. Frankfurt 1583.

Ed. p. in this form. Date given as 1584 BM and Bibl. Nat. Not Duv. or Ferg. The *Lectori* refers to previous collection now revised & augmented. Sudhoff (No. 198) gives date as 1583 and refers to the earliest collection, *Fasciculus Paracelsicae Medicinae . . .* Frankfurt, 1581. (Sudhoff No. 185.)

Prov. LIDDEL (?).

List of Paracelsian terms alphabetically arranged—no further text.

210. DORN, G. (Ed.) *Trevisanus de Chymico Miraculo quod lapidem philosophiae appellant. Dionys. Zacharius Gallus de eodem. Auctoritatibus variis principium huius artis Democriti, Gebri, Lullie, Villanovani, confirmati & illustrati. Per Gerardum Dorneum.*

C. Waldkirch, Basel 1600

[*Ed. p.* was P. Perna 1583. Duv. p. 69, and Ferg. I, 222; but not in Young collection].

Dedn. by Dorn, 1583. *Ad Lectorem* by Zacharius preceding his *opusculum*, which in turn is succeeded (cont. pagn.) by *Annotata quaedam ex Nicolo Flamello, autore Gallo.* Ferg. believes this to be one of the 'various authorities' mentioned in the title, and thinks that many commentators have raised a 'needless cloud of dust' over its inclusion. There is reason to believe that 'Nicholas Flamel' and his devoted wife were apocryphal: a pity!

211. DORTOMANNUS, NICOLAUS. . . . *libri duo de causis & effectibus Thermarum Belilucanarum parvo intervallo a Monspeliensi urbe distantium.*

Carolus Pesnot, Lyon 1579

Ed. p. (Both dedns. dated 1579; one is to the son of Admiral Coligny, whose attempted assassination may have precipitated the Massacre of Bartholomew (1572), in which he was one of the first to perish.)
Prov. BUTE.
The nature of 'hot' springs was naturally a problem of considerable interest.

DRYANDER, IOANNES. See No. 747.

212. DU CHESNE, JOSEPH (*syn.* Quercetanus). *Ad Iacobi Auberti Vindonis de Ortu et Causis Metallorum contra chymicos explicationem, Josephi Quercetani . . . brevis responsio. Eiusdem de exquisita mineralium, animalium, & vegetabilium medicamentorum Spagyrica praeparatione & usu perspicua tractatio.*

Io. Lertotius, Lyon 1575

Ed. p. (Ferg. II, 235).
Prov. *'Ex codicibus Johannis Clivei* [?] *pharmacopoei'.*
The work consists of (1) epistle, which includes a comparison of Aubertus with the Peripatetic philosopher Phormio of whom Hannibal said 'multos se deliros senes vidisse, sed qui magis quam ille deliraret, neminem' ('He had seen a good many old fools, but no one who drivelled more than he'). (Cicero *De Orat.* 2.75) (2) A reply to the 'slanderous letter in which Aubertus tried to overturn some remedies of the "Paracelsians", as he calls them'. (3) A reply to the tract on the origin of metals. The *De Mineralium Medicamentorum . . . usu* has sep. t.p. but cont. pagn. (Vol. I, p. 177.)

213. DU CHESNE, J. *Traitté de la Cure Generale et Particulière des Arcbusades avec l'Antidotaire Spagyrique pour preparer & composer les medicamens.*

Jean Lertout, Lyon 1576

Ed. p.
Prov. CARGILL.
Sigs. C iiii and (presumably) C v (unsigned) repeated near beginning.

The dedn. to the Duc d'Alençon is followed by *Ode de l'Autheur touchant les Misères de la France,* which gives a vivid picture of the shocking state of the country due to the civil wars of religion.

214. DU CHESNE, J. *Sclopetarius.*

Io. Lertotius, Lyon 1576

Ed. p. of the Latin version (of which it is a somewhat free translation) of No. 213. Frequently reprinted into the 17th cent. Several edns. in the Library.

215. DU CHESNE, J. *De priscorum philosophorum verae medicinae materia, praeparationis modo atque incurandis* [sic] *morbis praestantia. Deque simplicium & rerum signaturis tum externis tum internis, seu specificis, a priscis & Hermeticis philosophis multa cura singularique industria comparatis atque introductis, duo tractatus. His accesserunt eiusdem . . . de dogmaticorum medicorum legitima & restituta medicamentorum praeparatione libri duo. Itemque selecta quaedam consilia medica. . . .*

Io. Vignon, Geneva 1609

[*Ed. p.* S. Gervais 1603.]

This appeared in a French trans. as *Traicté de la Matière* . . . Paris 1626, containing some prelims. and following above title as far as 'signaturis'.

The *Praefatio* contains an interesting discussion of the relationship between the medical 'sects'.

216. DU CHESNE, J. . . . *ad Veritatem Hermeticae Medicinae ex Hippocratis veterumque decretis ac therapeusi, nec non vivae rerum anatomiae exegesi ipsiusque naturae luce stabiliendam, adversus cuiusdam Anonymi phantasmata responsio.*

Wolffgang Richter, Frankfurt 1605

Ferg. quotes a Paris edn. of 1604 as *Ed. p.* BM has both. A plea for the 'orthodoxy' of what he prefers to call the 'Hermetic' medicine. Among the 'Hermetics' whom England and Scotland supported he names Dee and Muffet. (See also No. 564.)

217. DÜRER, ALBRECHT. *Anderweysung der Messung mit dem Zirckel und Richtscheyt in Linien ebnen und gantzen Corporen durch Albrecht Dürer zusamen gezogen und durch inselbs (als er noch auff Erden war) an vil orten gebessert insonderheyt mit xxii Figuren gemert dieselbigen auch mit eygner Handt auffgerissen wie es dann eyn yder Werckmann erkennen wirdt. Nun aberzu nutz allen Kunst liebhabenden in Truck geben.*

(Col. Hiero. Formschneyder, Nürnberg 1538)

[*Ed. p.* Nürnberg 1525 'Underweyssung . . .'.]
Dedn. to Wilbolde [*sic*] Pirckheymer.
The figs. opposite col. show principles of linear perspective; earlier figures show similar end attained by chiaroscuro. Wilibald (Bilibald) Pirckheimer was the centre of the Nürnberg 'circle' at the end of the 15th century (Vol. I, p. 23). Dürer died 1528.

218. DÜRER, A. *De symmetria partium humanorum corporum libri quatuor e Germanica lingua in Latinam versi.*

Carolus Perier, Paris 1557

[*Ed. p.* 1532 Nürnberg]
The text is for the most part in both Latin and French, but the corresponding passages are often widely separated. Some fos. are in Latin only. A secn. *Ad Lectores* (presumably by printer) precedes *Liber tertius.* P. 115*v seq.* shows an interesting attempt by Dürer to build up a human figure out of cubic units.

219. DU PINAY, ANTOINE (*syn.* Pinaeus, Antonius). *Historia Plantarum, earum imagines, nomenclaturae, qualitates, & natale solum. Quibus accessere simplicium medicamentorum facultates, secundum locos & genera, ex Dioscoride.*
(*a*) (above title) Gabrielis Coterius, Lyon 1561.
(*b*) omits *quibus* and adds *Secunda editio.* Vidua G. Coterii, Lyon 1567.
  Prov. (*a*) Liddel, and some earlier (illegible).
    (*b*) 'Sum Richardi Wollastoni'.
The printer refers to a recent trans. by du Pinay of Mattioli's edn. of Dioscorides (No. 431). Arber has no ref. to Du Pinay except to this trans.

220. DURANTE, CASTORE. *De Bonitate et Vitio Alimentorum centuria. . . .*

Heirs of Barth. Cesano, Pisa 1565

Not BM or Bibl. Nat.
Prov. BUTE.
Simples listed with comments under stereotyped headings.

221. DURETUS, LUDOVICUS (= Ferrant, Louis). *Hippocratis Magnae Coacae Praenotiones . . . interprete & ennaratore L— D—.*

Jac. Dupuys, Paris 1588

Earliest BM 1658.
Dedn. by Io. Duretus, son of Ludovicus.
The text is in both Greek and Latin with lengthy notes and commentary. The *Praenotiones* is a supposititious work.

EDEN, RICHARD. See No. 178.

222. ELLINGER, ANDREAS. . . . *Hippocratis Aphorismorum* . . . *paraphrasis poetica.*

Andr. Wechel, Frankfurt 1579

Only edn.

The text actually consists of paraphrases of both *Aphorisms* and *Prognostic*, including the Latin selections made by Celsus. That the three parts constituted one publication is shown by the sequence of signatures.

223. ELYOT, THOMAS. *The Castel of Health. Corrected and in some places augmented by the fyrst author thereof, syr Thomas Elyot, knyght, the year of oure lord* 1561.

(Col. Thos. Marshe, London 1561)

[*Ed. p.* 1539 S.T.C. 7643. No. 223 is the 9th edn. cited by S.T.C.].
Prov. HENDERSON.

The 'Proheme' gives reasons for the writing of such a work by one who was not himself a physician—amplified on last page. Sir Arthur McNalty (*The Renaissance and its Influence on English Medicine*—Vicary Lecture 1945, p. 11 *seq.*) states that 'purpyls, meazels and small pockes' are mentioned in this form for the first time in a systematic work on medicine. He commends Elyot as an accurate observer, who though probably a layman, and writing in *English*, much to the disgust of the physicians, probably 'read Galen (too much perhaps!) with Linacre' (Vol. I, p. 59).

224. ENTZELT, CHRISTOPH (*syn.* Encelius). *De Re Metallica, hoc est de origine, varietate, & natura corporum metallicorum lapidum gemmarum, atque aliarum quae ex fodinis eruuntur rerum, ad medicinae usum deserventium li. iii.*

Christoph Egenolph. Franc [sic] [n.d.] [No. Col.]

BM cites 'Franc[ofurti 1551]' 'Franc[ofurti] 1557'.

From the absence of a date in No. 224 (and the date of the dedn. 1551), I infer *Ed. p.* (Duv. p. 192. Ferg. reports Young copy imperfect). Th. *Hist.* VI, 308-10 gives 1557 as *Ed. p.*, probably unaware of BM copy with inferred date 1551.

Dedn. by Egenolph to 'his friend, Philip Melanchthon' states that Entzelt wrote the work, not to dispute with Georgius Agricola, but for the advancement of philosophical studies. It is a useful source of vernacular names of ores, gems, etc., with (Duv.) an alchemical bias.

225. The four following items are here grouped in the order given in BM Old Catalogue. From the dates of dedns. they appear to have been published as four separate 'works'; but the choice of caps. and l.c. in the sigs. suggests that the four 'parts', together with their additions, were conceived as a single 'work'. The fact that the sigs. do not run continuously throughout serves to confirm the BM view of separate publication. Study of the Aberdeen copies was complicated by their having been bound, in the wrong order, in two volumes. No. 23 is bound in the same style (18th cent.?), gold stamped on spine, 'Erasti Opera', one being 'Tom 5'. They were acquired in the Forbes Collection and one has a pencilled note: '3 v[ols] 7/6' [!]

ERASTUS, T. (1) *Disputationum de Medicina nova Philippi Paracelsi pars prima, in qua, quae de remediis superstitiosis & magicis curationibus ille prodidit, praecipue examinantur* . . .

P. Perna, Basel n.d.

Only edn. BM (see below).

No col. but last fo. has (*r*) a fine portrait of Paracelsus (aetat 47), and (*v*) a libellous note on him by Conrad Gesner.

(2) *Disputationum de nova P— P— Medicina pars altera, in qua philosophiae Paracelsicae principia & elementa explorantur* . . . .

[Device of:] P. Perna, [Basel] 1572

Followed (sep. t.p. and pagn.) by *Explicatio quaestionis famosae illius utrum ex metallis ignobilibus aurum verum & naturale arte conflari possit* . . . *Declarantur in hac quaestione omnia ferme quae ad naturam, ortum, generationis modum, species, & materiam metallorum pertinent.*

[Device of:] P. Perna 1572

Also (sep. t.p. & pagn.) *Epistola de natura, materia, ortu atque usu lapidis sabulosi, qui in Palatinatu ad Rhenum reperitur.*

[Device of:] P. Perna, Basel 1572

Pars Altera, the *Explicatio*, and *Epistola* are treated by BM as one book.

(3) *Disputationum de nova P— P— Medicina pars tertia, in qua* . . . *verae medicinae assertio, & falsae, vel Paracelsicae confutatio continetur* ... *Cui accessit tractatio de causa morborum continente, eodem authore* . . . .

[Device of: P. Perna] [Col. P. Perna, Basel 1572]

The *Tractatus de Morborum causa continente* ('nunquam antea . . . editus') follows with sep. t.p. and pagn. The dedn. is 1572, the *explicit* of text 1561.

Only edn. BM.

(4) *Disputationum de nova P— P— pars quarta et ultima. In qua epilepsiae, elephantiasis seu leprae, hydropis, podagrae, & colici doloris vera curandi ratio demonstratur & Paracelsica—confutatur* . . . .

P. Perna, Basel 1573

Only edn. BM.

Dedn. to *consules* etc. of Ulm.

85

226. ERASTUS, THOMAS. *Disputatio de Putredine, in qua natura, differentiae, et causa putredinis ex Aristotele & ipsa rerum evidentia clare exponitur, consensio inter philosophos & medicos declaratur.* . . .

Sumpt. Oporinianorum, L. Ostenus, Basel 1580

Ed. p.

Prov. LIDDEL.

Followed by *Theses de Putredine . . . praeside . . . . D. Thoma Erasto . . . respondebit . . . Leo Wolfhardus . . . Anno Domini* 77 [*sic*] and *Disputatio de Febribus Putridis in qua tria de febribus paradoxa D. Laurentii Iuberti . . . excutiuntur a T— E—.*

Cont. pagn. For Joubert see No. 362.

227. ERASTUS, T. *De Astrologia divinatrice epistolae D. Thomae Erasti, iam olim ab eodem ad diversos scriptæ & in duos libros digestae ac nunc demum in gratiam veritatis studiosorum . . . aeditae opera & studio Joannis Jacobi Grynaei.*

Petrus Perna, Basel 1580

No earlier edn. BM or Bibl. Nat.—probably distributed in MS.

228. ERASTUS, T. *Disputatio de Auro Potabili, in qua accurate admodum disquiritur, num ex metallis, opera chemiae, concinnata pharmaca tute utiliterque bibi possint . . . Adiectum est ad calcem libri iudicium eiusdem de indicatione cometarum ex veris fundamentis & naturae principiis erutum.*

Off. Perna, C. Waldkirch, Basel

(Col. 1584).

Duv. p. 193 gives this as 2nd edn. *Ed. p.* 1578 (Bibl. Nat.—agreeing with dedn., though text is dated 1576 at end). BM, however, mentions an edn. of 1580, making this at least the third.

The section on comets has half-title *De Cometarum Significationibus . . .* and sep. pagn.

The dedn. (to Michael Conarsin) refers to a recent attempt by reputable physicians to consider the ways by which the purer and superior (*puriores et praestantiores*) parts of medicines could be separated from the impurer and baser (*impurioribus et ignobilioribus*). And he claims to know 'men who cheated others of an incredible amount of silver and gold for red syrup concocted out of the vilest ingredients'. The value of 'potable gold' was another *cause célèbre* at this time. (See Vol. I, p. 173 for significance.) The tract on comets is a sound account of Aristotelian theory, followed by a well argued case (based on 'artificial and natural signs') for rejecting the view that they portend disasters (see Vol. I, p. 122).

86

229. ERASTUS, T. *Varia Opuscula Medica . . . quae cum ipse studiosis communicare statuisset morte praeventus in lucem edere non potuit. . . .*

Io. Wechel, Frankfurt 1590

Ed. p. Dedn. to Frederic, son of Ludovic, Elector Palatine, by Jac. Castelvitreus, dated London, 1590.

Prov. Gordon. A second copy 'Jo. Gregory 1767'. A third copy also.

A collection of eleven short works, indexed in prelims.

230. ERASTUS, T. *. . . Disputationum & Epistolarum Medicinalium volumen . . . Nunc recens . . . editum, opera et studio Theophili Maderi. . . .*

Io. Wolphius, Typis Froschouer, Zürich 1595

Only edn. BM.

A second copy in the Forbes Collection—see No. 225

In his dedn. Wolphius refers to Erastus ('piae memoriae') whom he heard teaching at Basel and Heidelberg. *Epistola* vii, on the plague, contains the query, 'an seminaria pestis necessario eiusdem generis animante oriantur?' (See Vol. I, p. 278.)

See also Nos. 161, 729.

ESSLER, IOANNES. See No. 513

231. EROTIAN. *Vocum, quae apud Hippocratem sunt, collectio. Cum annotationibus Bartholomaei Eustachii. . . . Eiusdemque Eustachii libellus de Multitudine.*

L. A. Junta, Venice 1566

Only edn. in this form.

The *Lectori* contains an interesting account of the editor's search for the text which had been stolen from the Vatican Library, and of the concerted effort of several scholars to translate a mutilated text ultimately found bound with a volume of works by Hippokrates. The *De Multitudine* follows with cont. pagn. It is a commentary on Galen's work with the same title.

Erotian was a Greek medical commentator of 1st cent. A.D. See also Nos. 232, 236.

232. ESTIENNE, HENRI (the 2nd). *Dictionarium medicum, vel expositiones vocum medicinalium, ad verbum excerptae ex Hippocrate etc. Cum Latina interpretatione . . . Lexica duo in Hippocratem huic dictionario praefixa sunt, unum Erotiani, nunquam antea editum: alterum Galeni, multo emendatius quam ante excusum.*

Excudebat Henricus Stephanus illustris viri Huldrice Fuggeri typographus [n.p.] 1564.

87

This, the only edn. recorded in BM, is believed to have been printed in Geneva, whither his father, Robert, had to fly in 1552 to escape persecution by the Sorbonne. Putnam (*op. cit.* II, 52, 54) records the regrettable fact that in printing for Calvin the indictment against Servetus he was accessory to another's persecution.

Prov. 'Thomas Langton 1569'.

233. EUCHYON, DIODORUS. [*De Polychimia li. iv*] [Basel 1567?], t.p. and part of preface wanting: title taken from text heading; place and date, from BM.

Prov. LIDDEL.

Not in Ferg. or Duv.

The 'glossary' preceding text purports to have been compiled by 'Ioan. Garlandius Anglus'—for him see Ferg. I, 421 under 'Hortulanus'; but there seems to be a good deal of mystery about both 'Garland' and 'Euchyon' (see Sarton, *Intro.* III, p. 1575). Page 230—a good cut of a 'calcination furnace'.

234. EUCLID —. (*a*) *Optica & Catoptrica e Graeco versa per Ioannem Penam Regium Mathematicum.* . . .

Andr. Wechel, Paris 1557

Ed. p. But Conrad Dasypodius (No. 190) edited in the same year (Strassburg) a Greek text and Latin version of the *Catoptrica* alone. No. 234a was also bound up preceded by the Greek text under the following title *ΕΥΚΛΕΙΔΟΥ ΟΠΤΙΚΑ ΚΑΙ ΚΑΤΟΠΤΡΙΚΑ Euclidis Optica & Catoptrica nunquam antehac Graece edita* (No. 234b) above the same imprint as No. 234a

Prov. (No. 234a) Johnston.

No. 234b contains a long introduction by Pena which contains some discussion of the Copernican system as compared with the Aristotelian, and the bearing of the problem on the views attributed to Philolaos, Ekphantos, Seleukos, Aristarchos, Witelo, and others. This is followed by the postulates, including that which lays down that 'radios ab oculis emissos' (actually superfluous, since the *direction* of rays—to or from the eye—can make no geometrical difference). The *Catoptrics* (usually regarded as of later origin than Euclid's time) is defined as 'that part of optics which teaches the illusions (*fallacias*) of mirrors'. The introduction is addressed to Charles, Prince and Cardinal of Lorraine.

235. EUCLID. *The Elements.*

It is proposed to give a list of all editions prior to 1600 of which the Library possesses a copy. It may be as well to recall that this most influential work was first introduced to Western Europe in the Latin translation from the Arabic carried out by Adelard of Bath in Toledo about 1100. There were only two incunable editions of the Latin text. It is to be noted that many so-called 'editions' comprise only the enunciations of the theorems and vary considerably in regard to the number of the thirteen original books (the 14th and the 15th having being added by others) presented.

(a) *Incipit:* 'Praeclarissimus liber elementorum Euclidis per-spicacissimi in artem Geometrie incipit quam foelicissime . . . Punctus est cujus pars non est. . . .'

Col. *Opus Elementorum euclidis megarensis in geometriam artem in id quoque Campani perspicacissimi Commentationes finiunt. Erhardus Ratdolt . . .* Venice 1482, Hain 6693

Ed. p. a beautifully printed book with marginal diagrams—the difficulty in the printing of which, according to Ratdolt himself, was a bar to the early printing of the work. Note the confusion of Euclid of *Alexandria* with the earlier Euclid of *Megara*, a pupil of Socrates—a confusion probably started by Regiomontanus.

(b) *Euclidis Megarensis Geometricorum Elementorum li. xv, Campani Galli Transalpini in eosdem commentariorum li. xv. Theonis Alexan-drini Bartholomaeo Zamberto Veneto interprete, in 13 priores commentariorum li. XIIII. Hypsiclis Alexandrini in duos posteriores eodem B— Z— V— interprete commentariorum li. ii.*

Henri Estienne, Paris n.d. (1516 dedn.)

Prov. LIDDEL.

Ed. by Jacques Lefèvre of Etaples (Faber Stapulensis), this is probably the 5th edn., but the first to achieve anything approaching a definitive status. This copy is bound with several other works. Another copy without t.p. or col., but same registration, is bound alone.

(c) *EYΚΛΕΙΔΟY ΣΤΟΙΧΕΙΩΝ* βιβλ ιϲ *Adiosta Praefatiuncula in qua de disciplinis mathematicis nonnihil.* [Followed by the com-mentary of Proclus Diadochus.]

Io. Hervagius, Basel 1533

1st edn. of the Greek text.

Dedn. by the editor, Simon Grynaeus, to Bishop Cuthbert Tunstall (see No. 690). Both texts in Greek only.

(d) *Euclidis Megarensis . . . sex libri priores de Geometricis Principiis, Graeci & Latini. . . . Algebrae porro regulae, propter numerorum exempla, passim propositionibus adiecta, his libris praemissae sunt, eaedemque demonstratae, authore Ioanne Scheubelio, . . .*

Io. Hervagius, Basel [1550. Col.]

Dedn. to the Fugger brothers. The opening of the Algebraic secn. resembles closely the author's *Algebra* (see No. 625).

89

(e) *Jacobi Peletarii Cenomani in Euclidis Elementa Geometrica Demonstrationum libri sex.*

Io. Tornaesius & Gul. Gazeius, Lyon 1557

Prov. 'Ioannes Lyndesius Lutetiae 1578'.

Jacques Peletier (see No. 505) states that he 'has excluded proofs unsuitable to geometry, among which are those which rely on superposition of figures, as they are called'. He comments on the difficulties of contact between lines and circles, parallelism, etc. Appended are interesting *epistolae* to Fernel, Ronsard, Cardano, and Nonius.

(f) *Euclidis Elementorum libri XV Graece & Latine.* Widow of Gul. Cavellat, Paris 1598.

A later edn. of that published (as shown in dedn. and preface) in 1557. There is also a copy of the 1573 edn. (Ed. St. Gracilis). Prov. 'Thomae Rhoedi et amicorum' in his hand.

(g) *The Elements of Geometrie of the most auncient philosopher Euclide of Megara faithfully (now first) translated into the Englishe toung by H. Billingley, Citizen of London, whereunto are annexed certaine scholies. With a very fruitfull praeface made by M. I. Dee, specifying the chiefe mathematicall sciences, what they are & whereunto commodious: where, also, are disclosed certaine new secrets mathematicall & mechanicall, untill these our daies greatly missed.*

John Daye (Col. London 1570)

Ed. p. First English trans.

John Dee's 'Fruitfull Praeface' is a long discourse on the nature of mathematics, followed by a review of the arts and sciences based thereon—wordy, repetitive, over-nice, yet containing many historical allusions indicative of the intellectual climate of the Elizabethan age. The 'scholies' are in many cases of considerable length.

(h) *Euclidis Elementorum libri XV. Accessit XVI de solidorum regularium comparatione. Omnes . . . demonstrationibus accuratisque scholiis illustrati. Auctore Christophoro Clavio Bambergensi Societatis Jesu.*

Vincentius Accoltus, Rome 1574

The second volume has the title *Euclidis Posteriores libri sex a X ad XV. Accessit xvi de solidorum regularium comparatione*, Rome 1574.

There is also a copy with a slightly different t.p. Io. Bapt. Ciotti, Cologne 1591.

Dedn. to Emanuel Philibert.

Clavius was one of the leading mathematicians and commentators of 16th-17th cent. (No. 108).

(i) *Euclide Megarense . . . solo introduttore delle scientie mathematice . . . rassettato . . . per . . . Nicolò Tarteleo Brisciano seconde le due tradottioni . . . Coretto e ristampato.*

Heirs of Troian Navo, Venice 1586

Prov. Purchased 1958.

With an intro. and copious textual commentary.

236. EUSTACHIUS, BARTHOLOMAEUS. *Erotiani scriptoris vetustissimi vocum quae apud Hippocratem sunt, collectio. Cum annotationibus Bartholomaei Eustachii . . . Eiusdemque . . . libellus de Multitudine. Ed. p.*

Dedn. to Cardinal Julius Feltre.

FABER, STAPULENSIS. See LEFEVRE, J.

Juntae, Venice 1566

237. [FABRE, PIERRE JEAN. *Palladium Spagyricum*]

[Petrus Bosc, Toulouse 1624]

t.p. wanting; only a line col.—*le* 14 *Fevrier achevé d'imprimer* 1624. Title from dedn. (1623). Imprint from Duv. p. 201 who gives *Ed. p* as 1624. The h.t. suggests the edn. of 1638. The line col. may be copied from *Ed. p.* Ferg. I, 260 does not mention this edn.

Dedn. to Louis XIII. The *Lectori* provides a good *Apologia* for alchemy.

238. FABRE, P. J. *Myrothecium Spagyricum; sive pharmacopoea chymica occultis naturae arcanis, ex hermeticorum medicorum scriniis depromptis abunde illustrata.*

Petrus Bosc, Toulouse [Tectosagum] 1628

Ed. p. BM and Bibl. Nat.

Prov. P. A. Urquhart M.D. signs a reference to an index of this work in another book.

A not very convincing justification of 'spagyrist' remedies on rational grounds, since such abominations as the 'quinta essentia' of human blood, earthworms, etc., are included.

239. FABRICIUS, GEORGIUS. *De Metallicis Rebus ac Nominibus observationes variae & eruditae, ex schedis G— F— quibus ea potissimum explicantur, quae Georgius Agricola praeteriit.*

[Device of:] (Andreas?) Gesner, Zürich, 1565
(dedn. Autumn Equinox, 1566)

Only edn. BM.

Prov. BUTE.

Dedn. to Io. Kentmann (No. 365) by Iacobus Fabricius, the author's brother.

240. FABRICIUS (*syn.* Fabritius), Gulielmus (of Hilden). *Selectae Observationes chirurgicae quinque & viginti. Item de Gangraena et Spacelo tractatus methodicus.*

Gabriel Cartier (n.p.—Lyons?) 1598

Ed. p.

Dedn. by Io. Rheterius states that 'he chose and arranged the Observations and turned them into Latin' [from French].

De Gangraena sep. t.p. and date; cont. pagn.

Prov. Liddel.

Cuts of surgical instruments. Obs. XXV consists of 'objects in a dissected body worthy of observation'.

241. FABRICIUS, G. *Lithotomia Vesicae . . . Written first in High Dutch by G— F— H—. Afterwards augmented by the author, and first translated into Latin by his Scholler and Communer Henricus Schobingerus Sangalthensis and now done into English by N.C. for the generall good of this nation, and particular use of the Societe of Chirurgians. With better instruments than heretofore.*

John Norton, London 1640

'High Dutch' is, of course, what we call High *German*. The HG edn. is not BM or Bibl. Nat. A Latin edn. of 1628 is in BM.

Prov. Read.

Dedn. by Norton to 'the Worshipfull Companie of the Barber-Chirurgians'. Preface by Alexander Read, 'one of the Fellows of the famous medical College which is in London'.

242. FALE, T. *Horologiographia—The Art of Dialling.*

Kyngston, London 1626

[*Ed. p.* 1593 STC 10618. This edn. is 2nd cited in STC.]

Dedn. 'To my loving kinsman, Tho. Osborne', in which Fale says that the book had been 'in a manner perfected . . . these 7 years'. He refers to the figures having been 'graved' by M. Iod. Hondius, who also did those for 'the great globes set forth by M. Mullineux, and the maps of England for M. Camden's book'.

A second copy dated 1627, has on t.p. *v* 'Singulis artium mathematicarum studiosis in celeberrima Cantabrigiense Academia Thomas Falus eiusdem alumnus et verae matheseos studiosus exiguum hoc grati animi monumentum D.D. Anno 1593'.

A 'popular' work on construction of dials. Table of sines: 0°—90°.

243. FALLOPPIO, GABRIELE. . . . *Libelli duo; alter de ulceribus, alter de tumoribus praeter naturam. Nunc recens . . . editi.*

Donatus Bertellus, Venice 1593

Ed. p.

Prov. LIDDEL.

Dedn. by Bertellus to Io. Iac. Phucharus.

244. FALLOPPIO, G. *De Medicatis Aquis atque de Fossilibus . . . tractatus
. . . ab Andrea Marcolino Fanestri medico ipsius discipulo amantissimo
collectus. Accessit eiusdem Andreae duplex epistola.*

Ludovicus Avantius, Venice 1564

Ed. p.

The editor describes in dedn. and *Lectori* how the work was made up
from the author's lectures—the text is in fact dated Padua 1557.

Prov. BUTE.

Page 154 Antimony-alloy alleged to have been confused with
lead by 'Brasaulus' [Brasavolus? *q.v.*]

245. FALLOPPIO, G. (*a*) . . . *Opera quae adhuc extant omnia in unum
congesta & . . . nunc primum tali ordine excusa. . . .*

Heirs of Andr. Wechel, Frankfurt 1584

Dedn. by Felix Valgrisius, from Venice 1584, who speaks of having
been asked to collect all the extant works of F. but whether he has
succeeded he must leave to be judged by others.

Not in BM or Bibl. Nat. Edin. Univ. Lib. has a copy.

Prov. LIDDEL.

(*b*) Main heading same as (*a*) but the replacement of 'nunc
primum' by 'nunc denuo' indicates the existence of an earlier edn.
Different fount.

Heirs of Andr. Wechel, Frankfurt 1600

The dedn. by Valgrisius of (*a*) is replaced by a long and highly
interesting one on the iniquities of the chemists by Io. Crato, dated
Breslau 1585. (See Vol. I p. 257.)

This work is followed (as the main t.p. indicates) by a 'Tomus
Secundus' with sep. t.p. and note by printer.

The volume concludes with an Appendix, which has sep. t.p., pagn.,
and note by printer and is dated 1606.

246. FERNELIUS, IOANNES (*syn.* Fernel, Jean) . . . . *Monalosphaerium, partibus constans quatuor. Prima generalis horarii &
structuram & usum, in exquisitam monalosphaerii cognitionem
praemittit. Secunda, mobilium solennitatem, criticorumque dierum
rationes, multa brevitate complectitur. Tertia, quascunque ex motu primi
mobilis depromptas utilitates elargitur. Quarta, geometricam praxin
breviusculis demonstrationibus dilucidat.*

Simon de Colines, Paris 1526

(Col. Mar. 1527).

Ed. p.

Another copy bound with No. 612 (Prov. DAVID RAIT).

Both have the '1526' *t.p.* Sherrington (*The Endeavour of Jean Fernel*, Camb. 1946) pp. 187-9 records a '1527' variant.

Fine bordered t.p. by Oronce Fine (see No. 262 etc.). (Attribution by E.P. Goldschmidt, quoted by Sherrington, p. 188.)

Dedn. to Jacobus de Govea, later Principal of St. Barbe, from which it is dated 'Feb. 1526'.

247. FERNELIUS, IO. . . . *Cosmotheoria, libros duos complexa. Prior mundi totius & formam & compositionem, eius subinde partium (quae elementa & coelestia sunt corpora) situs & magnitudines orbium tandem motus quosvis solerter [sic] reserat. Posterior ex motibus siderum loca & passiones disquirit.*

Simon de Colines, Paris 1528 (Col. 1527 Feb.)

Ed.p. (Sherrington *op. cit.* p. 188). T.p. border similar to No. 246.

Dedn. to John III of Portugal, referring to the great tradition of Portuguese navigators and to Jacobus de Govea (No. 246). Dated 'Ex alma Parisiorum academia pridie Feb. MDXXVIII'.

On sig. Biii *r* he describes his measurement of an arc of meridian (Vol. I, p. 153) and refers to his recently published *Monalosphaerium*.

248. FERNELIUS, IO. . . . *de Proportionibus libri duo.* . . .

S. de Colines, Paris 1528

Ed. p. (Sherrington *op. cit.* p. 189). T.p. border similar to 246.

Dedn. dated 'Parisiis apud clarissimum divae Barbarae gymnasium ad calendas Novembres 1528' (Bound with No. 247).

249. FERNELIUS, IO. . . . *de naturali parte medicinae, libri septem.* . . .

Tornaesius & Gul. Gazeius, Lyon 1551

[*Ed. p.* S. de Colines, Paris 1542: Sherrington *op. cit.* p. 189] 3rd edn. recorded by Sherrington.

Prov. 'Hunc librum don. dedit Univ. Aberdon. bibliothecae C.S.S.', i.e. presented by Sir Charles Sherrington as a mark of gratitude for the placing at his disposal of No. 251. It was bought by him in Paris in 1935, and bears the names of two previous owners.

This work was renamed *Physiologiae li. VII* in the first collected edn. *Medicina*, A. Wechel, Paris 1554. (See Vol. I, p. 214.)

250. FERNELIUS, IO . . . . *de vacuandi ratione liber, quem vulgatiori nomine Practicam possumus inscribere.*

Gul. Rovillius, Lyon 1549

[*Ed. p.* Christ. Wechel, Paris 1545: Sherrington *op. cit.* p. 190.] This is the 5th edn. recorded by Sherrington.

This work, concerned mainly with the theory of venesection in relation to the 'humours', was from 1554 incorporated into the collected works as Book II of the *Therapeutice*.

251. FERNELIUS, IO. . . . *de abditis rerum causis libri duo.*

(*a*) *Ed. p.* Chris. Wechel, Paris 1548.

A variant t.p. is recorded by Sherrington, p. 191. This is the only copy in Great Britain (Sherrington, p. 191).

(*b*) Andr. Wechel, Frankfurt 1575.

(*c*) . . . *postremo ab ipso authore recogniti compluribusque in locis aucti.* . . . Andr. Wechel, Frankfurt 1581.
Prov. LIDDEL.

(*d*) As in (*c*). Thos. Soubron & Moyses des Prez, Lyon 1597.

252. FERNELIUS, IO. . . . *Universa Medicina tribus et viginti libris absoluta ab ipso quidem authore ante obitum diligenter recognita, & quatuor libris nunquam ante editis ad praxim [sic] tamen perquam necessariis aucta. Nunc autem studio . . . Guil. Plantii . . . elimata & in librum Therapeutices septimum . . . scholiis illustrata.*

(*a*) *Ed. p.* Andr. Wechel, Paris 1567 (Sherrington p. 195).
Prov. GORDON.

(*b*) Similar to edn. of Frankfurt, 1574, but contains No. 253 (see Sherrington, p. 196) Andr. Wechel, Frankfurt 1577.

(*c*) Andr. Wechel, Frankfurt 1581. Similar title, but *Postea autem studio* . . . . *Editio quarta.*
Prov. LIDDEL.

(*d*) Junta & Pa. Guittius, Lyon 1586.

The name of Paolo Guittio, the head printer, was included in the imprint, during this year only, after the death of Jeanne Giunta.
Prov. 'Ex. lib. Bullei'.

(*e*) . . . *Editio sexta. Qua nunc primum accedit vita auctoris ab eodem Plantio.* . . . .
(See Sherrington, p. 197.) Cl. Marnius & heirs of Io. Aubrius, Frankfurt, 1607.

The contents of this 'standard collection' changed gradually, but the *Ed. p.* contained: Preface; portrait of Fernel; intro. by Giullaume Plancy, friend, colleague, and biographer, of Fernel; *Physiologiae li. vii* (No. 249), *Pathologiae li. vii, Therapeutices universalis, seu medendi rationis li. vii* (No. 253), *De abditis rerum causis* (No. 251).

95

253. FERNELIUS, IO. . . . *Therapeutices universalis, seu medendi rationis, libri septem, quam totius medicinae tertiam fecit partem.* . . .

Andr. Wechel, Frankfurt 1581

Prov. LIDDEL. Several pages MS in his hand.

The *Therapeutice*, in seven books, had already been published in *Universa Medicina* (No. 252). In No. 253 it is followed by *Febrium curandarum methodus generalis* (No. 254); *De luis venereae curatione perfectissima liber* (*see* No. 254); and *Consilium epileptico praescriptum.*

This edn. is not recorded by Sherrington, who evidently did not know of it, since he fails to mention it in his list of works in which the *Consilium* was published (*op. cit.* p. 203).

254. FERNELIUS, IO. . . . *Febrium curandarum methodus generalis, nunquam antehac edita.*

Andr. Wechel, Frankfurt 1577

Ed. p. (Sherrington, p. 201).

The dedn. by Jean Lamy, the editor, to Jean Barjotte, Fernel's grandson, explains that the death of Plancy had delayed the publication of this work found among Fernel's papers after his death. The *Lues venerea* was also found and published separately 1579. Publication was necessary, as so frequently, to prevent injury to the author's reputation by garbled versions.

255. FERNELIUS, IO. *Medicinalium consiliorum . . . centuria. Ex adversariis quadringentarum consultationum eius selecta. Tertia editio. Priori non auctior solum sed longe correctior.*

Io. Wechel, Frankfurt 1593

Prov. LIDDEL. Also a second (damaged) copy.

Dedn. by Zacharias Palthenius, dated 'anno 93'. Also a second dedn. by Gul. Capellus to Iulianus Palmarius (No. 491), who had been a pupil of Fernel.

256. FERRIER, AUGER. *Liber de Somniis. Hippocratis de insomniis liber. Galeni liber de insomniis. Synesii liber de somniis.*

Io. Tornaesius, Lyon 1549

Ed. p. of this collection.

Dedn. to Robert Hay 'Remensem' ('of Rheims'), in which the author regrets the decay of the study of 'the most secret mysteries of nature; even those which we are apt to consider of least worth'. These 'mysteries', however, turn out to be nothing 'scientific', but the Sibylline books, auspices, *et hoc genus omne*!

The Greek originals are translated by J. C. Scaliger, Guinther of Andernach, and Marsilio Ficino respectively.

257. FERRIER, A. *Liber de diebus decretoriis secundum Pythagoricam doctrinam & astronomicam observationem.*

Io. Tornaesius, Lyon 1549

Ed. p.

In the dedn. to Cardinal Castillion, Ferrier rejoices in the benefit to morals and religion which will result from the Cardinal's designation as Archbishop [of Toulouse?]. He refers again to Robert Hay (No. 257). 'Equidem cum superioribus annis Roberti Hayi tui . . . consuetudine uterer familiariter videbar mihi pondus (quod ait Ecclesiasticus) ab humeris deponere'. The text contains a valuable outline of the doctrine of 'critical days'.

258. FERRIER, A. *Les jugements astronomiques sur les nativitez.*

Jean. de Tournes, Lyon 1550

Ed. p.

Dedn., to 'Catherine, Royne de France', refers to the Academy 'de vos ancestres' [the Medici.]

In the text instructions are given for the use of the Alphonsine and Regiomontane Tables. Englished in 1593 by Thos. Kelway—see Allen, p. 170.

See also No. 400.

259. FICINO, MARSILIO. [. . . *de Vita*].

(*a*) Sig a i wanting; a ii (unsigned) begins 'Prohemium Marsilii Ficini Florentini in librum De Vita . . .'.

The text ends '1489 *In agro Caregio*'.

Ed. p. of *De Triplica Vita* (*syn. De Vita*) Florence 1489 Hain 7065.

Prov. Hector Boece's auto. under col. There is no imprint. Copinger 2497 a (Vol. II, 2,309).

(*b*) *de Triplica Vita.* Venice 1495.

(*c*) . . . *de Vita libri tres, recens iam a mendis . . . vindicati . . . Quorum primus de studiosorum sanitate tuenda. Secundus de vita producenda. Tertius de vita coelitus comparanda. His accessit de ratione victus salubris opus nunc recens natum autore Gulielmo Insulano Menapio . . . . Epidemiarum antidotus . . . autore Marsilio.*

Barth. Westhemerus, Basel 1541

Prov. LIDDEL.

Cont. pagn. Numerous dedns.

FICINO, M. See under Plague Tracts, Nos. 752-3.

260. FIENUS, IOANNES. . . . *de Flatibus humanum corpus molestantibus commentarius novus.* . . .
Officina Sanctandreana, Heidelberg, 1592.
[*Ed. p.* Heidelberg 1589.]
Prov. BUTE.

A systematic work on 'spirit' (*spiritus = aer, ventus, flatus*) and its supposed consequences. Mainly based on classical sources. Numerous compound remedies are included.

261. FINCK, THOMAS. . . . *Geometriae rotundi libri xiiii.*
Seb. Henricpetrus, Basel (Col. 1583)
Ed. p.
Prov. LIDDEL.

The dedn. to Frederick II of Denmark exemplifies the importance of mathematics by reference to the report of the defeat of Philip V of Macedon as a result of his having had too short scaling ladders prepared for the assault of Melitaea (Polybius V, 97), and the same was said to have happened to the French before Milan. In the *Lectori* he calls by name upon many leading mathematicians.

Numerous cuts, including one showing the calculation for a ladder crossing a moat on to a wall; also the sextant level.

262. FINE, ORONCE [*syn.* Finaeus, Orontius]. . . . *Protomathesis: opus varium . . . nunc primum in lucem . . . emissum.*
(Col. Gerardus Morrhius & Io. Petrus) Paris 1532
Ed. p.
Prov. 'Theophili Stuart liber ex amicorum . . . Walter S. Gul (?) Rhaed'.

Dedn. to Francis I, who had made him 'liberalium disciplinarum professor regius'—perhaps the origin of 'Regius' professorships, but similar 'Chairs' were held under the Byzantine emperors. Fine's chair was in mathematics: there were others in languages, etc. The beautiful frontispiece of Urania surrounded by instruments, including the armillary sphere, is inscribed 'hanc author proprio pingebat marte figuram'; he is known to have done several t.p. margins for the Paris printers—see Fernel's early works.

The work, finely illustrated, includes geometry, arithmetic, cosmography, horology.

263. FINE, O. . . . *Quadrans astrolabicus omnibus Europae regionibus inserviens: ex recenti & emendata ipsius authoris recognitione in ampliorem ac longe fideliorem redactus descriptionem.*
Simon de Colines, Paris 1534

Ed. p. in this form: BM, also dedn. to Ludovicus Lassereus 'Regalis Collegii Navarrae provisori' referring to 'Planisphaerii (quod vocant Astrolabium)', and signed 'apud inclytam Parisiorum universitatem 1534'. The word 'universitas' is not common during the early 16th cent. For the astrolabe see Vol. I, p. 118.

264. FINE, O. . . . *de Arithmetica practica libri quatuor. Ab ipso authore . . . recogniti multisque accessionibus recens locupletati.*

Mich. Vascosanus, Paris 1555

[*Ed. p.* Paris 1535 Bibl. Nat. Earliest BM is 3rd (1542).]

Smith, p. 163, states that this is the arithmetical part of the *Protomathesis* (No. 262) extracted and revised. He takes a dim view of its originality. It is nevertheless very clearly arranged.

265. FINE, O. *Les Canons & documens tresamples touchant lusaige & practique des communs Almanachz, que l'on nomme Ephemerides. Briefue . . . introduction sur la iudiciaire astrologie pour scavoir prognostiquer des choses advenir . . . Le tout nouvellement & tresclerement redigé en langaige Francois.*

Simon de Colines, Paris 1543

Ed. p.

Earliest BM 1556.

Similar to Digges (No. 196), but lacking the diagrams and detailed tables.

266. FINE, O. . . . *In eos quos de mundi sphaera conscripsit libros, ac in planetarum theoricas, canon astronomicorum libri ii.*

Mich. Vascosanus, Paris 1553

Ed. p.

Not BM.

Clear line diagrams and tables. The hoary problems of the motion of the '8th sphere' and the eclipse of the '9th' sphere are discussed at the end. (See Vol. I, p. 114.)

See also No. 747.

267. FIORAVANTE, LEONARDO. *Del compendio de' secreti rationali . . . libri cinque.*

Giacomo Cornetti, Venice 1591

Ferg. I, 277 cites as *Ed. p.* that of Valgrisius, Venice 1564, but the dedn. to Marcantonio Colonna is dated Venice 1571. Not Duv.

2 copies.

The title comprises a list of contents, including '. . . Molte belletti chi usano le donne per apparer belle . . .'—evidently an early text of cosmetics.

268. FIRMICUS MATERNUS, JULIUS. . . . *ad Mavortium Lollianum astronomicωn lib. viii per Nicolaum Prucknerum astrologum nuper ab innumeris mendis vindicati. His accesserunt: Claudii Ptolemaei . . . Quadripartitum . . . lib. iiii. De inerrantium stellarum significationibus lib. i* ('per Nicolaum Leonicum a Graeco translata' added on p. 79). *Centiloquium eiusdem* ('Joviano Pontano interprete' added on p. 74) [and a number of other works by the Arabs and 'Chaldaeans']. *Postremo Othonis Brunfelsii de diffinitionibus [sic] & terminis astrologiae libellus isagogicus.*

Basel 1533

The earlier of Pruckner's two edns. *Ed. p.* of Latin text, Venice 1497.

Prov. LIDDEL.

The works of 'Ptolemy' (of the two, only the 'Quadripartitum' is believed to be genuine) follow with sep. pagn.

269. FOESIUS, ANUTIUS. *Oeconomia Hippocratis alphabeti serie distincta.* . . .

Heirs of Andr. Wechel, Frankfurt 1588

Ed. p.

A 'Hippocratic lexicon', as the author himself describes it.

FOESIUS A. See Hippocrates.

270. FONTANONUS, DIONYSIUS. (*a*) *De morborum internorum curatione libri quatuor, adiectis ab Ioanne Raenerio . . . causis & signis, ex Galeno . . . atque Aetio desumptis.*

Io. Frellonius, Lyon 1550

Not BM or Bibl. Nat. Earliest BM 1563.

(*b*) *De m— i— c— li— q— adiecta epistola de priscis & neotericis medicis, eorumque scriptis & curandi methodis in praesenti practica usurpatis.* . . .

Ant. de Harsy, Lyon 1574

Raenerius speaks of Fontanonus as having been his teacher at Montpellier.

271. FORESTUS, PETRUS. *Observationum et curationum medicinalium . . . libri . . .*

Off. Plantiniana, Franc. Raphelengius, Leiden (dates as under).

The small Roman numerals below refer to bound volumes; Arabic numerals to the years on the sep. t.p.'s. Titles are all of the form given above, but vary slightly in the listing of the *libri*. Some errors in 'gathering', etc., have been observed but are not recorded here.

(i) 1593 (dedn. 1584) Libri i & ii *De Febribus*, and 1591 (dedn. 1585) Libri. iii, iv, v, and 1591 (dedn. 1587) Libri vi, vii.

Prov. CARGILL, and 'Coll. Reg. Abred. ex dono Mgri. Jacobi Morisoni'. A second copy, LIDDEL.

(ii) 1590 . . . *Nempe octavus . . . nonus . . . decimus* (Liber x, new pagn.) Prov. LIDDEL.

(iii) 1591 Libri xi, xii, xiii, xiv, xv.

Prov. Two copies, one LIDDEL.

A third copy is bound with Libri xvi, xvii (1593), prov. CARGILL.

(iv) 1594 Liber xviii. 1595 Libri xix, xx and *de Scorbuto.* 1596 Libri xxi, xxii, xxiii. 1596 Libri xxiv, xxv and *de Scorbuto.*

Prov. CARGILL.

(v) 1596 Libri xxiv, xxv. 1597 Libri xxvi, xxvii. Also, Heurnius *De humana felicitate* (cont. pagn.).

Prov. as for (i).

(vi) 1596 Libri xxi, xxii, xxiii. 1596 Libri xxiv, xxv. Also, Heurnius *De humana felicitate* (cont. pagn.).

(vii) 1597 Libri xxvi, xxvii. 1599 Liber xxviii, sub-titled *De mulierum morbis.*

272. FOULLON, ABEL. *De holometri fabrica et usu, instrumento geometrico olim ab Abele Fullonio apud Henricum II Galliae regem invento nunc vero Io. Nicol. Stupani opera . . . sermone Latino ita explicato. Accessit Federici Delphini . . . disputatio de Aestu maris & Motu octavae spherae.*

P. Perna, Basel 1577

[*Ed. p.* Paris 1555 (French) Bibl. Nat. Earliest BM Paris 1561 (French). Latin edn. in neither.]

Prov. LIDDEL.

Secn. heading and sep. pagn. for the *De aestu;* secn. heading but cont. pagn. for the *De motu.* . . . Early running titles faulty.

The main text describes the earliest form of the surveying instrument, now known as the 'plane-table', which 'a quolibet vel mediocris ingenii homine intelligi possit . . . '. The *De fluxu et refluxu* correlates tidal movement with relative positions of sun and moon, and notes (p. 14) the peculiar case of 'mare nostrum' (Mediterranean). The *De Motu octavae sphaerae* provides a historical introduction to this evergreen problem. (See Vol. I, p. 114.)

273. FRACASTORO, GIROLAMO. . . *Lib. l de sympathia & antipathia rerum. De contagione & contagiosis morbis & eorum curatione libri tres.*

Gul. Gazeius, Lyon 1550

[*Ed. p.* Venice 1546.]

The *locus classicus* for the first systematic exposition of a theory of disease transmission by 'germs', concerning the importance of which authorities differ. (See Vol. I, p. 278.)

274. FRACASTORO, G. . . . *Opera omnia in unum proxime post illius mortem collecta.*

(*a*) Juntae 1555.

(*b*) 2nd edn. Venice 1574 (Col. 1573).

Prov. MELVIN.

(*c*) 3rd edn. Juntae, Venice 1584 (Col. MDLXXIIII.)

(*d*) Lyons 1591.

Prov. BUTE.

Ed. p.

Contents: *Vita. Homocentricorum sive de stellis . . . liber unus. De causis criticorum dierum libellus. De vini temperatura sententia. Syphilidis sive de Morbo Gallico libri tres. . . .* Also the works cited in No. 273 and some of purely literary interest. (For *Homocentricorum . . .* and *Syphilidis . . .* See Vol. I, p. 276 and p. 275 respectively.)

FRIGIMELEGA, FRANCESCO. See No. 753.

FUCHS, LEONHARD.

Owing to the confusion which seems to have arisen through the use of similar titles for the different works, strict chronological order has not been maintained.

275. FUCHS, L. *Paradoxorum medicinae libri tres, in quibus sane multa a nemine hactenus prodita Arabum aetatisque nostrae medicorum errata non tantum indicantur, sed & probatissimorum autorum scriptis firmissimisque rationibus ac argumentis confutantur . . . Obiter denique hic Sebastiano Montuo . . . respondetur eiusque annotatiunculae velut omnium frigidissimae prorsus exploduntur.*

Io. Bebelius, Basel 1535

Ed. p. BM and Bibl. Nat. but Biog. Univ. gives '1533', though this may apply only to the included *Annotatiunculae . . . per Leonhardum Fuchsium . . . collecta . . . Lugd.* 1533.

The dedn. to Huldrich, Duke of Würtemberg is dated 1534.

The critical character of the work is indicated by (e.g.) a discussion of the 'sinus seu cellulas' of the human uterus which, he says, Mondinus asserted 'praeter omnem veritatem' to be seven (fo. 109*v*); also 'our *Borago* does not differ from the *Buglossus* of the ancients' (fo. 33*r*.). (Genera of the modern family of flowering plants *Boraginaceae*.)

276. FUCHS, L. (*a*) [*De historia stirpium commentarii insignes . . .,
adiectis earundem vivis plusquam quingentis imaginibus, nunquam
antea ad naturae imitationem artificiosius effectis & expressis. . . .
Accessit succincta admodum vocum difficilium & obscurarum passim
in hoc opere occurrentium explicatio.*
Off. Isingrinana, Basel 1542.] Col. 'Palma Ising' device.
Ed. p. T.p. wanting—title copied from BM.
Prov. 'Est Gabrielis Humelbergii Medici 1542 Nov.' on ft.
   paper (see No. 641). This beautiful volume was recently
   acquired from the Library of the late Sir D'Arcy Thompson.
In 1531 Michael Isengrin married the daughter of Iohann Bebel and
became his partner. Bebel had been printing since *ca.* 1524 in association with Andreas Cratander and others, and had adopted the device
of a palm tree enclosing a printer's platen under which was usually
(but not in No. 276) a recumbent human figure. Isengrin continued to
use this device. (See C. W. Heckethorn *The Printers of Basle*, Lond.
1897, pp. 115 and 172.)

(*b*) Similar title—unillustrated.

Iac. Bogardus, Paris 1546
Glossed with English common names of plants in 17th cent. hand.

(*c*) Similar title, but different arrangement of the *adnotationes*
   following the several descriptions. The section 'forma' is
   illustrated by cuts (many crudely coloured in this copy).
Balthazar Arnolletus, Lyon 1551
Prov. JOHNSTON; also several other early MS. inscriptions.
(For description of this influential work see Vol. I, p. 192.)

277. FUCHS, L. . . . *In Hippocratis Coi . . . septem Aphorismorum libros
commentaria. Ab eodem authore nuper recastigata, adiectis adnotation-
ibus & locorum difficilium Galeni explicationibus.*
Seb. de Honoratis, Lyon 1558
[*Ed. p.* was presumably the work of 1545 with similar t.p. in
   Greek and Latin—BM.]
The dedn. includes a history of the commentaries on Hippocrates
and a ref. to Fuchs's 'answers to the calumnies of a certain envious
individual'—he was embroiled in many such cases and often replied
in unbridled terms. He refers to 'Ugo' (Benzi) and 'Foroliviensis'—
see Vol. I, p. 221.
Greek text with long commentaries *seriatim*.

278. FUCHS, L. *Methodus seu ratio compendiaria cognoscendi veram
solidamque medicinam ad Hippocratis & Galeni scripta recte*
103

intelligenda . . . emendata et quasi de novo . . . edita. Accesserunt huic de usitata huius temporis componendorum miscendorumque medicamentorum ratione libri tres . . . auctiores . . . eodem autore.

Iac. Dupuys, Paris 1550

Ed. p. in this form.

Prov. LIDDEL.

The *Epistola nuncupatoria* contains the phrase 'Physiologiae, hoc est tractationis de corporis humani natura'. This is an earlier use of the term, in something like the modern sense, than Fernel's (see No. 249 also Vol. I, p. 251). In his remarks on anatomy, Fuchs refers to 'charissimi amici D. Andreae Vesalii . . .' dated Tübingen 1548.

The whole work is of great importance in assessing the changing viewpoint in medicine.

The bibliography of this and similar works is difficult, and their arrangement in the catalogues of BM and Bibl. Nat. is confusing. From the Epistola (sig. a v *verso*) it appears probable that the work first appeared under the title *Compendiaria ac succincta admodum in medendi artem introductio* [Haganoae 1531 Bibl. Nat.], and in a revised and enlarged form at Strassburg 1535, Bibl. Nat. Bibl. Univ. (art. Fuchs) refers to a folio edn. [Basel 1541] with the title *Medendi Methodus seu ratio compendiaria.*

279. FUCHS, L. *Institutionum medicinae libri quinque; nunc denuo . . . recogniti. . . .*

Hiero. Gemusanus, Basel 1594

[Not BM nor Bibl. Nat. Earliest edn. traced is 2nd edn. Lyon, 1560 BM].

Prov. CARGILL.

Though based on No. 278 this is virtually a new book. In the dedn. Fuchs replies to critics who are complaining of the large number of editions of this work—'est enim haec, ni fallor, jam sextus'—and this dedn. is Tübingen 1565! The continuity of the edns. with No. 278 is shown by the repetition of Auger Ferrier's recommendation which appears in the latter.

Note the use of 'Institutes of Medicine'—which, either in this form or, as 'Institutes of Physic', was commonly used for the equivalent of 'Physiology' in something like its modern sense until at least the end of the 18th cent. A *graduated* urine glass is shown on p. 702.

280. FUCHS, L. (*a*) *De medendis singularum humani corporis partium a summo capite ad imos usque pedes passionibus ac febribus libri quatuor nunquam antea . . . editi. . . .*

Robert Winter, Basel 1539

The dedn. to Albert of Brandenburg states that he wrote it for the sake of those who 'put off by the difficulty and prolixity of the works of Galen and the other Greeks, go over to the Arabs and 'ineptissimam barbarorum medicorum turbam'.

(b) *De medendi methodo libri quatuor. Hippocratis Coi de Medicamentis purgantibus libellus iam recens editus.*

C. Neobarius, Paris 1539

The t.p. and dedn. only of this work are bound in a volume which otherwise consists of the *In Galeni de Usu partium libros epitome* of Theophilos Protospatharios, with imprint as above but 1540.

Neither (a) nor (b) is cited in BM, and only (b) by Bibl. Nat. The citation in BM '*De Curandi ratione lib. iv* [Tübingen ? 1539?] Imperfect; wanting title page', may refer to (a), since the dedn. is dated 'Tübingen 1539'.

281. FUCHS, L. (a) *De sanandis totius humani corporis eiusdemque partium tam internis quam externis malis libri quinque nunc primum . . . editi. . . . Obiter etiam in nuncupatoria epistola impudentissimum plagium Gualtheri Riffi Argentoracensis detegitur.*

Io Foucherius, Paris 1543

(b) Similar title.

Io. Frellonius, Lyon 1547

Prov. LIDDEL.

In the dedn. to (a) Fuchs refers to a work ('anno abhinc tertio') of which a new edn. had been called for. Comparison of this with No. 280(a) shows that No. 281(a) is an enlarged version of the former. Bibl. Nat. contains an edn. by Oporinus, Basel 1542; in the dedn. of No. 281(a) Fuchs refers to a request by Oporinus, so this may well be the *Ed. p.* of the enlarged version. Both these edns. contain the *De medicamentis purgatoriis* of Hippokrates 'nunquam prius . . . editus' (but see the t.p. of No. 280(b) and two extra chapters on fevers. The additional fifth 'book' is on tumours. Neither BM nor Bibl. Nat. cite (b). One or two fos. are wanting in this copy of (a).

282. FUCHS, L. *De curandi ratione libri octo causarum signorumque catalogum breviter continentes partim olim conscripti & nunc postremum recogniti, partim recens aucti adiecti.*

(Col. Io. Oporinus, Basel 1568)

This is a revised version of No. 281 together with three further 'books' on surgery—ulcers, wounds and fractures. The *Ed. p.* in this form was Lyon, 1548 (BM and dedn.).

The dedn. refers several times to 'medici nostrae Germaniae', 'who fell into error through adherence to the Arabs and more recent

105

writers instead of following Galen; nevertheless allowance must be made for the fact that Galen dissected apes and dogs'—a typical 16th cent. mixture of criticism and conservatism. Cuts of instruments for the extraction of 'tela' (pointed weapons), and of extension of fractured limbs, with refs. to Galen and Hippokrates. The 'Morbus Gallicus' is also called 'Hispanicus' and 'Neapolitanus'.

283. GABUCINIUS, HIERONYMUS. *De lumbricis alvum occupant-ibus, ac de ratione curandi eos qui ab illis infestantur, commentarius....*

(*a*) Io. Gryphius, Venice 1547.

(*b*) Gul. Rovillius, Lyon 1549.

Prov. LIDDEL.

(*a*) is the earliest in BM, but the dedn. is dated 1544.

A letter from Gentilis Arnulphus to the author contains the significant remark: 'What I chiefly admire is that you have not rated the authority of any one so high as to believe that you must agree with him without testing the facts' ('. . . quod nullius autoritatem tanti fecisti ut absque examine illi accedendum duxeris').

The text is based on a survey of classical sources; but the queries, under which the discussion is carried on, reveal a critical attitude.

284. GABUCINIUS, H. *De comitiali morbo libri iii.*

Aldus, Venice 1561

Ed. p.

Prov. 'Sum Guil. Charci'.

Epilepsy was called *morbus comitialis* by the Romans since its occurrence broke up the *comitia*—the name given to any general assembly of the citizens.

285. GAERINGUS, PHILIPPUS. *Fontium acidorum Pagi Spa et ferrati Tungarensis accurata descriptio, e Gallica Latina facta a Thoma Ryetio . . . Cuius et accesserunt in descriptionem & super natura & usu eorundem fontium observationes.*

Off. Henrici Hovii, Liége 1592

Ed. p.

Prov. BUTE. End papers contain Greek, English, and Latin, MS. referring to Fernel and Erastus. 'Thomas Rose' *in verso*.

The mineral spring at the Belgian village called 'Spa' (= *fons Spadanus*) gave its name to the whole class of such places ('Spaw' in 17th and 18th cent. English). On p. 58 is a ref. to Pliny (Nat. Hist. XXI, 12) 'who described the chalybeate ("ferratum") spring in the country of the Tungri, but did not recognise the acidity of the spring

PLATE III

Title-page of Guinther of Andernach's edition of Galen's text-book of anatomical procedures. The cadaver appears to be a 'wild' man (No. 287/2).

at Spa as some have wished to make out'. On p. 65 a 'certain noble Englishman' is referred to as having shown the great value of the waters in the treatment of complaints of the kidney.

GAIZO, ANTONIUS. See No. 169.

286. GALEATIUS DE SANCTA SOPHIA. *Opus medicinae practicae . . . antehoc nusquam impressum, G— de S— S— in Nonum Tractatum libri Rhaasis ad regem Almansorem de Curatione Morborum particularium huic seculo accommodatissimum. Libellus introductorius in Artem Parvam Galeni de principiis universalibus totius medicinae tam theoricae quam practicae ex doctrina Avicennae & aliorum philosophorum congestus. Omnibus ad Hippocraticam disciplinam anhelantibus . . . necessarius. Quae omnia . . . restituit . . . Georgius Kraut.*

Valentinus Kobian, Hague 1533

[There is a further t.p. at p. 93 where *Isagoges Iohannicii* is referred to: there appears to be no *text* of this since his name appears *passim* in the actual text.]

Presumably *Ed. p.* (not BM).

Incipit of Rhazes 'Quando rubedo et tensio & gravedo in facie . . .' (ThK, 546 refer to a 15th cent. MS). *Incipit* to Galeazzo's commentary 'In nomine domini & Dei misericordis, cuius nutu . . .'. ThK 329 state that this (without 'Domini et') is the *incipit* of Bartholomeo Galeazzo's commentary: for this family, associated with the 14th cent. Paduan medical school, see Castiglioni, p. 332.

287. GALEN, CLAUDIUS. No detailed study of the editions of the works of Galen has been attempted. Those printed before 1600 have been arranged in the chronological order *of their imprints.*

(1) *Galeni in librum Hippocratis de Victus Ratione in Morbis Acutis commentarius.* [All before a ii wanting. No registration; inferred Simon de Colines, since bound with three others from that House.]

(2) *(C)laudii (G)aleni (P)ergameni de Anatomicis Administrationibus libri novem Ioanne Guinterio Andernaco interprete.*

Simon de Colines, Paris 1531

Fine engraved t.p. showing an anatomy. (See Plate No. **IV.**)

(3) *Hippocratis Aphorismi, cum Galeni commentariis, Nicolao Leoniceno interprete. Item eiusdem Hippocratis Praedictiones cum Galeni etiam commentariis Laurentio Laurentiano interprete. . . .*

Simon de Colines, Paris 1532

Prov. 'Ioannes Hamyshoune [?] Pharmacopaeus'.

107

(4) *C— G— P— de Differentiis Febrium libri duo.* Simon de Colines, Paris 1535. (trans. Laurentianus, revd. Simon Thomas).

(5) *C— G— P— de Compositione Pharmacorum . . . libri decem,* Froben, Basel 1537. Io. Cornarius ed. and comm.

(6) *C— G— P— de Tuenda Valetudine secunda, libri sex.*
Balthas. Lasius, Basel 1538
Greek text.
Prov. LIDDEL; several pages of MS.

(7) *G— P— . . . Opera omnia* (2 pts.).
Io. Bebel, Basel 1538

(8) *De Plenitudine libellus, Ioanne Guinterio Andernaco interprete.* . . .
Chris, Wechel, Paris 1539
Prov. GREGORY.

(9) *C— G— P— in Hippocratis li. de Natura Hominis commentarius. Accesserunt Iacobi Sylvii . . . scholia.* . . .
Chris. Wechel, Paris 1540

(10) *C— G— P— . . . liber absolutissimus quem τεχνη ἰατρικη hoc est Artem Medicinalem, inscripsit.* . . .
Berth. Westhemerus, Basel 1541
Trans. Nicolaus Leonicenus and Io. Manardus.
Ed. (with comm.) Io. Agricola Ammonius.
Prov. LIDDEL.

(11) *De Usu Partium corporis humani libri xvii.*
Chris. Wechel, Paris 1543
Trans. Nic. Regius, revd. Iac. Sylvius and Martinus Gregorius.
Prov. GORDON.

(12) *C— G— P— de Temperamentis li. iii. De inaequali intemperie li.* [Wanting in this copy]. *Cum . . . scholiis marginalibus . . . per Iacobum Sylvium.*
Chris. Wechel, Paris 1545
(Trans. Thomas Linacre.)
Prov. LIDDEL (?)

(13) *C— G— methodi medendi, vel de morbis curandis li xiiii.*
G. Rovillius, Lyon 1546
Linacre's trans. ded. to Henry VIII. Also a letter from Gul. Budaeus (Guillaume Budé) to Thomas Lupset (a member of the Philip Sidney circle).

(14) *Oribasius de Musculorum Dissectione ex Galeno.*
Iac. Gazellus, Paris 1548
Trans. Paulus Crassus.
Prov. GORDON.

(15) *G. de Ossibus.*

Iac. Gazellus, Paris 1548

Trans. and scholia Ferd. Balamius.

Prov. GORDON.

(16) *Epitomes omnium G— P—operam universam illius viri doctrinam & methodum . . . continentis, sectio tertia.*

Hiero. Scotus, Venice 1548

Ed. Andreas Lacuna.

Prov. LIDDEL. (Binding stamped 'MPR 1567').

(17) *C— G— P— omnia tum quae antehac extabant, tum quae nunc primum inventa sunt opera.*

Io. Frellonius, Lyon [Cols. 1548-50]

Several secns. have t.p.'s. Ed. C. Gesner.

Prov. FORDYCE (owned by 'Ioan. Fordyce', brother of Sir William, and medical practitioner at Uppingham, Co. Rutland.

(18) *Epitome G— P— operum. . . .*

Mich. Isingrinus, Basel 1551

Ed. Andreas Lacuna.

(19) *. . . G— P— de Fasciis libellus a Vido Vidio Florentino Latinitate donatus congruisque iconibus illustratus.*

Half title only and no col. but printer's foreword makes it clear that he felt it ought to be bound with Lacuna's *Annotationes* (No. 373), which has here been done. Numerous cuts of showing methods of bandaging.

(20) *C— G— P— omnia quae extant in Latinum sermonem conversa. . . . His accedunt nunc primum Con. Gesneri praefatio & prolegomena tripartita de vita Galeni eiusque libris & interpretibus.*

H. Froben and N. Episcopius, Basel 1561-2

Prov. LIDDEL.

This magnificent edn., here bound in four volumes, contains *Isagogici libri, Septem classes, Libri extra ordinem classium, Spurii libri.* Profusely illustrated.

(21) *Theatrum Galeni.* Eusebius Episcopius and heirs of Nich.

Episcopius, Basel, 1568

A comprehensive index to Galen's works.

Prov. LIDDEL.

(22) *Opera omnia in Galeni libris . . . .*

Petrus Landry, Lyon 1587 (Col. 1586)

Prov. JAMES MORISON.

Ed. Thoma Rodericus a Veiga.

(23) [Jacopo da Forlì etc.] *Antiqua Thegni Galeni translatio cum commento Haly Rodoan. Nova eiusdem Galeni translatio cum expositione Jacobi Foroliviensis multotiens impressa. Expositio eiusdem J—super tribus libris Thegni Galeni.*

Octavianus Scotus, Venice 1519

Not BM ? *Ed. p.* in this form; of the commentary of J. da Forlì, Padua 1475; of Ali ibn Ridwan's 1487.

Prov. (i) Hector Boece, (ii) Balcarres, Edinburgi Feb. 3, 1636, (believed to be Lord Lindsay of Balcarres, a well-known alchemist).

Incipits: Text, 'In omnibus doctrinis. . . .'

Ridwan's commentary, 'Intendimus edere sermonem exponentem illud . . .'.

Of the great influence of these commentaries on this (the most popular in medieval times) work of Galen see the account in Vol I, p. 221.

See also Nos. 10, 72, 81, 115, 119, 121, 133, 169, 256, 277, 280, 373, 421, 527, 656, 679, 687.

288. GALLUCI, GIOVANNI PAOLO. *Theatrum Mundi et Temporis. . . . Ubi astrologiae principia cernuntur ad medicinam accommodata, geographiae ad navigationem, singulae stellae cum suis imaginibus item ad medicinam, & Dei opera cognoscenda & contemplanda, kalendarium Gregorianum ad divina officia, diesque festos celebrandos. . . . Nunc primum . . . editum.*

Io. Baptisma Somascus, Venice 1588

Ed. p.

This is a well-known 'collectors' piece', lavishly illustrated and supplied with beautiful working models (*volvelles*).

289. GALLUS, PASCHALIS. *Bibliotheca medica sive catalogus illorum qui ex professo artem medicam in hunc usque annum scriptis illustrarunt. Nempe quid scripserint, ubi, qua forma quove tempore scripta excusa, aut manuscripta habeantur. . . .*

C. Waldkirch, Basel 1590

Only edn.

Prov. LIDDEL; and a second copy, READ.

Probably the earliest systematic bibliography of purely medical works arranged in tables under various headings, e.g. on surgery, in French, etc. This involves a good deal of repetition—some of it confusing, owing to the appearance of the same writer under different names.

290. GANIVETUS, IOANNES. *Amicus medicorum . . . cum opusculo, quod inscribitur 'coeli enarrant'; & cum abbreviatione Abrahae*

Aveneezrae de Luminaribus & Diebus criticis. Quibus adiecimus Astrologiam Hippocratis. . . .
Rovillius & Barth. Fraenus, Lyon 1550

[*Ed. p.* Lyon 1496 (Hain 7467).]

Prov. READ.

The *epistola* of Gondisalvus of Toledo to his son, Antonius, defends astrology, and refers to Ganivetus (dated Lyon 1508). BM refers to the elder Gondisalvus as editor, hence 1496 is not the *Ed. p* of this form of the work.

291. GARETIUS, HENRICUS. *De Arthritidis praeservatione et curatione . . . medicorum consilia.*
Io. Wechel and P. Fischer, Frankfurt 1590

Ed. p. Only edn. BM 1592.

Prov. LIDDEL.

A collection of 'consilia' of Fernel, Vesalius, Sylvius and other less well known physicians.

292. GARIOPONTUS. . . . *Ad totius corporis aegritudines remediorum ΠΡΑΞΕΩΝ libri V. Eiusdem de febribus atque earum symptomatibus libri ii. Recens typis commissi & multis in locis suae integritati restituti.*
Henricus Petrus, Basel 1531

[*Ed. p.* Lyon 1526].

Prov. CHARLES SINGER (and an earlier).

This work was for long referred to as *Passionarius Galieni*. Several MSS are extant, one of which (11th cent.—ThK, 552) begins 'que vel quante sunt febrium diversitates', cf. secn. VI.2 of No. 292 'Quae vero vel quantae sunt febrium diversitates'.

Gariopontus seems to have been one of the leading physicians of the school of Salerno in XI-1, and this work marks the beginning of the 'Latinisation' (from Greek) of medical nomenclature (Castiglioni p. 302).

GEBER. See JABIR IBN HAJAN.

293. GEMMA, REINERUS (*syn.* 'Gemma Frisius'). *Arithmeticae practicae methodus facilis . . . iam recens ab ipso auctore emendata & multis in locis insigniter aucta.*
(Privilege to: Gregorius Bontius) Antwerp 1547

[*Ed. p.* Antwerp 1540] Smith, *R.A.* p. 200, lists four further edns. before that of Pelletier (No. 505) Paris 1545.

Prov. (i) 'Ex. lib. Andreas Aulii'. (ii) 'Arturi Jonstoni Scoti 1608'—the distinguished Latin poet, brother of William (see above, p. xviii).

Probably the most successful of the 'modern' arithmetics of the 16th cent. (See Vol. I, p. 88.)

294. GEMMA, R. *De radio astronomico & geometrico li. In quo multa quae ad Geographiam, Opticam, Geometriam, & Astronomiam utiliss. sunt, demonstrantur.*

G. Bontius, Antwerp, and Petrus Phalesius, Louvain (Col. Antwerp 1545)

Ed. p.

Prov. LIDDEL.

The text describes a kind of cross-staff with movable transversal, and contains ref. to Petrus Apianus's (No. 33) observations on comets (fo. 33r and v), also *Tabula gnomonica Georgii Peurbachii* (No. 513).

295. GEMMA, R. . . . *de astrolabo catholico liber quo . . . patentis instrumenti multiplex usus demonstratur & quicquid uspiam rerum mathematicarum tradi possit continetur.*

Io. Steelsius, Antwerp 1556

Ed. p.

Dedn. by Cornelius, son of Reinerus, to 'serenissimum Hispaniae, Angliae et Franciae Regem Philippum'.

The text comprises 81 chapters by R. Gemma on the orthographic projection by which the instrument is made; on its parts and uses (with a ref. to Vitruvius); and (p. 137) on the tails of comets (including a fig.). The remaining chapters are by Cornelius. A set of end-papers provide additional diagrams.

296. GEMMA, R. *Les principes d'astronomie et cosmographie avec l'usage du globe. Le tout composé en Latin pas Gemma Frizon & mis en langue Francois par M. Claude de Boissière D. Plus est adiousté l'usage de l'anneau astronomic par ledict Gemma Frizon. Et l'exposition de la mappemonde, composée par ledict de Boissière.*

Hierosme de Marnef et la Veufue de Guillaume Cavellat, Paris 1582

Ed. p.

Cont. pagn. with secn. headings. Contains interesting diagrams and maps ('anneau astronomique' fo. 68v). The '*abrégé*' of the *mappemonde* explains the nature of projection in a qualitative manner and deals with winds.

The dedn. to a woman (other than of a ruling family)—the 'Treshumaine Damoyselle Madamouselle Magdalene Brissonet'—is uncommon in 16th cent.

See also Nos. 33, 654 and 747.

297. GESNER, CONRAD. *Historia plantarum et vires ex Dioscoride, Paula Aegineta, Theophrasto, Plinio, & recentioribus Graecis iuxta elementorum ordinem . . . Adiecta ad marginem nomenclatura, qua singulas herbas officinae herbarii & vulgus Gallicum efferre solent.*

W. Lodovicus Tiletanus, Paris 1541

Ed. p. Unillustrated

From the dedn. this appears to have been intended as a 'Botanists' Pocket Book' especially adapted for use in the country. He says that he will 'send' in the near future, 'a book about the names of plants, in "our" [sc. German], French, Greek and Latin languages—just the names without reference to their virtues or their particular descriptions.' A sign of the times—presumably referring to his *Catalogus Plantarum, Latine, Graece, Germanice et Gallice.* Zürich 1542.

298. GESNER, C. (*Historiae Animalium* published in 5 vols. by Christopherus Froschoverus, Zürich 1551-87 BM).
The following volumes are in the library:

(*a*) *Lib. I. De Quadrupedibus viviparis.* 1551.
Two copies, of which (i) has coloured cuts, (ii) uncoloured, and also lacks the *Additiones & castigationes* and the colophon picture of the creation of Eve.

(*b*) *Liber II. De quadrupedibus oviparis.*
Two copies, (i) bound with 'a (ii)', dated 1554, with sep. t.p. and pagn. and appendix to *libri.* I and II. (ii) bound with *Lib.* V, but lacking t.p. and appendix.

(*c*) *Liber III qui est de avium natura,* 1555.
Two copies, one of which has coloured cuts.
Prov. Escutcheon and 'Ro. Keune 1584'.
Dedn. to Io. Iac. Fugger 'Mecaenas'.

(*d*) *Liber IIII qui est de piscium & aquatilium animantium natura . . . continentur in hoc volumine Gulielmi Rondeletii . . . & Petri Bellonii . . . de aquatilium singulis scripta. . . .* 1558.
Two copies, of which one has coloured plates.

This 'Liber' contains a letter from the English naturalist, William Turner (see Vol. I, p. 181), also a list of those who have helped the author in various ways, among whom may be mentioned 'Hieronymus Frobenius typographus Basileae clarissimus et de bonis literis quam optime meritus'; 'Io. Parkhurstus Anglus', 'Io. Fauconerus Anglus', and (most surprisingly) 'Theodorus Beza'! Gesner also notes that although there is nothing really new in Wotton's *De differentiis animalium* (No. 739) the grouping according to natural kinds is valuable.

(*e*) *Lib. V qui est de serpentium natura; adiecta est ad calcem scorpionis insecti historia a D. Casparo Wolphio Tigurino medico ex eiusdem Paralipomenis conscripta.*
One copy, bound (sep. t.p.) with Lib. II in '(a ii)'. 1587.

(*f*) *Icones avium omnes quae in historia avium C— G— describuntur, cum nomenclaturis singulorum Latinis, Italicis, Gallicis et Germanicis plerunque per certos ordines digestae.* 1555.
Bound with it (with sep. t.p.'s and pagn.) are similar sets for *Quadripeda* (1560) and *Animalia Aquatilia* (1560).

(*g*) (t.p. wanting—appears to be a duplicate of the corresponding part of (*f*), viz. *Nomenclator aquatilium animantium. Icones animantium aquatilium.* 1560.)
Prov. [Lord] 'Lumley'. English names of species added in MS.
For biblio. of the whole work see Van der Velde, A.J.J. *Rond Gesner's Historiae Animalium*-Koninklyke Vlaamse Academie voor Wetenschappen etc. Brussels 1951. The author includes No. 303, as 'Liber VI' though there is no warrant for this in the title.

299. [GESNER, C.] [*Thesaurus Euonymi Philiatri de Remediis Secretis liber physicus medicus, et partim etiam chymicus, nunc primum . . . editus*].

[Col. Andr. Gesner, Zürich 1554]
Ed. p.
'Euonymus Gesnerus Philiatros' in his dedn. to Nicolaus Zerchinta of Berne says that he has compiled the book mainly from others in various languages but has also added some observations of his own.

300. GESNER, C. *Euonymus . . . de remediis secretis, liber physicus, medicus, & partim etiam chymicus & oeconomicus in vinorum diversi saporis apparatu, medicis & pharmacopolis omnibus praecipue necessarius nunc primum . . . editus.*
[Only imprint is Froschover's device on t.p.]
BM cites Zürich 1554 as earliest under '*Euonymus*' (see below) and '*Liber Secundus*' 1569 under '*Gesner*'.
The dedn. of *Liber Secundus* in No. 300 is signed by Caspar Wolf and dated 'Tiguri, 1569'.
Sep. t.p. and pagn. for . . . *Liber secundus nunc primum opera & studio Caspari Wolphii . . . editus.*
In his dedn. to Petrus de Grantyre, Wolf speaks of various matters having been left to him by Gesner to be finished.
The titles are misleading. From Gesner's preface referring to the 'chymistica ars . . .' onwards there is constant reference to matters of

interest to the historian of chemistry, e.g. II, 156 'De oleo vitrioli . . .
ex Valerio Cordo . . .' and the fig. on fo. 179*v*—'spacium vacuum'.

301. GESNER, C. *Sanitatis tuendae praecepta cum aliis tum literarum studosis*
[sic] *hominibus & iis qui minus exercentur cognitu necessaria.*
Contra luxum conviviorum. Contra notas astrologicas ephemeridum de
secandis venis.

Iacobus Gesner, Zürich [1556—Conrad's dedn.]

Prov. Liddel.

A collection of excerpts from classical authors and others, followed
by original brief warnings—especially concerning venesection.

302. GESNER, C. *Cassii Iatrosophistae Naturales et Medicinales Quaestiones*
LXXXXIIII circa hominis naturam & morbos aliquot Conrado
Gesnero . . . interprete nunc primum editae. Eaedem Graece, longe
quam antehac castigatiores, cum scholiis quibusdam. His accedit
catalogus medicamentorum simplicium & parabilium quae veneno
adversantur . . . authore Antonio Schnebergero. . . .

(Col. Iac. Gesner, Zürich) (1562 Pref.)

Ed. p.

In the dedn. Conrad states that the Greek book (printed in Paris)
reached him in 1541, a Latin trn. (by Iunius Hornanus) appearing in
the same year. His further comments cast a light on Gesner's character
& scholarship. The *Catalogus* has sep. t.p. and pagn; also dedn.
(Cracow 1561).

303. GESNER, C. (*et al.*) *De omni rerum fossilium genere, gemmis, lapidibus,*
metallis et huiusmodi, libri aliquot, plerique nunc primum editi.

. . . . Iac. Gesner, Zürich 1565-6

Ed. p.

Many of the works have sep. t.p.'s and pagn. but the *errata* at the end
of the volume indicates that they were issued ('for the most part for
the first time') in one volume. The concluding work—Gesner's
De rerum fossilium . . . figuris & similitudinibus liber—is of special importance and contains numerous cuts of fossils.

304. GESNER, C. [ed. Caspar Wolf]. *Epistolarum medicinalium C—*
G— . . . libri iii. His accesserunt eiusdem Aconiti primi Dioscoridis
asseveratio & de Oxymelitis Elleborati utriusque descriptione & usu
libellus. Omnia nunc primum per C— W— data.

Christ. Froschoverus, Zürich 1577

Ed. p.

Prov. Liddel.

The 'additions' have sep. t.p.

305. GESNER, C. [trans. George Baker]. *The Newe Jewell of Health,*
wherein is contayned the most excellent secretes of phisicke & philosophie
devided into fower bookes. In the which are the best approved remedies
for the diseases as well inwarde as outwarde of all the partes of man's
bodie, treating very amplye of all dystillations of waters, of oyles,
balmes, quintessences with the extraction of artificiall saltes, the
use & preparation of antimonie & potable gold. Gathered out of
the best and most approved authors by that excellent doctor Gesnerus.
Also the pictures, and maner to make the furnaces, and other instruments
thereunto belonging. Faithfully corrected and published in Englishe, by
George Baker, chirurgian.

Henrie Denham, London 1576

Only edn. BM
Prov. Fordyce ('John Fordyce' t.p.)
Fully illustrated with figs. of glass cutters, water baths, sand baths,
etc. The preface makes an interesting claim for the use of the English
language, namely, that Hippocrates had written in Greek, Celsus in
Latin, and Avicenna in Arabic, for their countrymen respectively.
It includes also the names of many practitioners in the City of London.
It is often described as a 'translation' of No. 300: the correspondence
is not, however, very close.
See also Nos. 67, 175, 749.

GHINI, LUCA. See No. 424 and Vol. I, p. 194.

306. GIACCHINI, LEONARDO (*syn.* Jacchini, Leonardo). . .
opuscula . . . nempe: Praecognoscendi methodus. De rationali curandi
arte. De acutorum morborum curatione. Quaestiones naturales.

[No imprint, but device of P. Perna] Basel 1563

Bound with the following work, and despite its different date,
t.p., and pagn., cited as a single work BM. Text includes refs. to the
use of urine as an indicator (p. 7) and to the precautions to be taken in
making the observations (p. 9). Among the *Quaestiones* is 'why gnats do
not attack all men equally with their bites' (p. 133). Cont. pagn. with
secn. headings.

In Nonum Librum Rasis . . . de partium morbis . . . commentaria.
Opera . . . Hieronymi Donzellini . . . emendata ac perpolita.

P. Perna, Basel 1564

Ed. p.
Prov. LIDDEL.
Also a sep. vol. of the *In Nonum Librum Rasis . . .* Carolus Pesnot,
Lyon 1577.

The book of ar-Razi was a very popular text for commentary (cf. No. 708). This one does not cite the text but consists of a course of lectures ('Coepimus superiore anno ornatiss. iuvenes . . .') whose connection with ar-Razi seems rather tenuous.

307. GILBERT, WILLIAM. *De magnete, magneticisque corporibus, et de magno magnete tellure; physiologia nova, plurimis & argumentis & experimentis demonstrata.*

Petrus Short, London 1600

Ed. p. (Two copies.)
(See Vol. I, p. 164.)

308. GILBERTUS ANGLICUS. *Compendium medicine . . . tam morborum universalium quam particularium nondum* [sic] *medicis sed et cyrurgicis utilissimum.*

Iacobus Saccon for V. de Portonariis, Lyon 1510

Only edn.
Prov. FORDYCE, and another copy.
Ed. Michael de Capella. *Incipit:* (Prologue) 'Liber morborum tum universalium . . .'. (Text) 'A morbis universalibus propositi [nostri] intentio est inchoare. Cf. ThK 1.
See fo. ccclxii *de regimine transfretantium* (treatment of sea-sickness).

309. GOEBELIUS, SEVERINUS. *De Succino libri duo . . ., horum prior liber continet piam commonefactionem, de passione, resurrectione, ac beneficiis Christi quae in historia Succini depinguntur. Posterior veram de origine succini addidit sententiam.* [No imprint, but cont. registrn. with previous work—No. 239] Zürich, 1565.

Only edn. BM (Dedn. 1558.)
A curious work in which the 'history' of amber and other 'bituminous' substances is used as the basis of a devotional and moral lesson. The opinions of many well-known men are included.

310. GOEUROT, JEHAN. *Summaire très singulier de toute medicine et cyrurgie. Speciallement contre toutes maladies survenantes quotidiennement au corps humain . . . Item régime singulier contre la peste . . .* [Sev. fos. wanting—no imprint exc. in Col. to *Le Traicté des Urines* which follows].

Alain Lotrian and Denis Janot, Paris 153-?

Only edn. with this title in BM is Lyon 1544. Many edns. Bibl. Nat. mostly undated.
Le traicté des urines, lequel traicte de leurs couleurs et ce quelles peuvent signifier par lesquelles urines on peust congnoystre facilement toutes les maladies

qui peuvent estre au corps de l'homme et de la femme et dou ils procedent. Approuve par plusieurs medecins comme Avicenne, Razis, Isaac, Ypocras, et Gilles. Et plusieurs austres assez expers en lard [sic!] *de medecine.*

311. GORRAEUS, IOANNES. *Definitionum medicarum libri xxiiii literis Graecis distincti, ab authore ante obitum recogniti . . . adaucti & nunc denuo . . . editi.*

Andr. Wechel, Frankfurt 1578

[*Ed. p.* presumably 1564 from dedn. (Not BM.) New dedn. by the author's son for the 1578 edn.]

Two copies.

Comprises an explanation, in Latin, of Greek medical terms. Io. Gorraeus the elder was the translator of Nikandros (No. 476).

312. GRATAROLUS, GULIELMUS. *De literatorum & eorum qui magistratibus funguntur conservanda praeservandaque valetudine, illorum praecipue qui in aetate consistentiae, vel non longe ab ea sunt, compendium, cum ex probatioribus autoribus, tum ex ratione ac fideli experientia concinnatum.*

(*a*) Henricus Petrus (Col. Basel 1555).

Ed. p. (earliest BM Paris 1562).

(*b*) Andr. Wechel, Frankfurt 1591.

(*c*) Io. Wechel and P. Fischer, Frankfurt 1591.

Prov. (*a*) 'Patrick Anderson gave it to John Blackburne'. Peter Blackburne's name also appears. 'Joannes Hariel' on cover.

Dedn. to his relative, Franciscus, gives information about other members of the Gratarolo family.

The text is concerned with diet and regimen.

313. GRATAROLUS, G. (Ed.) *Ioannis de Rupescissa qui vixit ante cccxx annos, de consideratione quintae essentiae rerum omnium, ... Accessere Arnaldi de Villanova, Epistola de Sanguine humano destillato.* Raymundii Lullii, *Ars operativa, & alia quaedam.* Michaelis Savanarolae *libellus . . . de Aqua Vitae, nunc valde correctior quam ante annos LXX editus. . . .*

C. Waldkirch, Basel 1597

Ferg. II, 305 quotes an edn. with no imprint but '*epistola* 1561': the contents are, however, not identical. He cites a work with similar imprint, pagn. & contents, but not under 'Gratarolo'. Duv. p. 54 cites an edn. of 1561, but not No. 313, which, however, is cited by BM.

Prov. LIDDEL, and a second copy.

In dedn. Gratarolus says that he had had Io. de Rupescissa's work lying by him for some time, and accused Ulstadius of having lifted

118

pages of it into his *Coelum Philosophorum* (No. 694). He says that Trithemius (No. 688) recorded Ioannes as having been alive in 1340.

The book also contains two tracts by Gratarolo, *Modus faciendi quintam essentiam simplicem* (containing instructions for obtaining 'aqua ardens' by seven distillations of the 'best wine you can find') and *De viribus, usu, ac mistione aquae ardentis*. (See Vol. I, p. 171.)

314. GRATAROLUS, G. (Ed.). *Verae alchimiae artisque metallicae citra aenigmata doctrina certusque modus, scriptis tum novis tum veteribus, nunc primum . . . maiori ex parte editis.*

H. Petrus & P. Perna, Basel 1561

Ed. p.

In his dedn. Gratarolo describes how he got possession of writings old and new. New pagn. after p. 244

315. GRATAROLUS, G. (Ed.). *Alchemiae, quam vocant, artisque metallicae doctrina certusque modus, scriptis tum novis tum veteribus duobus his voluminibus comprehensus.*

P. Perna, Basel 1572

and a presumed Vol. II, according to contents on fo.):(4v of Vol. I) lacking t.p. or any identification marks except index entitled *Partis Secundae*.

Ferg. I, 341 states that these two vols. are an abridgment of the folio version (No. 316). There are two copies of 'Vol. II' both of which start at sig. aa, suggesting absence of prelims. One has the imprint of Brestonius, Leiden 1588.

316. GRATAROLUS, G. (Ed.). *Auriferae artis, quam chemiam vocant, antiquissimi authores sive Turba Philosophorum.*

P. Perna, Basel 1572

Ferg. I, 51. gives a good deal of information about the various combinations and translations.

There is no ref. to Gratarolus on t.p. but the printer in his *Lectori* refers to the success of the previous collection of works 'opera Gulielmi Grattaroli Medici collectum . . . ante annos aliquot' and to his decision to issue a further collection. The *Turba* itself occupies pp. 1-151, and is followed by *Aenigmata, Exercitationes*, etc. referring to the *Turba*.

317. *Calendarium Gregorianum Perpetuum.*

Chris. Plantin, Antwerp 1583

A tract headed 'Gregorius Episcopus servus servorum dei ad perpetuam rei memoriam . . .' giving account of the circumstances and nature of the task laid upon him by the Council of Trent for the reform of the calendar. (See Vol. I, p. 117.)

218. GREVINUS, JACOBUS (*syn* Grévin). . . . *de venenis libri duo, Gallice primum ab eo scripti, et a multis hactenus Latini* [sic] *desiderati & nunc tandem opera & labore Hieremiae Martii . . . in Latinum sermonem . . . conversi. Quibus adiunctus est praeterea eiusdem auctoris de Antimonio tractatus, eodem interprete.*

Chris. Plantin, Antwerp 1571

Ed. p. Latin trn.
Original French, Antwerp 1568.
Prov. READ.

Numerous cuts of poisonous animals and of plants used as antidotes—see esp. *De Torpedine*, p. 130.

Dedn. by Martius to Maximilian II speaks of the 'turbulence of the times and the state of affliction of the whole Christian polity', and includes a list of cases of poisoning. Dated Augsburg, 1569.

The work is mainly a paraphrase of the poems of Nikandros (No. 476).

319. GROSSETESTE, ROBERT. *Libellus Linconiensis de phisicis lineis angulis et figuris pfr* [i.e. per] *quas omnes acciones naturales complentur* [whole title in Rom. Caps.]

'Nurenberge' 1503

Only edn. BM. There were no incunable edns. of Grosseteste's works (Hain) except of the Commentary on the *Posterior Analytics* of Aristotle 1494 and 1497.

Incipit: 'Utilitas considerationis linearum . . .'. ThK, 746 quote no MS.

The editor, Andreas Stiborius Boius, refers to the rarity of the work, and laments the fact that 'it is little known owing to the idleness of the men of our times'.

The cut on t.p. shows the sun's rays, reflection by convex speculum, and (symbolically) refraction. The work is a *locus classicus* of the medieval concern with 'perspective' (No. 503) as an illustration of 'physical action'. It deals with the relation between the illuminating power (*virtus*) and inclination of the rays, and demonstrates deductively that light is more weakened by reflection than by refraction since in the former it has receded more from the 'straight' path. Sig. a ii *v* 'Sed natura operat meliori modo quo potest, ergo melius operat super lineam rectam', i.e. rectilinear propagation is the 'best' way in which nature could act.

320. GUAINERIUS, ANTONIUS [*syn.* Guarnerius, Guaynerus etc.]
(*a*) *Practica . . . cum additionibus . . . Ioannis Falconis . . . List of*

sections: *De egritudinibus capitis* . . . *De Peste* . . . *De Balneis, Antidotarium, Rhazel de Pestilentia.*

Const. Fradin, (Col. Lyon 1517)

Falco's 'additions' start at sig. lii *v* and appear to end at sig. lxv *r*. '*Rhazel*' not found.

Col. '. . . Et aureum est volumen: et iuvenibus ad opus practicum noviter accedentibus maxime utile. Et . . . emendati [*sic*] per . . . Hieronymum Faventinum . . . . Habes additiones Ioannis Falconis . . . hactenusque non impressas. . . .

(*b*) . . . *Opus praeclarum ad praxim non mediocriter necessarium cum Ioannis Falconis nonnullis . . . adiunctis . . .*[contents as in (*a*)].

Const. Fradin, (Col. Lyon 1525)

Hain (8097) states that the *Ed. p.* of the *Practica* (Venice 1497) was based on *De medicina tractatus*, Pavia 1481. Copinger (2802) reports that parts were separately issued in 1474.

Col. slightly different from (*a*): 'emendati' corrected to 'emendatum'.

321. GUIBERTUS, NICOLAUS. *Alchymia ratione et experientia ita demum viriliter impugnata & expugnata una cum suis fallaciis & deliramentis quibus homines imbubinarat ut nunquam imposterum* [sic] *se erigere valeat*. . . . *Item de balsamo, eiusque lachrymae quod* OPOBALSAMUM *dicitur, natura viribus & facultatibus admirandis.*

L. Zetznerus, Strassburg 1603

Ed. p. Ferg. I, 353. Duv. p. 272.

An 'exposure' of the follies of alchemy, accompanied by verse compositions by those of like opinions, e.g. J. C. Scaliger.

322. GUIDO DE CAULIACO (*syn.* Guy de Chauliac). *Chirurgia Guidonis de Cauliaco, addita recepta aquae balnei de porecta per egregium . . . Thuram de Castellio . . . edita, nuperrime impressa & . . . emendata & aucta.*

(*a*) Vincentius de Portonariis, Lyon 1537.

Ed. p.

Prov. 'Simion Henrysoun'.

(*b*) Gaspar a Portonariis, Lyon 1559.

(*c*) Sebastian Honoratus, Lyon 1572.

Prov. HENDERSON.

(*d*) *La Grande Chirurgerie* de M. Guy de Chauliac . . . par M. Laurens Joubert . . . Etienne Michel (Col. Etienne Brignol) Lyon 1579.

ThK quote MS with *Incipit* 'Postquam prius gratias egero deo . . .' Erfurt, 1417.

The preface contains a short account of the march of surgery, with the well-known reference to No. 358 as 'una fatua rosa anglicana'. Also a passage not so well known as it deserves to be. . . . 'Scientiae enim per additamenta fiunt, non enim est possibile eundem incipere & finire. Pueri enim sumus in collo Gigantis, quia videre possumus quicquid Gigas & aliquantulum plus . . .' (sig. a 2 *r*). 'The sciences are created by constant additions; for it is not possible for the same person to begin and complete them. We are as boys on the shoulders of a giant and for this reason are able to see what the giant sees and a little further . . .' . . . sentiments not traditionally associated with the Middle Ages! Nevertheless the comparison was made by John of Salisbury.

323. GUINTERUS, IOANNES *von Andernach* (*syn.* Guint(h)er). . . . *Commentarius de balneis & aquis medicatis in tres dialogos distinctus.*

(Col. Theo. Rihelius, Strassburg 1565
Only edn.

The dedn. to the Archbishop of Trier refers to the works of Agricola and Reinerus Solenander (No. 650).

The text (p. 68) mentions the (still extant) warm springs of Wiesbaden and the 'acid' springs at Ems, Niederlahnstein, Andernach, and Daun. See also No. 400.

324. GUINTINI, FRANCESCO (*syn.* Junctinus, F.). *Commentaria in Sphaeram Ioannis de Sacrobosco . . . Omnia iudicio S.R. Ecclesiae submissa sunta* [sic].

Philippus Tinghius, Lyon 1578
Ed. p. 1577—dedn. of No. 324 dated 1577.

The unusual *submissa* is accounted for by the fact that Guintini is described as 'Sacrae Theologiae Doctor'.

This work provides documentary evidence for the persistence of a undamentally medieval approach to physics throughout the greater part of the 16th cent. The General Introduction (on the seven Liberal Arts) indicates the sources and scope of Io. de Sacrobosco's book and the reasons for its popularity. Among the sources Guintini mentions 'Thebit', though in fact Ioannes makes no reference to the latter's theory of 'trepidation' (see Vol. I, p. 114.)

325. HAJEK, TADEAS [*syn.* Thaddeus Hagecius ab Hayck]. *Epistola ad Martinum Mylium. In qua examinatur sententia Michaelis Moestlini & Helisaei Roeslin de cometa anni* 1577. *Ac simul etiam asseritur contra profanas & Epicureas quorundam opiniones, qui cometas nihil significare contendunt.*

Ambrosius Fritsch, Görlitz 1580

Not in BM or Bibl. Nat., which, however, cite the same author's *Descriptio Cometae* . . . 1577 . . . Prague 1578.

Explicit 'Pragae ex aedibus nostris', followed by reply from Michael Maestlin (the teacher of Ioannes Kepler, see. No. 418) 'Gorlicii . . . anno 1578'.

HALL(E), JOHN. See No. 375.

326. HAJEK, TH. *Apodixis physica et mathematica de cometis tum in genere tum in primis de eo qui proxime elapso anno LXXX in confinio fere Mercurii & Veneris effulsit & plus minus LXXVI dies duravit.* . . .

[Col. A. Fritsch, Görlitz 1581]

Not BM or Bibl. Nat.

Prov. LIDDEL.

Cut on t.p. shows path of comet through the signs.

In a preliminary epistle to Hajek, Andreas Dudith (to whom the tract is dedicated) refers to Tycho Brahe, Wittich, Gemma Frisius, Maestlin, and also to the opinion of 'Dn. Henrici Savile Angli . . . praestantissimi iuvenis . . .' and says '. . . nobiles & doctrina ac moribus praestantes adolescentes, Anglos, Dn. Robertum Sidnaeium ac Dn. Henricum Nevellum, officiose ex me ut salutes, valde te rogo'. dated Breslau, 1581.

Dedn. dated Prague, 1581.

HANDSON, RALPH. See No. 526.

327. HARPUR, JOHN ['J.H.' t.p.] *The Jewell of Arithmetick or the explanation of a new arithmeticall table, whose portraiture is here demonstrated.*

Kyngston, London 1617

Ed. p. S.T.C. 12796.

The epistle to the reader takes more than two pages to explain the necessity of avoiding the epithet, 'tedious'!

The *Jewell* is virtually a printed abacus frame, using cardinals instead of units.

328. HARRIOT, THOMAS. *Artis analyticae praxis ad aequationes algebraicas nova expedita & generali methodo resolvendas.*

Robt. Barker (& heirs of John Bill) London 1631

Only edn.

T.p. bears a kind of dedn. to Henry Percy, Earl of Northumberland, who ordered the publication of the work ten years after the author's death. The work is included here since Harriot was about forty at the turn of the century.

Unsigned *Praefatio ad Analystas.* (Vol. I, p. 93.)

329. HASAN IBN HASAN [*Syn* Ibn Haitham; Al Hazen]. *Opticae thesaurus . . . libri septem nunc primum editi. Eiusdem liber de crepusculis & nubium ascensionibus. Item Vitellonis Thuringopolini libri X. Omnes instaurati, figuris illustrati, & aucti, adiectis etiam in Alhazenum commentariis a Fiderico Risnero.*

Eusebius and heirs of Nich. Episcopius, Basel 1572

Only Edn.

Prov. Gregory and a second copy.

Sep. t.p.'s and pagn, but second t.p. undated and only one col.

Incipit: Of Alhazen—'Invenimus quod visus . . .'. ThK, 366 give 'Invenimus (quod) visum quando . . .' for three MS of 13th and 14th cent.

Cf. *De crepusculis*—'Ostendere volo in hoc tractatu quod sit crepusculum . . .'. ThK, 477 give 'Ostendam autem . . .'. The trans. is said to be that of Gerard of Cremona.

T.p. of Witelo's work follows closely that of the main t.p. but adds 'infinitisque erroribus, quibus antea scatebant, expurgati' and includes cut of burning mirrors. The original dedn. is to Frater Gulielmus de Morbeta, knowledge of whose date (*ca.* 1269) furnishes the only clue to that of Witelo.

Incipit: Of intro. 'Praesens itaque negotium. . . .'

Of text, 'Definitiones, quae vero per modum principiorum . . .' as in ThK, who mention 14th cent. MSS.

The volume is ded. to Catherine de'Medici. Both works are of highest importance for the sources of medieval 'perspective'. Al-Hazen, p. 6, shows a longitudinal section of the human eye, including its component membranes revealed as continuous with those of the brain.

330. HENERUS, RENATUS. *Adversus Iacobi Sylvii Depulsionum anatomicarum calumnias pro Andrea Vesalio apologia. In qua praecipue totius negotii anatomici pene controversiae breviter explicantur. . . . Additus est & ipse Iacobi Sylvii Depulsionum libellus quo aequus lector commodius de omnibus iudicium ferre possit* [no printer]. Venice 1555.

Ed. p.

Sep. half-title for *Depulsiones;* cont. pagn.

In his dedn. Henerus states that he was in Paris when Sylvius let loose his virulent attack on Vesalius.

Both works are recorded by Cushing, *Biobibliography of Andreas Vesalius,* but without comment. (Vol. I. p. 227.)

331. HERILACUS, PAMPHILIUS. . . . *Aquarum natura et facultates per quinque libros digesta: vinorum et aquarum effectuum invicem comparatorum tractatus. De arthritide et podagra consilium.* . . .

For Io. Bapt. Ciotti, Cologne 1591

Only edn.

Prov. BUTE.

Sep. secn. heading, dedn. & pagn. for *De arthritide.* . . .

The preface is of interest by reason of its comments on the method of Medicine by 'definition, composition and resolution' (see Vol. I, p. 220). The *Periocha* shows Aristotelian dichotomous classification carried to absurd lengths.

332. HERON OF ALEXANDRIA. *Spiritalium liber a Federico Commandino . . . ex Graeco nuper in Latinum conversus.*

(*a*) [No printer]. Urbino, 1575.

Ed. p. (Privilege of Gregory XIII).

Prov. Io. Chalmers, Principal of King's College Aberdeen.

(*b*) Aegidius Gorbinus, Paris 1583.

Prov. 'Sum Samuelis Kechelii'.

The *Lectori* states that Commandinus died before the translation was printed, which would account for 'anything . . . which may not smack of the diligence and outstanding erudition of the translator'.

Numerous cuts showing double-action pump, wind organ, aëlopile, syphons, etc. Commandino also produced the first definitive translation of Archimedes (No. 39).

333. HERVET, ISRAEL [*syn.* Harvetus]. *Demonstratio veritatis doctrinae chymicae adversus Ioan. Riolani comparationem veteris medicinae cum nova, Hippocraticae cum Hermetica, Dogmaticae cum Spagyrica.* . . .

Typis Wechelianis—Marnius and heirs of Aubrius, Hanover 1605.

Ed. p. Ferg. I, 367.

Dedn. 'War was already declared by Riolan on the chemists twenty years ago as appears from his writings . . .'.

Valuable for diverse uses of the word 'alchemy'. Letters to du Chesne and Riolan at end. The views of Paracelsus on the origin of 'invisible diseases' are noteworthy (p. 11).

334. HÉRY, THIERRY DE. *La methode curatoire de la maladie venerienne vulgairement appellée grosse vairolle & de la diversité des ses symptomes.*

Thibaut Payan, Lyon 1568

[Only edn. in Bibl. Nat. is Paris 1552.]

Dedn. 'à la Republique Française. Aux lecturs [*sic*] de bon vouloir salut . . .'.
The text opens with the affirmation of the necessity for a 'partie theorique & speculations des choses universelles . . . partie en la pratique . . .'.

335. HEURNIUS, IOANNES. *Praxis medicinae nova ratio . . . recognita & emendata ab auctore. . . .*

Off. Plant. Raphelengius, Leiden 1590

Ed. p.
Prov. LIDDEL.
Cuts of stills, baths, clysters, etc. Rather an unusual form of *Practica*.

336. HEURNIUS, IO. *De febribus liber.*

Raphelengius, Leiden 1598

Ed. p.
Prov. LIDDEL (?)

337. HEURNIUS, IO. *. . . de morbis pectoris liber editus post obitum auctoris ab eius filio Othone Heurnio.*

Raphelengius, n.p. 1602

Not BM. Catalogued under 'van Heurne' Bibl. Nat.—vol. not available.
Prov. LIDDEL (?)
Dedn. from Leiden to Maurice of Nassau by Otho 'parentis nomine'.

338. HEURNIUS, IO. *. . . de morbis qui in singulis partibus humani capitis insidere consueverunt. Hic artificiosa methodo . . . morborum ideae, causae . . . graphice depinguntur.*

Raphelengius, Leiden 1594

Not BM.
Prov. LIDDEL.
The preface explains that this is to be the first of a series 'if God allows me the leisure'. Epistle from Jos. Scaliger and dedn. to 'Bataviae et Westphrisiae Ordinibus . . . . Ex Academia vestra Leyda in Batavis 1593': The University of Leiden was in fact instituted by the Estates as a 'war memorial' in 1575.
Apparently 'morbis capitis' included *incubi* and *succubi*!

339. HEURNIUS, IO. *De morbis oculorum aurium nasi dentium et oris. Liber editus post mortem. . . .*

Raphelengius, n.p 1602.

Not BM.
Prov. LIDDEL.
See also No. 271.

340. HILL, THOMAS. *The Schoole of Skil: containing two bookes, the first of the sphere of heaven, of the starres, of their orbes & of the earth etc. The second, of the sphericall elements, of the celestial circles, and of their uses, etc.* . . .

W. Jaggard, London 1599

Only edn. STC 13502.

The printer (*To the Reader*) writes concerning the decease of the author that 'the government (God be blessed) hath a long time (now these 40 years) bin so peaceable that students never had more libertie to looke into learning of any profession. . . . Lastly the meanes otherwise, as well out of the universities, as in them, have been and are so good, . . . that I dare be bold to say, England may compare with any nation for numbers of learned men, & for variety in professions.' A striking comparison with the state of France (see No. 213); but the '40 years' included that of the Armada, showing how effective England's 'sure shield' was to internal affairs even then.

See Johnson, p. 183, for numerous other works by Hill.

341. HIPPIUS, M. FABIANUS. *ΨΥΧΟΛΟΓΙΑ Physica, sive de corpore animato libri quatuor, toti ex Aristotele desumti, morborum saltem doctrinis ex medicis scriptis adiecta. Quibus multa physiognomonica & chiromantica passim . . . inseruntur.*

Wolfgang Richter for Io. Spiessius, Frankfurt 1600

Not BM. Only edn. Bibl. Nat.

Written in form of question and answer.

342. HIPPOCRATES. No detailed study of editions of the works of Hippocrates has been attempted. Those printed before 1600 have been arranged in the chronological order of their imprints.

(1) *Antiqua Hippocratis translatio supra septem sectiones Aphorismorum una cum . . . Galeni commentatione. Nova eiusdem Hippocratis translatio supra easdem . . . per Theodorum Gazam habita cum Iacobi Foroliviensis Marsiliique expositionibus. Questiones eorundem J— & M— supra predictas septem sectiones.* . . .

Octavianus Scotus, Venice 1519

The '*antiqua translatio*' is prefaced by Constantinus Africanus.

(2) *Omnia Opera Hippocratis* [repeated in Greek.]

Andr. Torresanus and heirs of Aldus Manutius, Venice 1526

Ed. p. Entirely in Gk. except the privilege of Clement VII.

(3) *Hippocratis . . . opera: quibus maxima ex parte annorum circiter duo millia Latina caruit lingua: Graeci vero & Arabes & prisci*

nostri medici, plurimis tamen utilibus praetermissis, scripta sua illustrarunt: nunc tandem per M. Fabium Rhavennatem, Gulielmum Copum Basiliensem, Nicolaum Leonicenum & Andream Brentium . . . Latinitate donata, ac iam primum . . . aedita. . . .

And. Cratander, Basel 1526

Prov. 'Io. Gregorie'.

End papers wanting.

(4) *Hippocratis Aphorismi, cum Galeni commentariis, Nicolao Leoniceno interprete. Praedictiones, cum G— commentariis, Laurentio Laurentiano interprete.*

Simon Sylvius [n.p.] 1527

Prov. 'This book perteinis to the librarie of New Abd. . . . M. Thomas Cargill 1585'.

(5) *. . . ΕΠΙΔΗΜΙΩΝ liber sextus a Leonharto Fuchsio . . . latinitate donatus . . . auctus.*

Io. Bebel and Mich. Isingrinus, Basel 1537

Prov. GORDON.

(6) *Hippocratis de Haemorrhoidibus libellus. Item Galeni de Locis affectis.*

[Col. Thoma Platterus, Basel 1540]

Prov. FORDYCE.

(7) *Aphorismorum Hippocratis Sectiones VII ex Franc. Rabelaisi recognitione. Quibus ex Ant. Musae commentariis adiecimus & octavam & quaedam alia. . . .*

Seb. Gryphius, Lyon 1543

Ed. p. 1532 Bibl. Nat.

Prov. HENDERSON.

The 'eighth' section was created by Brasavolus by joining those Aphorisms purposely omitted by Galen with those in the Greek codices which were held to be spurious. The collection also contains the *Regimen in Acute Diseases, Human Nature* (of doubtful authenticity), and *Praesagia*, of Hippocrates, and the *Medical Art* of Galen.

(8) *Hippocratis . . . opera quae hactenus ad nos extant omnia. Per Janum Cornarium, . . . Latina lingua conscripta. Recens per nos . . . castigata. . . .*

Hiero. Scotus, Venice 1546

[*Ed. p.* of this edn. Basel 1538.]

(9) *Sententiae omnes ac verba quae in . . . Hippocratis Aphorismis continentur iam recens . . . digesta. Auctoribus . . . Bartholomaeo & Petro Reatiniis.*

Balthas. Constantinus, Venice 1555

(10) *Hippocratis Aphorismi digesti . . . Cum expositione singulis Aphorismis ex Galeno supposita. Eiusdem . . . Praenotionum libri tres cum explanatione eodem ex fonte hausta. Insigniores aliquot sententiae selectae ex libris A— C— Celsi. . . .*
(*a*) Io. Tornaesius and Gul. Gazeius, Lyon 1555.
Prov. Numerous including 'Georgius Pavus Pharmacopola.'
(*b*) Io. Tornaesius, Lyon 1580.
Prov. LIDDEL.
The author's name, Ioannes Butinus, does not appear on t.p. but only in the dedn. of (*a*).

(11) . . . *Hippocratis . . . opera quibus addidimus commentaria Ioan. Marinelli.*

Io. Valgrisius, Venice 1575
Additional notes by Io. Culmanus.

(12) *Hippocratis . . . viginti duo commentarii tabulis illustrati . . . Theod. Zwinger Bas. studio et conatu.*

Imp. Episcopiorum, Basel 1579
Prov. Two copies: 'Ioan Craigius D. Med.' and Gordon.
Greek text and Latin trans., marginal refs., tabular summaries and analyses. A classification of arts and sciences includes 'Botanica'—an early example of this usage.

(13) . . . *Prolegomena et Prognosticorum libri tres. Cum . . . commentariis Iohannis Heurnii.*

Off. Plant. Raphelengius, Leiden 1597
Prov. READ.

(14) . . . *Hippocratis . . . opera omnia quae extant . . . Nunc recens Latina interpretatione & annotatibus illustrata. Anutio Foesio authore.*

Heirs of Andr. Wechel, Frankfurt 1595
Prov. WILLIAM GORDON, Mediciner of K.C. 1632. And another copy.
Francis Adams of Banchory (*Genuine Works of Hippocrates*, Sydenham Society, London 1849, p. 31) held the view that the works of no ancient author owe more to the exertions of a single individual than those of Hippocrates do to the labours of Foës. Only in regard to the authenticity of the several works did he regard Foës as following Galen too slavishly. (For Francis Adams, see C. Singer *Bull. Hist. Med.* **12** (1942), 1.)
See also Nos. 72, 81, 115, 119, 133, 169, 221, 222, 232, 256, 277, 280, 281, 400, 447, 483, 636, 700, 705, 738.

HOHENFELD, MARCUS VON. See No. 419

343. HOLLERUS, IACOBUS. *Hippocratis . . . Praesagia . . . cum inter-*
pretatione & commentariis Iacobi Holleri . . . nunc primum Desiderii
Iacobi . . . opera . . . editis & eiusdem . . . commentariorum ad
idem opus libri tredecim.

Gul. Rovillius, Lyon 1576

Prov. GREGORY.

344. HOLLERUS, I. . . . *de Morbis internis lib. ii. Illustrati . . . eiusdem*
authoris scholiis . . . non antea excusis; deinde Ludovici Dureti . . .
in eundem adversariis & Anthonii Valetii . . . exercitationibus
luculentis. Eiusdem de Febribus; [*de*] *Peste, de Remediis* κατατοπως
in Galeni libros; [*de*] *Materia chirurgica. Omnia nunc . . . castigata.*

For Petrus Landry, Lyon 1578

Not BM or Bibl. Nat. An edn. with different title but including
some of the above contents, Venice 1562. Edn. with same title, Paris
1571.

The printer states that a reprint of Hollerus's works, previously
printed by the former's brother, had been called for. Valetius, in a
further *Lectori*, indulges in a diatribe against 'innovations', and refers
to an edn. of four years earlier and the consideration of a third. The
commentary of Duretus occurs *passim* and includes a secn. of *Rara*
quaedam fo. 296*v*. The four secns. have sep. pagn.; the *Praefatio* to the
Chirurgica has a good assessment of the history of the subject.

345. HOLLERUS, J. *In Aphorismos Hippocratis commentarii septem.*
Illustrati scholiis . . . per Ioan. Liebautium . . . Et nunc primum
impensis Heinrici Osthausi bibliopolae Lipsensis in Germania
correctius editi.

No name or place 1597

Ed. p. Paris 1582. Bibl. Nat.
Prov. LIDDEL.

346. HOLLERUS, J. *Ad libros Gal. de Compositione medicamentorum. . . .*

Carolus Macaeus, Paris 1577

Ed. p. ? [Earliest BM 1589, Bibl. Nat. 1603.]
Followed by (no. t.p. but new pagn.) *De morbis internis liber II,*
febrium historiam & curationem continens. Antonii Valetii opera . . .
auctior ac integrior factus & scholiis . . . illustratus [Lib. I
wanting.]
[*Ed. p.* 1571 BM.]
Followed by (cont. pagn.) *De peste libellus. Ant. Valetii scholiis illus-*
tratus [Not BM.].
See also No. 667.

347. HOPTON, THOMAS. *Speculum Topographicum or the Topographicall Glasse, containing the use of the T— G—, theodolitus, plaine table, & circumferentor. With many rules of geometry, astronomy, topography, perspective, and hydrography.*

Simon Waterson, London 1611

Only edn. STC. 13783.

In his *To the Reader* Hopton refers to his *Geodeticall Staffe* and to Gemma Frisius's astrolabe, 'the mater whereof being excellent and a most arte-like projectment resembling the true lineaments of the sphere and is now made perpetually famous with the addition of the rete of our ingenious countryman M. I. Blagrave'.

348. HORSTIUS, JACOBUS. . . . *de Aureo Dente maxillari pueri silesii, primum utrum eius generatio naturalis fuerit necne; deinde an digna eius interpretatio dari queat. Et de noctambulonum natura . . . denuo auctus liber.*

Mich. Lantzenberger for Vol. Voegelinus, Leipzig 1595

Only edn. BM.

Prov. A little scoring similar to Liddel's—he is known to have taken a leading part in the controversy about this boy alleged to have been born with a golden tooth. The arguments on both sides shed an interesting light on the 'medical science' of the period.

Ded. to Rudolph II from Helmstadt where Horstius was Liddel's colleague.

The work on sleep walking follows with cont. pagn. Intercalated are a few pages (*Regni Turcici catastrophe*) by Antonius Torquatus and some letters which passed between Horstius and Conrad Paedopater.

349. HUNICHIUS, CHRISTOPHERUS. *Archimedis opinio de Arenae Numero explicata & geometrice examinata.* . . .

A. Lamberg, Leipzig 1593

Not BM or Bibl. Nat.

Prov. 'Suo Georgio Rumbaum mittit K.L.V. Helmaestadium'.

Bound with numerous pamphlets, many bearing Liddel's auto.

A thesis to be disputed at Leipzig, the proposer being also *praeses.*

350. HUBER, SAMUEL. *De peste themata sacra tria in quibus potissimum horum opinio confutatur, qui illam & contagiosam & fugiendam esse asserunt.*

[No printer] Basel 1584

Ed. p. (Dedn. 1584) Not BM or Bibl. Nat.

Dedn. from Burgdorf, of whose church Huber was minister, contains the statements 'clandestina ferrula Domini castigamur' and 'timent aliqui pestilentiae contagionem'.

The text compares the medical and theological reactions to the plague. The conservative temper of the author is indicated by the sentence 'hae sunt nigrae nives & Copernici rationes' (sig. B 5 r).

For rival theories of cause of plague—contagion, miasma, or 'Divine arrow'—see Hirst, F. *The Conquest of Plague* Oxford 1953 also Vol. I, Ch. XIV.

HUMELBERG, GABRIEL. See No. 641.

351. HUSAIN IBN ABD ALLAH [*syn.* Ibn Sina, Avicenna]. (*a*) *Liber canonis, de medicinis cordialibus, et cantica cum castigationibus Andreae Alpagi . . . una cum eiusdem nominum Arabicorum interpretatione.*

Juntae, Venice 1544

(Preceded by—same imprint—*Interpretatio Arabicorum nominum . . .* and *Avicennae vita a Nicolao Massa . . . latinitate donata nusquam antea excusa.*)

Fine cuts of reduction of dislocations, etc.

(*b*) [New t.p.:—] . . . *postea vero Benedictus Rinius . . . lucubrationibus decoraverat. . . .*

Juntae, Venice 1562

Prov. GORDON.

This contains, in addition to the matter of (*a*), the *De syrupo acetoso.*

(*c*) [New t.p. in red and black, and a second t.p. expanding the title:] . . . *A Ioanne Costaeo & Ioanne Paulo Mongio annotationibus . . . illustratus.* [With additions by Fabius Paulinus and ed. by Fabritius Raspanus.]

Juntae, Venice 1596

Two copies.

The 1st t.p. has an engraved border showing instruments, stills, trephining etc. A full-page line fig. of the integuments etc. of the eye shown in sagittal section is included.

All the above are based on the first trn. from the Arabic into Latin by Gerard of Cremona. *Ed. p.* of Canon given in BM as having been from the press of the Brethren of the Common Life, near Cologne 1465-78. GW 3114 gives 'Strassburg vor 1473'.

(*d*) *Avicennae . . . Libri in re medica omnes qui hactenus ad nos pervenere. Id est, Libri Canonis quinque. De Viribus Cordis. De removendis nocumentis in regimine sanitatis. De Sirupo Acetoso. Et Cantica.*

Omnia novissime . . . a Ioanne Paulo Mongio . . . & Ioanne Costaeo . . . recognita. Accessere . . . annotationes.
                                          V. Valgrisius, Venice 1564
Canon based on Gerard's trn. and Alpagus's corrections. Bks. 1-3 cont. pagn. and 4-5. Some sep. t.p.s and dedns.

352. HUSAIN IBN ABD ALLAH (*syn.* Avicenna) *Dinus in chirurgia cum tractatu eiusdem de ponderibus et mensuris necnon de emplastris et unguentis. Additi sunt insuper Gentilis de Fulgineo super tractatu de lepra: atque Gentilis de Florentia super tractatibus de dislocatione et fracturis commentarii. Omnia nuper . . . recognita.*
                                          Juntae, Venice 1544
     Prov. GORDON—much marginal glossing.
     Prologue 'Clarissimi artium et medicinae doctoris magistri Dini de Florentia expositio super tertia & quarta Fen quarti canonis Avicennae & super parte quinte [*sic*] feliciter incipit'.
     Incipit: 'Iam locuti sumus in libro primo de apostematibus & speciebus. . . .' ThK, 308 quote 15th cent. MSS of Dinus de Garbo and an *Ed. p.*, Venice 1499 as well.
     Make up of text: (i) The tracts of Dinus break off at fo. 119v A, and are taken up by Gentilis de Florentia *'de dislocationibus & fracturis' quos Dinus non exposuit . . . (explicit* fo. 147r B). (ii) *Dinus de Ponderibus* fo. 147v. (iii) *Dinus Compilatio Emplastrorum* fo. 148v B. (Note interpolation between 54r B and 58r B of Gentilis de Fulgineo *'De lepra Avicenne'* after *'et quia Dinus non exposuit tractatum de lepra ideo ad maiorem perfectionem libri huius placuit ponere expositionem G— de F—'.*

HUTTEN, ULRICH VON. See No. 501.

JABIR IBN AFFLAH See No. 34.

353. JABIR IBN HAIJAN (*syn.* Geber). (*a*) . . . *de Alchemia libri tres*
                                          Io. Grieninger. (Col.) Strassburg 1529
     Earliest BM similar title and imprint 1531.
     Prov. ARBUTHNOT.
Cut of apparatus on t.p. and others in text.

     (*b*) . . . *Summa Perfectionis Magisterii in sua natura ex Bibliothecae Vaticanae exemplari undecunque emendatissimo nuper edita cum quorundam capitulorum, vasorum, & fornacum in volumine alias mendosissime impresso omissorum, librique Investigationis Magisterii, & Testamenti eiusdem Geberis, ac aurei Trium Verborum libelli, et Avicennae . . . Mineralium additione. . . .*
     (Col.) 'Apud Petrum Schoeffer Germanum Maguntinum', Venice 1542.

(2nd Col.) Apud Dominum Ioannem Baptismam Pederzanum Brixiensem Anno 1542.

Not BM. Duv. p. 238, Ferg. I, 342, cite this and other edns. These are not merely different edns. of the same work. The incipit of the text is 'Totam nostram scientiam quam ex libris [dictis in (*b*)] antiquiorum abbreviavimus . . .'. The explicit of the *libri tres* of (*a*) is '. . . quae materia novam et corruptam dedit formam', which occurs at the end of the *Liber Investigationis magisterii* following four 'libri' in (*b*). The whole work is of course under suspicion of having been compiled about four centuries after Jabir's death (815?): no Arabic MS has ever been found. (See Vol. I, p. 168.)

354. INDAGINE, IOANNES AB. *Chiromantia.*

(Col. Io. Schottus, Strassburg, 1534)

Contents:—*Physiognomia ex aspectu membrorum hominis. Periaxiomata, de faciebus signorum. Canones astrologici, de indiciis aegritudinum. Astrologia naturalis. Complexionum noticia, iuxta dominium planetarum.*

[*Ed. p.* (slightly different title), Strassburg, 1522.]
Prov. LIDDEL.
Many cuts.

355. INGRASSIA, IOANNES PHILIPPUS. [Collection]
Angelus Patesius, Venice 1568.
Contents: *Quaestio de purgatione per medicamentum atque . . . de sanguinis missione, an sexta morbi die fieri possint. Quia occasione de omnibus etiam aliis diebus determinatur in quibus praecipue purgare vel sanguinem mittere possumus.*
[*Ed. p.* 1558 from dedn.]
. . . Ducis Terraenovae casus enarratio & curatio e quibus tum penetrantis in thorace vulneris, tum fistulae curandae methodus elucescit.
[*Ed. p.* 1562.? Dedn. undated but contains a letter from Eustachius 1562.]
Quaestio utrum victus a principio ad statum usque procedere debeat subtiliando an (ut multi perpetuo observant) potius in grossando.
[*Ed. p.* 1568 ? No dedn. or t.p. but sep. pagn. and registrn.]
Quod veterinaria medicina formaliter una eademque cum nobiliore hominis medicina sit, materiae duntaxat nobilitate differens. Ex quo veterinarii quoque medici non minus quam nobiles illi hominum medici ad regiam protomedicatus officii iurisdictionem pertineant.
[*Ed. p.* 1564 Dedn. undated, but text ends 'Panhormi 1564' Sep. t.p., pagn. and registrn.]

Ed. p. of collection from dedn. and 'Omnia nunc primum in unum corpus redacta'. t.p.

Prov. READ.

The question 'whether veterinary medicine is formally the same as human, merely differing in respect of the "nobility" of its material' has a modern ring, and technically implies an advanced attitude to comparative anatomy and physiology—sciences hardly recognised before the 17th cent.

See also No. 755.

356. IO[H]ANNES HISPALENSIS. *Epitome totius astrologiae con-scripta . . . ante annos quadringentos, ac nunc primum . . . edita. Cum praefatione Ioachimi Helleri . . . contra astrologiae adversarios.*

Io. Montanus & Ulricius Neuber, Nürnberg 1548

Only edn.

Incipit: 'Zodiacus dividitur in duodecim signa principaliter . . . ' ThK, 784, 14th cent. MS.

Explicit: 'Est itaque exordium ipsarum mansionum secundum tabulas quae sunt factae in motu solis a circulo recto a sedecim gradibus Arietis hoc tempore 1142 annorum Christi. Tabula viginti octo mansionum lunae, ad annum Christi 1142 completum'. [Tables are given to 1']

Heller was a friend of Melanchthon. See Allen, *Star Crossed Renaissance*, Durham, N. Carolina 1941, pp. 65-6.

357. IOANNES ISAACUS (*syn.* Isaac of Holland). . . . *Opera mineralia sive de Lapide Philosophico, omnia duobus libris comprehensa. Nunquam antehac edita, ac nunc primum ex optimis manuscriptis Teutonicis exemplaribus . . . in Latinum sermonem translata a P.M.G.*

Richardus Schilders, Middelburg 1600

Ed. p. of Latin. No German edn. BM.

Prov. LIDDEL.

Ferg. mentions two 'Hollands' but their difference and dates are uncertain. Dedn. signed 'L.D. 1600'.

Text alchemico-symbolic. See also No. 507.

IOANNES DE RUPESCISSA. See No. 313

358. JOHN OF GADDESDEN (*syn.* Ioannes Anglicus). (*a*) *Rosa Anglica practica medicine a capite ad pedes.*

Io. Antonius Birreta, Pavia 1492

Ed. p. Hain 1108.

Prov. CARGILL.

Incipit: (intro.) 'Sicut dicit Galienus primo de ingenio sanitatis . . .'. ThK, 274.

(*b*) *Ioannes Angli Praxis medica,* '*Rosa Anglica*' *dicta quatuor libris distincta . . . recens edita opera ac studio . . . Philippi Schoppfii. . . .*

Mich. Mangerus, Augsburg 1595

Prov. LIDDEL—two vols. bound in one, with cont. pagn., and a second copy in two sep. vols.

The late reprinting of this mediaeval work is difficult to explain. Guy de Chauliac dismissed it as 'Rosa fatua' (No. 322); the commendation (quoted on t.p. *v* from Gesner's Bibliography) of Symphorien Champier (No. 143 etc.) is somewhat half-hearted: 'John, of the English nation, a man greatly experienced in medicine, in philosophy and in secular affairs not unlearned, wrote certain medical works whose reading is not altogether to be disparaged.' The dedn. by Schopfius [*sic*], having extolled the Greek and 'Latin' physicians (who rather unexpectedly include Constantinus Africanus), rails at those (unnamed) who 'scatter on the crowd either ravings or goodness only knows what dreams concocted in their own heads, and in their empty subtleties . . . spread a snare for the inexperienced'. John is presumably recommended on the grounds that there is no justification for preferring the moderns because they have added to knowledge or corrected the ancients; since it is easier to make additions than to draw up a set of ordered precepts!

The text (p. 78) mentions among the extrinsic causes of paralysis 'contact with a certain fish called "Tarcon", referred to by Avicenna'. On p. 644 is a list of *somnifera* for the relief of pain in sciatica, gout and 'arthritis' (which probably includes rheumatism).

359. [IOANNES DE KETHAM]. [*Fasciculus Medicinae*] t.p. wanting. From *explicit:* Io. and Gregorius de Gregoriis, Venice MCCCC [*sic*. cf. Hain 9776]

> *Explicit fasciculus medicine in quo continentur Primo, iudicia urinarum cum suis accidentiis. Secundo, tractatus de flobotomia. Tertio, de cyrogia. Quarto, de matrice mulierum et impregnatione. Quinto, concilia . . . contra epidimiam. Sexto de anothomia Mundini totius corporis humani. Septimo, de egritudinibus puerorum.*

[*Ed. p.* Venice 1493 (Italian)—Copinger 3449.]
Prov. 'Ex libris Ioannis Pacoli 1623.'

Make-up: 1st fo. blank. 2nd fo. *r*, fine cut of Petrus de Montagnana, books, patients with urine containers; *v*, physicians examining urine. 3rd fo. *r*, urine chart; *v*, 'Incipit fasciculus medicine compositus per . . . Ioannem de Ketham Alamanum . . . annectuntur multi alii tractatus per diversos . . . doctores . . . Necnon Anathomia Mundini . . .' The '*Quartus* tractatus' precedes the '*Tertius*'.

Numerous cuts, e.g. wound-man, signs of zodiac in relation to phlebotomy, anatomy demonstration.

For a critical account see *The* Fasciculus Medicinae *of Iohannes de Ketham* by Charles Singer and Karl Sudhoff, Milan 1927.

360. IORDANUS, THOMA. *Luis novae in Moravia exortae descriptio.* . . .
And. Wechel, Frankfurt 1580

Ed. p.

Prov. LIDDEL.

Bound in rubricated parchment MS stuffed with several pages of MS in German script. Dedn. to Julius Alexandrinus of Trent contains biographical details of his contemporaries; with a reply dated 1579.

361. IORDANUS, T. *De Aquis medicatis Moraviae, commentariolus.*
Heirs of Andr. Wechel, Frankfurt 1586

Only edn.

Dedn. to his fellow countrymen, also to Hinekkus, Baron of Waldstein, contains an historical (?) glorification of the Marcobrunni.

The text contains a list of springs, and is of interest in giving details of 'examination of water by distillation' (p. 11, 23, 27) with actual weights on p. 37. There is a letter (dated 1571) from Aldrovandi (p. 115) referring to the receipt of 'mumia' (see Vol. I, p. 297 n).

IORDANUS NEMORARIUS, See No. 381.

362. JOUBERT, LAURENT . . . . *de Peste, liber unus.* . . . *Accesserunt duo tractatus, unus de Quartana febre, alter de Paralysi in quibus . . . quaestiones aliquot explicantur eodem authore.*
Io. Frellonius, Lyon 1567

Ed. p. Bibl. Nat. Not BM.

Prov. (i) 'R. Harveii 1583'. (ii) Bute. Also a second copy.

The two tracts have sep. pagn., secn. headings, and dedns.

In addition to the dedn. to Henricus Stapedius, Joubert prefaces the main text 'prudentissimis quibusque suae sanitatis studiosis . . .', which begins 'Quum ante biennium pestis has regiones invadere coepisset . . .'. Refs. to 'bubones, carbunculi & tumores praeter naturam' indicate that it was bubonic plague.

363. JOUBERT, L. . . . *Medicinae practicae priores libri tres. Editio tertia ab ipso autore recognita & tertia fere parte adaucta. Accessit eiusdem isagoge therapeutices methodi. De affectibus pilorum & cutis, praesertim capitis & de cephalalgia, tractatus unus. De affectibus internis partium thoracis, tractatus alter.*

137

(*a*) Antonius de Harsy, Lyon 1577.

Not BM. Monsr. P. Josserand (see No. 495) kindly confirmed the absence of any edn. in Bibl. Nat., but states that the Library of the Faculté de Médecine has a copy of each of the edns.—1572 (*Ed. p.* —dedn. of No. 363 dated 1572) and 1575. He points out that 'ce Joubert ne devait pas être très bien considéré de la Faculté en raison du titre de son ouvrage le plus connu—Erreurs populaires au fait de la médecine—' which may account for the rarity of his works even in France!

> Prov. Name obliterated, but ft. cover *v* inscribed 'Arteriae pulsant non propter spiritum a corde immissum sed quia . . . habent cum corpore cordis 106 Cap Ani'. This fits the context of p. 106 of Primrose's *Animadversiones* . . . but is not a quotation. Its sense is identical (except for '*spiritum*') with Harvey's statement.

> (*b*) Carolus Pesnot, Lyon 1577.
> Not BM. Not Bibl. Nat.
> Prov. Two copies, one Fordyce. The other '. . . ex dono Mgri Iacobi Morisoni', with earlier inscription, 'This book pertaines to me William Lindsay'.
> (*b*) lacks some of the poems present among the prelims. in (*a*), otherwise similar.

The second copy of (*b*) wants t.p. and the whole gathering of sig.* (viz. 8 fos.), also the *Accessiones*, which in the first copy are comprised under a single sep. t.p. and cont. pagn. Both copies have a col. after the main text 'Excudebat Petrus Roussin Lugduni 1575'. The t.p. and dedn. of the *Accessiones* are dated 1577 and 1576 respectively.

In the *Lectori* (presumably by the printer) it is explained that Joubert had abandoned the traditional arrangement of the *Practica* . . . 'a capite in calcem'—in order to help those studying medicine and to avoid needless repetitions. For this method there was the precedent of Galen— 'excepto uno Hippocrate medicorum omnium facile principem—'.

The dedn. of the *Accessiones* by Io. Guichardus (the younger; his father, Petrus is named on sig. 5) claims that the study of Hippocrates and Galen is no longer sufficient since 'permulti morbi novi & antehac inauditi miseris his cetemporibus magno omnium malo obrepserunt, quorum nulla sit usquam mentio apud Graecos & Arabes—' (' A whole host of new and previously unheard of diseases have appeared in these unfortunate days to the great hurt of all; and of these there is nowhere a mention in the works of the Greeks and Arabs').

364. JOUBERT, L. *Segonde partie des Erreurs Populaires et propos vulgaires touchant la medecine & le regime de santé refutés . . . avec des*

catalogues de plusieurs autres erreurs . . . qui n'ont été mancionnés an la
premiere & segonde edition de la premiere partie. Item deus autres
petis traites . . . du maime auteur. Le tout corrigé & augmenté par
l'auteur. . . .

For Abel l'Angelier, Paris 1580.

Ed. *p.*

Dedn. by Berthelemy Cabrol (who had them printed) defending the practice of writing in French, and referring to Charles Etienne and Jean Liébaut.

For an account of Joubert's versatile activities see P. J. Amoreux, *Notice historique . . . sur Laurent Joubert*, Montpellier 1814.

365. KENTMANNUS, IOHANNES. *Calculorum qui in corpore ac membris hominum innascuntur, genera XII, depicta descriptaque, cum historiis singulorum admirandis . . . .*

[Device of:] Andreas [?] Gesner, Zürich 1565

Only edn. BM.

Prov. BUTE.

In his dedn. to Conrad Gesner Kentmann refers to letters and to Gesner's books on stones and gems, in which he had seen pictures of stones growing in the earth; so he is sending him examples of similar ones growing in human bodies.

Numerous cuts of calculi with accounts of their origin, with references to Peucer (No. 512), 'Gabriel Valopius (No. 343)', Valerius Cordus (No. 174) and various princes.

366. [KERTZENMACHER, PETRUS]. *Alchimia, das ist alle Farben, Wasser, Olea, Salia, unnd alumina damit man alle corpora, spiritus, unnd calces prepariert, sublimiert unnd fixiert zubereyten. Unnd wie man dise ding nutze auff dasz sol unnd luna werden möge. Auch von soluiern unnd scheydung aller Metall polierung allerhandt edelgestein fürtrefflichen wassern zum etzen scheyden unnd soluiern. Unnd zuletzt wie die gifftige dämpff zu verhüten ein kürtzen bericht etc.*

Frankfurt 1574 (Col. heirs of Chr. Egenolff)

Earliest in BM, but both Ferg. and Duv. cite earlier edns. going back to 1534. Ferg. I, 19 quotes the Young Colln. copy of 1613 which corresponds almost exactly to No. 366. BM copy wants pp. 9-24.

Sig. a i *r* gives a glossary of selected Latin words, e.g. 'Sol' = 'Goldt.' Sig. a ii *r* 'Petrus Kertzenmacher etwan Burger zu Meyntz ein berhümpter Alchimist Wünscht dem Leser alles guts . . .'.

A work of great interest—the First Book being a straightforward text of preparative chemistry, the Second having a more alchemistical flavour, the last dealing with poisonous vapours, where is to be found (fo. 76, r) an admirably observed description of carbon monoxide poisoning, part of which is here quoted: 'Derselbig Rauch oder Dampff ist gifftig, und sonderlich so die Kolen etwas feucht sein, dann dieser Dampffe beschwert das Haupt unnd Brust, nemlich, so einer lang darbey ist und man zu lang darbey verhart, so wirt ihm das Gesicht betrübt, also dasz einen beduncket es sey grün und blauw und dergleichen vor seinen Augen oder Fliegen im lufft. Dieser Dampff machet auch schwer unnatürliche Schlaff unnd zu Zeiten schwer Glieder . . .' (punctuation added). ('This smoke or vapour is poisonous, particularly if the coal is rather damp, in which case the vapour makes the head and breast feel heavy. If it persists or any one stays in it too long, his sight will become so dimmed that things appear green, blue, etc. before his eyes or as if there were flies in the air. This vapour also produces deep unnatural sleep and at times heavy limbs. . . .') (I am indebted to my colleague, Dr. B. J. Kenworthy, for the meaning of the rather obscurely expressed middle section.)

367. KHALAF-IBN-ABBAS (*syn.* Albucasis, Alsaharavius). *Liber theoricae necnon practicae Alsaharavii in prisco Arabum medicorum conventu facile principis: qui vulgo Acararius dicitur.*

(Col. Sigismund & Marcus Wirsung, Augsburg 1519) Not BM. Only edn. Bibl. Nat.

Prov. GORDON.

Fine cut of medical consultation t.p. Good early examples of privileges by Leo X 'per universum Christianum orbem' and Maximilian I 'per universum imperium' under penalty of excommunication and fine. (Plate I.)

Incipit: 'Finis intentionis medicinae est in sanis sanitatem conservandi speculatio. . . .'

A complete treatise of medicine, including gynaecology and pediatrics. Albucasis, who flourished in Cordoba in the khalifate of Al-Hakam II (A.D. 961-76), was even more famous as a surgeon (No. 368).

368. KHALAF-IBN-ABBAS. *Albucasis chirurgicorum omnium primarii lib.tres. I. De cauterio cum igne & medicinis acutis . . . cum instrumentorum delimatione. II. De sectione & perforatione phlebotomia, & ventosis. De vulneribus & extractione sagittarum & caeteris similibus. Cum formis instrumentorum. III. De restauratione & curatione dislocationis membrorum. Cum typis item instrumentorum.*

Io. Schottus, Strassburg 1532

This finely illustrated work was apparently printed only as a 'supplement' to that of 'Horatianus' (Priscianus) No. 538 with which it is bound and cont. paged.

140

PLATE IV

LIBER THEORICAE

NECNON PRACTICAE ALSAHARAVII IN PRI=
fco Arabum Medicorum conuentu facile principis:qui vulgo
Açararius dicitur:iam fumma diligentia & cura
depromptus in lucem,

Cum priuilegio fummi Pontificis
et Imperatoris Romani.

Title-page of the treatise on medicine by Albucasis
(Khalif-ibn-Abbas), No. 367, inscribed 'Book belonging
to King's College, Aberdeen, the gift of Francis Gordon'.

Not BM; only edn. Bibl. Nat.

Incipit (of *Praefatio*): 'Postquam complevi vobis, o filii, librum . . .'
(as in ThK 498); (of text): 'Sed ante quidem quam . . .' (ThK 243
'Et ante . . .)'. 'Explicit liber chirurgiae quem transtulit Magister
Gerardus Cremonensis . . .' of which 13th cent. MS and *Ed. p.* of
Venice, 1499 are extant.

The cuts in this work, especially that of the human skeleton, are
outstanding for this period.

See also No. 740.

369. [KHUNRATH, CONRAD] *Medulla destillatoria et medica, das
ist warhafftiger eigentlicher gründtlicher Bericht wie man den Spiritum
Vini durch Mittel seines hinter ihme verlassenen Saltzes item die
Perlen Corallen deszgleichen alle andere Olitelen aus den Crescentibus
als Fruchten refinen und anderen Sachen mehr zum Auro potabile und
andern Arcanen dienstlich künstlich destilliren nachmals in Quintam
Essentiam zur hochsten Exaltation bringen soll. Item etzlicher
herlicher Wundt Balsam Stichpflaster und guldene Wasser Praepara-
tiones . . . Durch C[onrad] C[unrath] L[ipsiensem.]*

Nicolaus Wegener, Schlesswig (1594 from dedn.)

Ed. p.

Prov. LIDDEL.

The title above corresponds to that of Ferg. I, 461 presumed by him
to be *Ed. p.* The title in Duv. corresponds to the third entry in Ferg.

Dedn. to 'Bürgermeistern und Rathmannen der Keyserlichen
Reiches und Freyen Stedten Lübeck, Lüneburg, und Magdeburg . . .'
announces his intention to publish '. . . Arcana, preparations of nobler
things, metals and minerals . . .'.

The text is medical in intention, free from symbolism and obscurity,
but the 'cures' must have relied largely on faith, e.g. fo. 1*r* 'Spirit of
urine' for dissolving gold and curing gout! On the other hand, the
instructions for making 'rechten Spiritum Vini' and for testing it (by
burning) to show its freedom from 'phlegmatischen substanz' is clearly
described.

Conrad is to be distinguished from (his brother?) Heinrich Khunrath.

370. KOEBEL, IOANNES. *Astrolabii declaratio eiusdemque usus mire
iucundus utilis ac necessarius: verum etiam mechanicis quibusdam
opificibus non parum commodus: a Io. . . . K . . . facilioribus formulis
nuper aucta. . . . Cui accessit Isagogicon in Astrologiam Iudiciariam.*

Gul. Cavellat, Paris 1552

Earliest BM and Bibl. Nat. is Mainz 1535 (without *Isagogicon*),
but the title quoted by each ends 'nuper aucta'.

L   141

Numerous cuts of parts of astrolabe. Fo 26*r* (printed '16') has list of 'various instruments of mathematicians, with their several uses'.

371. KUEFNER, IOANNES. *Pharmacopolitarion saluberrima synthetorum pharmacorum in officinis medicamentariis passim promercalium symmicta, ad medibiles quoscunque morbos curandos apprime conducibilia promens.* . . .

Alex. Weissenhorn, Ingolstadt 1542

Date only on t.p. Col. also has device in Greek, Latin and Hebrew.
Ed. p. (from dedn.).
Prov. 'Edoardus Lapworthus Magdalensis 1601'.

372. L'AGNEAU, DAVID. *Harmonia seu consensus philosophorum chemicorum* . . . *in ordinem digestus & a nemine alio hac methodo distributus.* . . .

Cl. Morellus, Paris 1611

Earliest BM, and dedn. dated 1611; also Duv. p. 334. But Ferg. II, 84, mentions, though he admits that he had not seen, an octavo edn. of 1601.

L'Agneau wrote several *medical* works, including two on phlebotomy. This contains a table of alchemical operations with synonyms.

373. LAGUNA, ANDRES DE [*syn* Lacuna, Andrea]. *Annotationes in Galeni interpretes: quibus varii loci, in quos hactenus impegerunt lectores, & explicantur &* . . . *restituuntur.* . . .

Gul. Rovillius, Lyon 1553

Ed. p. in this form; but the dedn. implies that it was included in an *Epitome*, 1548.

In the dedn. he refers to 'superioribus annis, quum Lutetiae Parisiorum politiores disciplinae inciperent iam renasci & florescere . . .' dated Venice 1548.

The text is a collection of annotations from Cornarius, Linacre, Nicolaus Leonicenus, etc.

374. LALAMANT, JEAN (*syn.* Lalamantus). *Externarum fere omnium et praecipuarum gentium anni ratio & cum Romano collatio.*

Crispinus [n.p.] 1571

Comparative chronology.
Only edn. BM gives Geneva as place of printing.

LANDUS, BASSIANUS. See No. 753.

142

375. [LANFRANCHI, GUIDO]. *A most excellent and learned woorke of chirurgerie, called Chirurgia Parva Lanfranci, Lanfranke of Mylane his briefe: reduced from dyvers translations to our vulgar or usuall frase, & now first published in the Englyshe prynte by John Halle Chirurgien. Who hath thereto necessarily annexed a table as well of the names of diseases and simples. . . . And in the ende a compendious worke of Anatomie more utile and profitable than any heretofore in the Englyshe tongue publyshed. An historicall expostulation also against the beastly abusers, both of chyrurgerie and phisicke in our tyme. . . .*

Thos. Marshe, London 1565

Ed. p. STC 15192. Lanfranc's *Chirurgia Parva*, Louvain 1481 (Copinger 3484).

Dedn. to the 'Maisters, Wardens & consequently to all the whole company & brotherhood of Chirurgiens of London . . .' includes a protest against 'pedlars' of surgery. Refs. to texts employed; to Recorde (No. 554), William Turner (Vol. I, p. 181), Vigo (No. 718), and 'Wm. Cunningham, doctor of Physicke, for his so many learned lectures which he red unto you in our halle . . .'.

Epistles from Cunningham '. . . unto the professors of Chirurgerie, salutations'; 'Thomas Gale, maister in chirurgerie, unto . . . John Hall . . .', justifying the author's qualifications for the task of translation. The *Anatomie* has sep. t.p. and pagn.; the Expostulation, secn. heading and sep. pagn. The text contains many historical refs. e.g. to 'Henricus de Ermunda villa' (Henri de Mondeville?) (*Anatomie* sig. L i r); and Fernel's 'new division' of fevers, which is described on sig. A i r.

376. LANGHAM, WILLIAM. *The Garden of Health conteyning the sundry rare and hidden virtues and properties of all kindes of simples and plants. . . .*

[No printer] London '1579'

Ed. p. STC 15195, where it is stated that the correct date is 1597.

The astrological character of the work is exemplified by *Lunaria* or *Moonwort*—'Falling sickness, drink it when luna is in virgo in the wane' (p. 344).

377. LANGIUS, IOANNES. . . . *Epistolarum medicinalium volumen tripartitum, denuo recognitum, & . . . auctum . . . .*

Heirs of Andr. Wechel, Frankfurt 1589

Ed. p. in this form, based on edn. of 1557 BM.

143

Prov. (i) 'Liber Ioannis Henrichii Dassiliensis [pastoris?] ad D. Jacobum . . . Hamburgi'. (ii) 'Duncanus Liddelius. Emptus Hamburgi Anno 96'.

Dedn. by Nich Reusner, 'iuriconsultus', dated 'Jenae ex Academia . . . 1589'.

Castiglioni p. 443 emphasises that Langius (1485-1565) 'insisted on direct study of the classics', which is borne out by this book, though quotations from his contemporaries are included. Ref. to wounds caused by missiles driven by gunpowder (p. 38).

378. LANSBERGER, PHILIPPUS VON. *Triangulorum geometriae libri quatuor* [etc.].

Raphelengius, Off. Plant. Leiden 1591

Ed. p.

Includes tables of sines, tangents and secants. See Johnson.

379. DU LAURENS, ANDRÉ (*syn.* Laurentius, Andreas). *A discourse of the preservation of the sight: of melancholike diseases; of rheumes, and of old age.*

F. Kingston, London 1599

Ed. p. of trans. (by Richard Surphlet), STC 7304.

Earliest French orig. 'Reveuz de nouveau . . .' 1597 BM and Bibl. Nat. Several later edns. and trans.

Prov. READ—characteristic scoring and glossing.

In his *To the Reader* Surflet refers to '. . . the lamentable time and miserable daies that are come upon us in this last and weakest age of the world . . .'. So much for the Elizabethan Age!

The author of the original apologises to the reader for sometimes 'grappling with Aristotle' and for writing in French.

The text includes a good observation on the optic nerve (p. 25) and (p. 68) a ref. to the harm (as well as good) that may be done in blood-letting from the 'store-house of nature'.

380. LAVATER, LUDOVICUS. *De spectris, lemuribus, et magnis atque insolitis fragoribus, variisque praesagitionibus quae plerunque obitum hominum, magnas clades, mutationesque imperiorum praecedunt, liber unus.*

Anchora Crispiniana, Geneva 1570

Earliest BM Geneva 1575.

Prov. LIDDEL.

In his dedn. the author explains the purpose of the book, namely, to combat undue credulity and superstition, while claiming the authority of scripture for the reality of some spectres—both good and bad.

381. LEFÈVRE, JACQUES, D'ETAPLES (*syn.* Iacobus Faber Stapulensis). (*a*) *In hoc libro contenta epitome compendiosaque introductio in libros Arithmeticos divi Severini Boetii adiecto familiari commentario dilucidata. Praxis numerandi certis quisbusdam regulis constricta. Introductio in Geometriam sex libris distincta: primus de magnitudinibus & earum circumstantiis; secundus de consequentibus contiguis & continuis; tertius de punctis; quartus de lineis; quintus de superficiebus; sextus de corporibus; Liber de quadratura circuli. Liber de cubicatione sphere. Perspectiva introductio. Insuper astronomicon.*

(Col. fo. xlviii: Henricus Stephanus 'absolutum in almo Parisiorum studio 1503'. 'Emissum' 1510.)

Bound with: (*b*) *In hoc opere contenta Arithmetica decem libris demonstrata. Musica libris demonstrata quatuor. Epitome in libros Arithmeticos divi Severini Boetii. Rithmimachie ludus qui et pugna numerorum appellatur. Haec secundaria superiorum operum aeditio venalis habetur Parisii in officina Henrici Stephani . . .* (Col. '. . . 1514 absolutumque reddidit eodem anno . . .'

(*a*) contains (i) Iosse Clichtove's commentary on Lefèvre's *Epitome* and Clichtove's *Praxis Numerandi*. There are no geometrical parts.

(*b*) contains Lefèvre's (i) edn. of text and commentary of the *Elementa Arithmetica* of Iordanus Nemorarius (13th cent.), (ii) *Elementa Musicalia*, (iii) *Epitome* of Boethius's *Arithmetica*, (iv) the mathematical game 'Rithmimachia'.

382. LEFÈVRE, J.

[8 fos. wanting inc. t.p.] Contents: (All l.c. letters refer to parts of No. 381) (1) b (i) less 7 fos.; (2) b (ii)—sig. f iii wanting, but replaced by MS in 16th C. hand; (3) additional MS in same hand on sig. h vii; (4) b (iii); (5) b (iv). Col. as in No. 381. Sigs. a ii *r* & *v* and part of a iii *r* are supplied in MS

Prov. 'Liber domini David [?] de Monymusk'.

No. 381 and No. 382 constitute 'sample' collections of works in current use in the early 16th cent. for the arithmetic and music of the *quadrivium*. Their popularity is shown by the numerous printings (see Smith, pp. 30 and 65). An edn. on similar lines is cited by BM for 1495.

383. LEMNIUS, LEVINUS. *De Miraculis occultis naturae libri iiii. Item de vita cum animi et corporis incolumitate recte instituenda. . . . iam postremum emendati & aliquot capitibus aucti: hic vero nunquam ante hac editus.*

(*a*) Guil. Simon, Antwerp 1567 [Col. excudebat Chris. Plantin].
(*b*) Chris. Plantin, Antwerp 1581.

(c) Andr. Wechel, Frankfurt 1590.

(d) . . . de gli occulti miracoli . . . Venice 1560.

Ed. p. Antwerp 1559.

Orig, dedn. by Lemnius 'Illus. Potentis. Suecorum, Gothorum, Vandalorumque Regi ac Levoniae Domino ERICO eius nominis decimoquarto'. It will be seen that the Swedes were in strange company at that time! Plantin deds. his edn. of 1581 to Ortelius 'Geographio Regio . . .'.

The *Paraenesis sive exhortatio ad vitam optime instituendam* . . . follows with secn. heading and cont. pagn. It was not included in (a)—the title quoted is therefore that of (b).

384. LEMNIUS, LEVINUS. *Della Complessione del corpo humano libri due. . . . Nuovamente di Latino in Volgare tradotti & stampati.*

Dom. Nicolino, Venice 1564

Only edn. BM.

Prov. BUTE.

385. LEMNIUS, L. *De astrologia liber unus in quo obiter indicatur quid illa veri quid ficti falsique habeat & quatenus arti sit habenda fides: in quo denique multae rerum physicarum abditae amoenissimaeque causae explicantur. . . .*

(No imprint. Dedn. 1553).

Bound (with half-title and cont. pagn.) with No. 386.

Ed. p. Antwerp 1554. *Biog. Univ.*

Prov. LIDDEL.

Not as original as might be inferred from the title—mainly quotations from classical authors.

386. LEMNIUS, L. *Similitudinum ac parabolarum quae in bibliis ex herbis atque arboribus desumuntur dilucida explicatio.*

Io. Wechel & Petrus Fischer, Frankfurt 1591

[*Ed. p.* Antwerp 1569 *Biog. Univ.*] Earliest BM, Frankfurt 1596, but see No. 387.

The dedn. is dated 1566. See Ferg. II, 23.

The t.p. also names *De gemmis aliquot* . . . *auctore Francisco Rueo* and *De astrologia Levini Lemnii.*

Prov. LIDDEL (unusual auto.).

387. LEMNIUS, L. *Herbarum atque arborum quae in bibliis passim obviae sunt & ex quibus sacri vates similitudines desumunt* . . . *dilucida explicatio. . . .*

. . . Gul. Simon, Antwerp 1566

Ed. p. BM.

Prov. LIDDEL.

388. LEMOSIUS, LUDOVICUS. *De optima praedicandi ratione libri sex. Item Iudicii operum magni Hippocratis liber unus. . . .*

Robt. Meietus, Venice 1592

Only edn. BM.

The dedn. to Eustachius Rudius is explicitly, and the *Lectori* implicitly, by the printer, since in the latter he speaks of the 'Libellus' of Lemosius having come into his hands.

The *Iudicium* is of some biographical interest.

LEONICENO, NICOLO. See No. 748.

389. LEONUS, DOMINICUS. (*a*) *Ars medendi humanos. . . .*

Io. Rossius, Bologna 1583

Ed. p.

(*b*) [As above] *nunc primum in Germania ab innumeris quibus prior editio scatebat mendis . . . vindicata.*

Heirs of A. Wechel, Frankfurt 1597

This work is of interest in that each secn. is preceded by an anatomical description. There is also a long secn. on the care of the teeth and gums.

390. LEOWITZ, CYPRIANUS (*syn.* Leovitius). *Ephemeridum novum atque insigne opus ab anno Domini* 1556 *usque in* 1606 *accuratissime supputatum cui praeter alia omnia in caeteris editionibus addi solita etiam haec accesserunt.* [List too long to quote.]

Th. Ulhardus, Augsburg 1557

Only edn. BM.

Prov. 'Ex dono Io. Gordon D. Medicinae', and in another hand 'E Libraria suppellectile Ja. Gordon . . . Anno 65'.

Dedn., to the Elector Palatine Otho Heinrich, refers to the latter's aid to the 'Academy' of Heidelberg, to numerous authorities (e.g. Schöner's *Tabulae Resolutae*, but not the Prutenic Tables), also to Georg. and Huldrich Fugger of Roseberg [*sic*]. Dated Augsburg 1557. The *Lectori* by Andrea Rosetus refers to the Fuggers of Rosenberg. See Th. *Hist.* VI, iii *seq.*

391. LEOWITZ, C. *De coniunctionibus magnis.*

Thos. Vautrouiller, London 1573

S.T.C. 15484 cites it as a sep. work. No. 391 is bound with No. 191, having sep. t.p.

392. LIBAVIUS, A. *Neoparacelsica, in quibus vetus medicina defenditur adversus* τερετισματα *tum Georgii Amwald, cuius liber de Panacea excutitur . . . tum Iohannis Gramani . . . qui omnes medicos acerbissima charta est insectatus.*

Io. Saur for P. Kopff, Frankfurt 1594

Only edn. BM. Not in Duv. or Ferg.

Prov. 'Prestonii . . . 1595'.

The title is somewhat misleading: the whole volume consists of two sections with sep. t.p. and pagn. but same imprint. The first section, ded. to the 'Senate' of Rotenburg, of which Libavius was archiater, and with a foreword by Io. Hartmann Beyer to Bernhard Stieber urging publication, consists of a detailed attack on Amwald's *Panacea*, followed by a work entitled *Liber Antigramaniorum* taking the form of about 550 pp. of diatribe against Paracelsus and all his rout; its interest lies in the numerous witnesses called for the prosecution. With cont. pagn. and half-title follows *Anatome tractatus Neoparacelsici de Pharmaco Cathartico*.

The second section is entitled *Tractatus duo physici; prior de Impostoria Vulnerum per unguentum armarium sanatione Paracelsicis usitata . . . Posterior, de Cruentatione Cadaverum in iusta* [sic] *caede factorum praesente qui occidisse creditur . . . His accessit epistola de examine Panaceae Amvaldinae. . . .*

The first two were proposed as theses to be defended at Jena and are here published with added scholia. The added *Epistola* is addressed to Beyer.

393. LIBAVIUS, ANDREAS. *Rerum chymicarum epistolica forma ad philosophos et medicos quosdam in Germania excellentes descriptarum liber primus. . . .*

Io. Saurius for Petrus Kopfius, Frankfurt 1595

Only edn. BM. Duv. quotes three parts, Ferg. (**II**,32) only two.

The dedn. refers to many contemporaries, to those of the recent past, and also 'veteres'. He deplores the fact that no chemist is held in popular esteem 'unless he follows the principles and footsteps of the utterly corrupt (*vitiosissimi*) Paracelsus'. Further refs. in *Praefatio* include Dr. Dee and Brahe's Uraniborg (see Vol. I, p. 121).

The text of letters addressed to famous contemporaries (e.g. Caspar Bauhin, p. 240) is preceded by an index. The purpose of these letters is to expound the application of chemistry to Hippocratic medicine. Libavius shows himself to be well versed in the 'history' of the subject.

(Followed by—sep. t.p. and pagn.) *Rerum C— E— F— ad medicos praestantes Germaniae conscriptarum, liber secundus. Continens operationes chymicas artificum praeceptis naturae documentis & experientia declaratas additis . . . problematis.* (Imprint as for Lib. 1).

The *Epistola* of Bernhardus Stiberus to Libavius mentions (sig. ff 3*v*) 'Horstius Helmstadensis'—he was a colleague of Duncan Liddel.

The text is more 'practical' than that of *Lib.* i, but still smacks of the schools.

394. LIBAVIUS, A. *D.O.M.A. Alchemia . . . opera e dispersis passim optimorum autorum veterum & recentium exemplis potissimum . . . in integrum corpus redacta. Accesserunt tractatus nunnulli Physici Chymici item methodice ab eodem autore explicati . . . Sunt etiam in chymicis eiusdem D. Libavii epistolis, iam ante impressis, multa huic operi lucem allatura.*

Io. Saurius for P. Kopfius, Frankfurt 1597

Ed. p. Duv. p. 356. Ferg. I, 31 same collation.

T.p. *v* gives list of contents of second part, which has sep. t.p. and pagn. following the Col. of first part, but seems always to have been issued with it.

The second part comprises:

D.O.M.A. Commentationum metallicarum libri quatuor de natura metallorum, Mercurio philosophorum, Azotho, et Lapide seu tinctura physicorum conficienda. E rerum natura, experientia, et autorum praestantium fide studio & labore A— L— depromti & expositi, more veteris philosophiae cum perspicuitate evidente [same imprint as above] [Half titles to parts] followed by (half-title only & cont. pagn.):

D.O.M.A. Ars probandi mineralia libris duibus compraehensa & ex plurium artium, potissimum Chymiae & Physicae, concursu, experientiaque optimorum artificum, in primis Georgii Agricolae Modestini Fachsii, opera & studio A— L— . . . .

(These contents correspond to those on main t.p. *v.*)

Also (half-title & cont. pagn.):

D.O.M.A. De iudicio aquarum, mineralium, et horum quae cum illis inveniuntur, libri tres physici chymici, ex rerum natura artibus cognatis, & praestantium auctorum commentariis . . . selecti . . . studio A—L—.

395. LIBAVIUS, A. *D.O.M.A. Singularum A— L— pars quarta et ultima continens historiam & investigationem fontis medicati ad Tubarim sub Rotemburgo. Libros batrachiorum duos de natura, usu, & chymia ranarum utriusque generis. Sectiones duas historiae & confutationis Panaceae, violentia Ambaldi iterum extortae & non tantum re gesta & rationibus, sed & experimentis infelicis curae penitus confectae. Et tandem de Aethiopibus & senibus duplicibus tractariunculum. . . . Nunc primum. . . prodita.*

Io. Saurius for P. Kopfius, Frankfurt 1601

The four parts, of which this is the last, were issued in 3 vols. (BM) 1599-1601. All are mentioned by Ferg. (none in Young Collection), the first two only by Duv.

Cont. pagn. with half-titles after the first secn.

396. LIBAVIUS, A. *D.O.M.A. Alchymia . . . recognita, emendata et aucta tum dogmatibus & experimentis nonnullis; tum commentario medico physico chymico qui exornatus est variis instrumentorum chymicorum picturis partim aliunde translatis partim plane novis. In gratiam eorum qui arcanorum naturalium cupidi ea absque involucris elementarium & aenigmaticarum sordium intueri gaudent. Praemissa defensione artis: opposita censurae Parisianae. . . .*

Io. Saurius for P. Kopfius, Frankfurt 1606

Prov. READ.

This is not just a '2nd edn.' of No. 394, but contains an intercalated addition of 402 pp. of text. The make-up is as follows: Text of *Alchymia* 1st edn. part 1, followed by (*a*) *Commentariorum, pars prima* with sep. t.p., dedn. and pagn. (*b*) *Commentariorum, pars secunda*—essentially Part 2 of the 1st edn. but with new t.p. and wanting the dedn. The half-title of *De lapide philosophorum* is here followed by three additional fos. portraying, in four fine cuts, the 'mythology' of alchemy, with explanatory legends and tables.

The t.p. of (*a*) is *D.O.M.A. Commentariorum Alchymiae . . . Pars prima, sex libris declarata. Continens explicationem observationum chymicarum priore artis libro comprehensarum adiectis fornacum et aliorum vasorum figuris partim ex impressis antehac autoribus partim aliunde acceptis & ex latibulis officinarum productis. Praemissa est defensio alchemiae et refutatio objectionum ex censura Scholae Parisiensis, quae licet videri nolit hanc Alchemiam, sed Quercetani damnasse, nimis tamen frigide de arte sentit, eaque proponit, quae in ludibrium & ignominiam artis simpliciter possunt converti, nec sonant aliter.* On p. 16 of this part (in the course of Defensio) he mentions a Scotsman 'who to-day in Belgium and Germany and other places is said to be making gold out of lead', but he is unwilling to give any opinion about this report.

397. LIBAVIUS, A. *D.O.M.A. Alchymia triumphans de iniusta in se Collegii Galenici spurii in Academia Parisiensi censura; et Ioannis Riolani maniographia, falsi convicta & funditus eversa. Opus Hermeticum vere didacticum (ne ἐριστικον saltem putetur) solida explicans chymiatriae Hippocraticae fundamenta. De quinta essentia, magno perfectoque lapidis magisterio, principiis extractis, oleis, aquis, salibus, elixyribus & . . . . Studio A— L—.*

Io. Saurius for P. Kopfius, Frankfurt 1607

The *Lectori* sets forth the origin of the dispute and the nature of the attacks Libavius wishes to combat. For detailed study see Vol. I, ch. XIII.

398. LIDDEL, DUNCAN. *Themata de Melancholia, . . . praeside Francisco Parcovio. . . .*

Iac. Lucius, Helmstadt 1596

Liddel's own thesis, on account of which he was granted the degree of Doctor of Medicine.

Note: There are in the Library several bound volumes of theses delivered at Helmstadt under Liddel's presidency. Many of these theses were published in the years immediately preceding 1600. It was thought more appropriate to include them all in a memoir on Liddel's library, whose publication is planned.

399. LIÉBAUT, JEAN. *Quatres livres de secrets de médecine et de la philosophie chimique. Faicts Francois par M. Jean Liébaut.* . . . .

Benoist Rigaud, Lyon 1593 (Col. Heirs of Pierre Roussin)

Duv. p. 358 gives evidence of *Ed. p.* now lost. Earliest Bibl. Nat. is Paris 1579, but dedn. is dated 1573. Ferg. I, 36 points out that the work is in fact a trans. of Gesner's *De secretis remediis*—2nd part (See No. 299 and 300).

The dedn. describes (fo. 4) the function of chemistry, and (fo. 5*v*) its relation to the medicine of Hippocrates and Galen.

400. LIÉBAUT, J. (ed.). *Thesaurus sanitatis paratu facilis selectus ex variis authoribus.* . . . .

Iac. du Puys, Paris 1577

Only edn. BM.

Contents: *Hippocratis, sive Polybii Hippocratis discipuli, de salubri victu idiotarum libellus. Iacobi Sylvii* . . . *de victus ratione paratu facili ac salubri pauperum scholasticorum libellus egregius. Synopsis remediorum paratu facilium ad quosvis morbos* . . . . *Thesaurus pauperum vulgo inscriptus a Petro Hispano* . . . *primum editus* . . . . [*Iacobus Sylvius*] . . . *Libellus de peste & febri pestilenti.* [*Guinther of Andernach*], *De victus et medicinae ratione* . . . *cum alio, tum pestilentiae tempore observanda, commentarius.* [Liébaut], *De praecavendis curandisque venenis.* [Auger Ferrier], *De pudendagra sive Lue Hispanica* [running title—*De Morbo Gallico.*] *Quod chyna et apios diversae res sint, adiecto utriusque radicis usu.*

The *Praefatio* of the book of diet for poor scholars has the pregnant remark that 'it is easier to resist than to cure diseases'—hence the writing of the book.

See also No. 364.

401. LINDHOUT, HENRICUS A. (*a*) *Introductio in physicam iudiciariam in qua* . . . *vera astrologiae fundamenta & rerum humanarum consensus cum superioribus atque divinis aperte demonstratur. Item in quo vera ac legitima praesagiendi methodus statuitur & quam multae lateant in genethliacae Arabum doctrina vanitates involutae. Contra calumniatores artis astrologiae, eosque qui praedictionem penitus nullam ferunt vel ultra fas hominis moliuntur.* . . . .

151

Off. Binderiana per Philippum de Ohr (Col. 'imp. Abraham Kertzeri') Hamburg 1598.
Only edn. BM [but see (b)].
Dedn. 'Ad Elizabetam Angliae Reginam invictissimam . . .' is a good example of late renaissance verbiage.
On p. 189 the 'naturalistic' explanation which had been put forward for the plague (1598) in Hamburg is 'answered'.

(b) *Speculum Astrologiae* . . .[title similar to above] *Addita est de Astrologiae praestantia et utilitate M. Gotardi Arthus . . . Praefatio.*
Wolfgang Richterus for Conradius Meulius, Frankfurt 1608
This is essentially the same work as (a) with an additional secn. on astrology, and the dedn. to Elizabeth dropped.
These two works (neither mentioned by Allen, *Star Crossed Renaissance*) are interesting in respect of the numerous nativities of famous men (cf. Cardano, No. 126) and the quaint admixture of theology (e.g. fo. 16) showing the nine choirs beyond the *primum mobile* as distinguished by (pseudo) Dionysius.
Only edn. in this form BM.

402. L'OBEL, MATTHIAS DE (*syn.* Lobelius) and Pena, Petrus.
Nova Stirpium Adversaria. Perfacilis vestigatio luculentaque accessio ad priscorum, praesertim Dioscoridis & recentiorum, materiam medicam . . . Quibus accessit appendix cum indice variarum linguarum . . . Additis Gulielmi Rondeletii aliquot remediorum formulis, nunquam antehac . . . editis.

Chris. Plantin, Antwerp 1576
The dedn. is dated London 1570. According to Colin Clair (*Christopher Plantin*, London 1960, pp. 117-8) Plantin bought 800 copies of the *Ed .p.* printed by Purfoot (see No. 404) and replaced the title page by the one here cited.
The share of Petrus Pena in the joint authorship has never been determined—he is at best a shadowy figure (Arber, p. 90).

403. L'OBEL, M. DE. *Plantarum seu stirpium historia . . . cui annexum est Adversariorum volumen.*

Chris. Plantin, Antwerp 1576
Ed. p. in this form, but it was based on the *Stirpium adversaria nova*, Purfoot, London 1570 (Col. 1571).
A comparison of Nos. 402 and 403 shows that there are many more and larger cuts in the later edn. Only one cryptogam—'Alga marina'—was found in the earlier. The order of treatment, however, appears to be the same, namely, according to the structure of the *leaves*, an arrangement (originated by L'Obel) which brought out the 'natural' relationships

corresponding to the division into mono- and di-cotyledons. There are, however, some strange bedfellows: e.g. 'Pseudomelanthium' (*syn. Gitago* of Tragus—similar if not identical to *Lychnis githago*) and 'melampyrum'—an obvious Legume—occur in the midst of a number of grasses. No. 403 contains 671 pp. instead of the 646 stated by Clair.

404. L'OBEL, M. DE. . . . *In G. Rondeletii . . . methodicam pharmaceuticam officinam animadversiones quibus depravata & mutilata ex authoris mente corriguntur & restaurantur. Accesserunt auctaria: In Antidotaria vulgata censurae . . . simplicium medicamentorum explicationes. Adversariorumque volumen eorumque pars altera & illustramenta quibus ambigua enodantur. Cum Ludovici Myrei Pharmcopolae Reginei paragraphis utiliss.*

Thos. Purfoot, London 1605

Not BM, nor Arber.

Dedn. to Edward, Baron Zouche, whose herb garden at Hackney was laid out by L'Obel. Following the *Tabula* (of contents) there is a second dedn. to 'Clariss. et honorabilibus inclyti Collegii Medici Londinensis . . . praesidi, professoribus & praefectis.' The privilege (of James I) is dated 'Grynwich 1605 Aula Regia Nostra'.

On p. 156 there is a Col., following which are what appear to be the sheets of the original text (compared with No. 403), but with new prelims. and the index placed *before* the text. The appendix is excluded, the first page of this being replaced (sig. Qq) by a collection *Rariorum aliquot naturae miraculorum recentiones* [sic] including the 'plantae quoddam rudimentum' of the 'Insula Portlandia', the 'Concha Anatifera' (i.e. 'Barnacle Goose') and 'Lithoxyla', described on pp. 654-5 of the earlier *Observationes*. The *Adversariorum altera pars* comprises sigs. Q q 2 *r*—V v 6 *v* (the effective t.p. being a table of contents Q q 2 *r*), and ends with a work on Balsams having sep. secn. heading and dedn. to Clusius (No. 3), the running title being, however, continued. The volume concludes with *G. Rondeletii Tractatus de Hydrope nunquam antehac in lucem editus*, and *Eiusdem G— R— Elephantiasis nova methodica curandi ratio.* . . . Rondelet (No. 581) was L'Obel's teacher at Montpellier.

405. ED. LOKERT, GEORGE. *Quaestiones et decisiones physicales insignium virorum Alberti de Saxonia, in octo libros Physicorum* [etc.], *Thimonis in quatuor libros Meteororum* [etc.], *Buridani in Aristotelis lib. de Sensu & Sensato* [etc.] . . . *Recognitae rursus & aemendatae . . . Magistri Georgii Lokert Scoti a quo sunt tractatus proportionum additi.*

Badius Ascensius & Conradus Resch [Col:—] Paris 1518
[*Ed. p.* 1516 (Bibl. Nat.)]

The *Epistola nuncupatoria* by Lokert to his pupils at the Collège Montaigu (1516) and, later, Collège de Sorbon is of great interest—see Vol. I, p. 81 n.

The text consists of the vigorous criticism of the works of Aristotle which took place at Paris in (XIV-1) under the stimulus of the nominalism of William of Ockham.

406. LOMMIUS, IODOCUS. *Commentarii de sanitate tuenda in primum librum de Re Medica Aurel. Cornelii Celsi.* . . .

A. M. Bergagne, Louvain 1558

Ed. p.

Prov. ALEXANDER [?]; IACOBUS CLARK; A. LUMISDANE; PATRICK DUN.

In his dedn. to William of Nassau (Tournai 1557) Lommius states that Celsus is read by few, and by fewer understood. A strange comment on this is that Lommius's work was reprinted at least three times in the 18th cent. The *Praefatio* deals at length with the place of Celsus in the history of medicine.

The text consists of excerpts from Celsus with much longer comments. Refs. to plague at the end.

407. LOMMIUS, I. *Medicinalium Observationum libri tres.*

Gul. Sylvius, Antwerp 1560

Ed. p. (reprinted as late as 1745 and, in English, 1747).

Prov. BUTE.

408. LONICERUS, ADAMUS. *[Bo]tan[icon] plantaru[m] historiae, cum earun[dem] [a]d vivum artific[i]ose expressis iconibus, tomi duo. . . . Addita sunt animantium terrestrium, volatilium, & aquatilium brevis descriptio. Item de gemmarum, metallorum, succorumque concretorum vera cognitione, delectu & viribus. Praeterea: de stillatitiorum liquorum ratione & instrumentorum ad eam artem praeparatione, . . . tractatio. Postremo onomasticon, quo variae plantarum nomenclaturae ex diversis linguis, item voces, quarum frequens in descriptionibus usus est, explicantur. Omnia . . . recognita.*

Heirs of Christ. Egenolph, Frankfurt 1565

Based on a 2-vol. work, Frankfurt 1551 BM.

The dedn. (Frankfurt 1551) draws attention to the care of gardens, etc., referred to in the works of Varro, Palladius, Columella, etc.

The copy is incomplete, breaking off at fo. 216v, though the index refers to fo. 352 at least. T.p. (somewhat mutilated) has (in this copy) coloured cut showing garden, woodland, sick-chamber, multiple still.

An attractive example of the semi-popular works in which Egenolph and his son-in-law, Lonicer, specialised; but see Arber, p. 72, referring to the piracy of the illustrations from L. Fuchs, etc.

409. LORITUS, HENRICUS (*syn.* Glareanus). . . . *de Geographia, liber unus ab ipso authore . . . recognitus.*

G. Rikart, Paris 1542. (Col. 'excudebat Io. Lodivicus Tiletanus 1542').

[*Ed. p.* Basel 1527—edn. of 1542 not BM.]

In the dedn. Loritus says that 'among the liberal arts . . . in my opinion the chief place must be given to Geography'. This view is characteristic of the renaissance modification of the medieval *studium generale* (cf. No. 149). He goes on to say that he undertook the task of writing this work owing to the unsuitability of Ptolemy's *Geography*— at least for beginners.

In the text there is a world map printed by the Rickarts in 1542, in which North America is still represented only by 'Terra de Cuba' separated from the South; fo. 35 *r* actually refers to 'two islands'. Generally, his descriptions are based on Ptolemy, but he notes that 'Hybernia . . . is now called Irlandt', and that the island of Britain is divided into two kingdoms, 'Anglia and Scotia'.

410. LORITUS, H. *De VI Ar[i]thmeticae Pr[a]cticae speciebus.*

Io. Faber Emmeus, Freiburg-in-Breisgau, 1539

Probably *Ed. p.* Earliest BM is Paris 1558.

In the dedn. he claims to have heard Erasmus say frequently that nothing is more harmful to great authors than the epitomes made by bad ones. He refers to his own audiences in Paris. A very elementary text for the Latin schools.

411. LOWE, PETER. *A discourse of the whole art of Chyurgerie, wherein is exactly set down the definitions, causes, accidents, prognostications, & cures of all sorts of diseases both in generall & particular, which at any time heretofore have beene practised by any Chyurgion; according to the opinion of all the ancient professors of that science . . . . Compiled by Peter Lowe, Scottishman, Doctor in the Faculty of Chyurgerie at Paris and ordinary Chyrurgion to the French King of Navarre. Whereunto is added the rule of making remedies which chyrurgions doe commonly use: with the Presages of divine Hippocrates. 3rd edn. corrected & much amended.*

Thos. Purfoot, London 1634

[*Ed. p.* '. . . the whole course of Chirurgerie . . .' London 1597, S.T.C. 16870.]

The general dedn. (1612- 2nd edn.) refers to the earlier attempt. This is followed by a 'professional' dedn. to Gilbert Primrose, 'Sergeant Chirurgion' to James I; James Harvie, 'chiefe Chirurgion' to the Queen; and to the 'Worshipful Company of Chirurgions in London and Edenborough'. Lowe was the founder (under a Charter of James VI) of the Faculty of Physicians and Surgeons in Glasgow, 1599, which was unsuccessfully opposed by the University. See p. 391 for a good description of the kinds of leeches, with a suggestion as to the nature of contagion.

(Genl. ref. J. Finlayson *Account of the Life and Works of Maister Peter Lowe*. Glasgow 1889).

LUCAR, CYPRIAN. See No. 673.

412. LUISINI, ALOYSIUS. *De Morbo Gallico omnia quae extant apud omnes medicos cuiuscunque nationis, qui vel integris libris vel quoquo alio modo huius affectus curationem methodice aut empirice tradiderunt . . . conquisita . . . erroribus expurgata & in unum . . . corpus redacta. In quo de Ligno Indico, Salsa Perillia Radice Chynae, Argento vivo caeterisque rebus omnibus ad huius luis profligationem inventis, . . . tractatio habetur. . . . Tomus prior.*

Iordanus Zilettus, Venice 1565

Ed. p.

Prov. GORDON.

In his *Lectori* Zilettus refers to the editorship of Luisini and promises a second volume 'before you've finished the first' (of 736 pp.!). Actually it appeared in 1566-7, according to BM.

The text consists of extracts from well known works and includes (p. 292 f.) a good review of the use of *argentum vivum* (mercury)—a bone of contention for many years.

413. LULL, RAIMONDO. (*Syn.* Lullius, Raymond Lully). *De Secretis Naturae sive Quinta Essentia libri duo. His accesserunt Alberti Magni . . . De Mineralibus & rebus metallicis libri quinque. Quae omnia . . . repurgata . . . recens publicata sunt per Gualtherum H. Ryff.*

[Col. Balth. Beck, Strassburg] 1541

[*Ed. p.* Augsburg 1518; 2 imperfect copies of No. 413, the first to include Albert's *De Mineralibus.*]

Prov. LIDDEL (?)

Incipit: (*Praefatio*) Liber secretorum naturae, seu quintae essentiae . . . (Text) 'Iam vero ut rem ipsam . . .' (ThK give 'Deus gloriose, cum [virtute] tue sublimis . . .' no date).

Incipit of Albert's work 'De commixtione et coagulatione, similiter & congelatione . . .' as in ThK, 172 b (14th cent.).

Of these two works, the former is almost certainly supposititious, the second almost certainly correctly ascribed. But they are both of great interest for the scholar of the early sixteenth century. For refs. to Lull see Vol. I, p. 175 and p. 299.

414. LULL, R. *Opera ea quae ad adinventam ab ipso artem universalem, scientiarum artiumque omnium brevi compendio . . . pertinent . . . .*

L. Zetzner, Strassburg 1598

Ed p. in this form; of *Ars Universa*, Venice 1480 (Hain 10320). Prov. GREGORY.

The dedn. by Zetzner refers to commentaries by Giordano Bruno and Cornelius Agrippa, which are included, with cont. pagn. Bound with *Clavis Artis Lullianae . . . Opera & studio Iohannis Henrici Alstedi;* Same imprint 1609.

415. LULL, R. *Tractatus . . . de conservatione vitae. Item liber secretorum seu quintae essentiae qui doctrinam eius extractionis & applicationis ad corpus humanum & ad opera mirabilia totius artis medicae facienda, nec non ad metallorum transmutationem, instituit; estque speculum & imago omnium librorum super his tractantium. Nunc primum . . . editus.*

L. Zetzner, Strassburg 1616

Only edn. BM.

Incipit of *De Conservatione . . .* 'Intendimus componere rem admirabilem, tactam ab Hippocrate, Galeno, et Avicenna . . .' (ThK, 358).

Incipit of *Liber secretorum . . .(Praef.)* 'Deus gloriose, cum tuae sublimis bonitatis . . .' (ThK, 191). (Cap. 1.) 'Ordimur namque tu fili quod necesse. . . .' (ThK, 475—15th cent. MS).

416. LULL, R. *Secreta secretorum R— L— et Hermetis philosophorum in libros tres divisa. . . . Cum opusculo S. Thomae Aquinatis De esse et essentia mineralium & Cornelii Alvetani . . . De conficiendo divino elixire libellus.*

Gosuinus Colinus, Cologne 1592

Only edn. BM.
Prov. LIDDEL.

In an unsigned *Lectori* we are told (presumably by the printer) that these three MSS 'came into his hands', and, 'thinking them worth reading', he decided to submit them to the judgment of the learned.

Incipits: I (of *Liber primus artis R— L— operandi quintam essentiam:* 'Intendo componere rem mirabilem Hippocratis, Galeni & Haly, & Avicennae . . .' (cf. No. 415).

II (of *Liber secundus de septem herbis* . . .) 'Blasius Africus discipulus N. studio Atheniensi epiloico . . .'

III (of *Liber tertius* . . . *ars operativa*) 'Summe necessarium est, quod antequam propines. . . .'

I can not trace identical *incipits* for II and III in ThK. But one almost identical with that of the work by 'St. Thomas Aquinas' is cited by them for a work by Thomas Capellanus.

See also No. 313.

417. MACER. *De herbarum virtutibus* . . . *cum succincta* . . . *Georgii Pictorii* . . . *expositione antea nunquam* . . . *edita. Adhaerentibus graduum compendiosa tabula* . . . *ut quisque* . . . *percipiat omnium morborum* . . . *medicinalem curam. Cum carmine de Herba quadam exotica, cuius nomen mulier est rixosa eodem G— P* . . . *authore.*

(Col. Henricus Petrus, 1559) Basel

Ed. p. in this form. (See below.)

Prov. two copies, one Melvin, the second with MS notes.

Incipit: 'Herbarum varias dicturus carmine vires . . .'. ThK 289 (13th cent. Zürich) has 'quasdam' for 'varias'. Scholia and rather crude cuts.

A third volume in the Library begins at sig. a ii, *in verso* of which '*Guillermi queroaldi in Macri floridi medici et poete excellentissimi interpretatiunculas prefatio* . . .'. Text and running commentary—'*finis*' sig. v vii *in verso*. No. Col.

Sig. v viii presumably wanting.

Prov. 'Ex codicibus Iohannis Fraser' (end papers). Cuts crudely smeared with green wash. Glossed in bold (17th cent.?) hand.

I compared this volume with an edition in BM having the same collation but including t.p. and sig. v viii. The latter has title *Herbarum vires macer tibi carmine dicet* with a cut showing herbalist with books and vessels. There is no col. BM gives 'Paris? 1515?'

Choulant (*Handbuch der Bücherkunde für die Ältere Medizin*, Leipzig 1841, p. 240) describes several separate edns. from Paris 1510-17, whose make-up is the same except for differences of paper, abbreviations, etc. The Aberdeen volume appears to be identical with the BM except for such differences. Choulant states that the edn. of 1511 (Paris) was the first to contain both Gueroaldus's commentary and the woodcuts: the *Ed. p.* of the poem was Naples, 1477.

It is doubtful whether Aemilius Macer (d. BC 16) ever wrote a herbal. The version commonly attributed to him was by Odo of Meung (XI). See also No. 551.

418. MAESTLIN, MICHAELES. *Consideratio & observatio cometae aetherei astronomica qui anno MDLXXX mensibus octobri, novembri* *et decembri in alto aethere apparuit. Item descriptio terribilium aliquot & portensorum* [sic] *chasmatum quae his annis MDLXXX & MDLXXXI confecta sunt.* . . .

Iac. Mylius, Heidelberg 1581

Only edn. BM [A work referring to the comet of 1577 was published at Tübingen 1578].

Prov. LIDDEL.

T.p. has cut showing comet's path, with portrait of Maestlin *in verso*.

419. MAESTLIN, M. (*praeses*). *Disputatio de eclipsibus solis et lunae quam, divina favente gratia,* . . . *defendere conabitur Marcus ab Hohenfeld.* . . .

G. Gruppenbach, Tübingen 1596

Not BM under Hohenfeld, Marcus.

420. MAGINI, GIOVANNI ANTONIO [t.p. wanting—no col.]. [*Novae coelestium orbium theoricae congruentes cum observationibus Nicolai Copernici*—(from BM).] Venice, 1589.

Ed. p.

This copy has been carelessly gathered e.g. the first fo. of the *Praefatio* succeeds the remaining two; and the first fo. of corrections should precede the *Praefatio*. Sig. a 4 wanting.

Lectori: '. . . ideoque uno omnium voto exoptetur aliquis qui coelestes motus cum postremis observationibus congruentes non ad absurdas hypotheses, quales Copernicus confinxit, sed ad eas quae a vero minus abhorrere videantur, referat ac revocet' ('and to that end there is a general desire for some one who would systematise the motions of the heavenly bodies, in agreement with the latest observations, not according to the absurd hypotheses such as Copernicus contrived, but according to such as should seem less repugnant to truth'). The *Praefatio* lays down two kinds of motions, namely, diurnal and 'opposite', 'since it is necessary that the individual motions of the secondary movables, which to us appear non-uniform and unordered, should thus be in agreement with [*conciliare*] an eternal uniformity and most constant order in such manner as the greatest perfection and excellence of the heavenly bodies demands'. The sustained influence of Platonic and medieval (e.g. Grosseteste, following Aristotle) thought is here apparent.

Finely illustrated with numerous figs.

421. MANARDUS, IOANNES (*a*) . . . *Epistolarum medicinalium lib. duodeviginti. Hi partim* . . . *ab ipso autore iam recens castigati sunt, partim iam primum* . . . *aeduntur. Eiusdem annotationes & censura in medicamina simplicia & composita Mesue.*

Io. Bebel (device only) Basel 1535

Revised edn. of a work first printed at Ferrara, 1521.

Prov. GORDON—the MS index appears to be in his hand.

The 'Palma Beb' device is in the less common form, having no human figure beneath the platen (No. 276).

Dedn. by Coelius Calcagninus, this edn. having been printed after the death of Manardus.

(b) . . . *Epistolarum medicinalium libri XX.* . . .

Godefridus and Marcellus Beringorum fratres, Lyon 1549
Similar to (a) but having two extra *libri*.

[*Ed. p.* in this form, Venice 1542.]

The *Annotationes* to Mesue, dedicated 'iuvenibus medicinae studiosis' by Manardus from Ferrara 1521, open with a list of omissions from Mesue (No. 740). On the last page are *Castigationes . . . quas post obitum Manardi habuimus.*

(c) [Same title as (b).]

Io. F. Camotius (Col. Io. Grypheus excudebat) Venice 1557
Prov. LIDDEL (?).

This collection well illustrates the importance of the study of *medical* works for the understanding of the emergence of the natural sciences in the 16th cent.; e.g. on the source of the variation of terrestrial temperatures. (See also Vol. I, p. 220.)

Of interest also are (c, p. 432) the letter to Montanus on Galen's *De differentiis pulsarum*, with diagrams; and (a, p. 314) the ref. to the Cracovian Academy and the use of mercury in the treatment of the *Morbus Gallicus* (see also No. 412).

See also No. 754.

422. MANISCALCO, AGOSTINO COLOMBRE. *Della natura de'cavalli, et del modo di medicare le loro infermità libri iii.*

Francesco Fagiani, Venice 1561

Not BM.

Prov. BUTE.

Includes some descriptive anatomy (e.g. gut, nerves) without diagrams. Aristotelian in style.

423. MARANTA, BARTHOLOMEUS. *Methodi cognoscendorum simplicium libri tres.*

Off. Erasmiana Vincentii Valgrisii, Venice 1559

Only edn. BM.

Prov. (LIDDEL ?)

In a letter Falloppio (No. 243) tells the author that he has read the work 'animi voluptate'. In other prelims. the author gives interesting bibliographical data, and states that he has followed Ruellius's trn. of Dioscorides (No. 199c).

424. MARQUARDUS, IOANNES. *Practica theorica empirica morborum interiorum e praelectionibus Io—M—collecta, cui duo tractatus de lue Venerea accesserunt; unus Lucae Ghini . . . alter Hieronymi Capivaccii. . . . Omnia denuo recusa & . . . recognita.*

Bernardus Albinus, Speyer 1592

Ed. p. Speyer, 1589.
Prov. LIDDEL (?)
Dedn. of whole vol. by printer, of Capivacci's work by Philip Schopf 'Durlaci 1590', so presumably added after *Ed. p.* Sep t.p. for this, otherwise cont. pagn. and secn. headings only.
See p. 84 for use of mercury.

425. MARTIN, GREGORY. *Commentatiuncula in libri qui inscribitur de chymicorum cum Aristotelicis et Galenicis consensu ac dissensu caput XI. Quod est de principiis chymicorum. Tractationem quaestionis an sal, sulphur & mercurius sint prima perfecte mixta & reliquorum perfecte mixtorum principia . . . continens. . . .*

Frid. Hartmannus, Frankfurt 1621

Not BM.
Ferg. II, 81 cites this edn., but says that the work was 'published in 1619'. The dedn. is, however, dated 1621.
Prov. READ—glossed in his hand, and bound in contemporary cover with inscription.
Biographical details in dedn.

426. MASSA, NICOLAUS. *Anatomiae liber introductorius in quo quam plurimae partes, actiones, atque utilitates humani corporis nunc primum manifestantur quae a caeteris tam veteribus quam recentioribus hucusque praetermissa fuerant.*

Off. Stellae Iordani Zilleti, Venice 1559

Ed. p. 1536 BM. No other cited. There is, however, no indication that No. 426 is a later edn.—the text is headed '*nunc primum . . . aeditus*'; see, however, dedn.
Prov. 'Jo. Gregory 1769'.
The dedn. to Paul III tells how, being in Venice in former years, Massa took part in anatomies, and particularly in the company of Hiero. Marcellus, to whom he made a promise that he would write the book (dated Venice 1536).
Of special interest in relation to the question of Vesalius's claim to be a pioneer. The work has been strangely neglected—it is not mentioned by Castiglioni. (See Vol. I, p. 231.)
See also No. 753.

427. MASSARIA, ALEXANDER. . . . *De abusu medicamentorum vesicantium disputatio secunda apologetica ad librum Herculis Saxoniae De Phoenigmis.*

Geo. Graecus, Vicenza 1593

Ed p. of *Disputatio secunda; prima* was published Padua 1591 BM. Dedn. to 'Professores bonarum artium Patavii 1593' where Massaria was Professor of Practical Medicine. The license to print contains a ref. to 'x regulam *indicis librorum prohibitorum* a sacra Synodo Tridentina' (Council of Trent).

For the work of Hercules Saxonia see No. 615.

428. MASSARIA, A. *Disputationes duae: altera de scopis mittendi sanguinem cum generaliter, tum speciatim in febribus; altera de purgatione principio morborum.*

Franc. de Franciscis, Venice 1588. (Col. 1589)

Ed. p.

429. MASSARIA, A. *Praelectiones de morbis mulierum, conceptus & partus editae in gratiam studiosorum medic. ut quasi prodromum haberent omnium totius corporis humani morborum Θεραπειας suo tempore edendae.*

Abrahamus Lambergus, Leipzig 1600

Ed. p.

The work is introduced by Henricus Osthauser. The text opens with the statement that the lectures were delivered in 1591.

430. MASSARIA, A. *Tractatus quatuor. Quorum est I de Peste* [1579]. II *De affectibus renum & vesicae.* III *De pulsibus* [1603]. IV *De urinis, cui est annexum pro exemplo consilium de febre catarrhali cum totius macie mesenterii obstructione moestitia & vigiliis. Nunc tandem post auctoris obitum ab haeredibus reperti, caeterisque eiusdem operibus adiuncti.*

Matt. Becker, Frankfurt 1608

Ed. p. in this form. Dates in [ ] are those of individual tracts, of which the first is of special interest in opening with a systematic account of the plague in various parts of Italy in 1575 (Vol. I, p. 268.)

431. MATTIOLI, PIETRO ANDREA (*a*) . . . *Commentarii secundo aucti in libros sex Pedacii Dioscoridis Anazarbei de Medica Materia. Adiectis quam plurimis plantarum & animalium imaginibus, quae in priore editione non habentur, eodem auctore. His accessit eiusdem Apologia adversus Amathum Lusitanum: quin & censura in eiusdem enarrationes.*

Off. Valgrisiana, Venice 1559

[*Ed. p.* of Latin trans. 1554; of the orig. Italian 1544.]

In the Latin dedn. to the Emperor Ferdinand (Gorizia 1554) Mattioli, having referred to the labours of Barbaro, Leoniceno, Ruel, Fuchs, and others to bring back the study of *materia medica* into its pristine light after dispersing the fog of the previous century (!), explains that he was so appalled by the errors of the Italian pharmacists ('qui non satis Latinam linguam sciunt') that he brought out his own work (greatly extending that of Dioskurides) in the vernacular. The success of the venture is indicated by the number of Italian editions (four, he thinks) before the first in Latin.

Prov. JAMES FRAZER.

The *Apologia* has sep. t.p. dedn., and pagn.

(*b*) [Similar title] . . . *iam denuo ab ipso autore recogniti et locis plus mille aucti. Adiectis plantarum & animalium iconibus supra priores editiones longe pluribus ad vivum delineatis. Accesserunt quoque ad margines Graeci contextus quam plurimi ex antiquissimis codicibus desumpti qui Dioscoridis ipsius depravatam lectionem restituunt.*

Off. Valgrisiana, Venice 1569

New dedn. to Maximilian II and the Electors Ferdinand and Charles, in which the help of Busbecq (Arber, p. 9) is acknowledged, dated Prague 1565. Arber mentions a previous edn. of the enlarged work in Czech, Prague 1562.

Prov. FORDYCE.

(*c*) Off. Valgrisiana, Venice 1570

Same t.p. and pagn.; but following p. 956 is *De ratione distillandi aquas ex omnibus plantis et quomodo genuini odores in ipsis aquis conservari possint*—with half-title and pages unnumbered.

Not BM.

Prov. JOHNSTON.

(*d*) Felix Valgrisius, Venice 1583

Greatly enlarged, different pagn., much finer cuts—'adiectis magnis ac novis . . . iconibus'. Same prelims., but 'Candidis Lectoribus' by both Valgrisius and Borgarutius (Venice 1565) added. *De ratione distillandi* . . . follows p. 772; half-title, fine cuts of stills, pages unnumbered. The whole divided into two parts with sep. t.p. and pagn.

Mattioli's great *Commentaries* are much more than the name implies: they are outstanding for their scholarship and illustrations (though these are perhaps a little 'fussy' as compared with those in Fuchs—No. 276); also as a source of new knowledge. See Arber, p. 92 *seq.*

432. MATTIOLI, P. A. *Epistolarum medicinalium libri quinque.*

(*a*) Off. Geo. Melantrichius for V. Valgrisius, Prague 1561.

Ed. p.

Prov. (i) 'Prestonii 1580', (ii) 'Georgius pacoli? pharmacopoeus',

(iii) 'A. Mure M.D.'

(*b*) Caes. Farina, Lyon 1564. (Same title; but privileges and list of contents omitted—pirated?)

Prov. Scored in red with a few glosses—probably READ.

The work contains a portrait of Mattioli, letter to Jul. Alexandrinus referring to Mattioli's *Dioscorides*, and replies from correspondents. The text includes discussion of metallic remedies (p. 16)—and in a letter from Io. Viscerus (p. 395) occurs the following request: '. . . If you have any senna pods [*semen Senae*] and any other rarer seeds which you may think worth sending me, and which I may pass on to my relative, Dr. Fuchs. . . .'

433. MATTIOLI, P. A. *De simplicium medicamentorum facultatibus secundum locos & genera. Accesserunt quoque praefationes quaedam huic opusculo admodum necessariae.*

Gul. Rovillius, Lyon 1571

[*Ed. p.* 1569.]

Two copies.

Prov. LIDDEL, 'Rostockii 1589'. MS notes on end-papers.

In his *Lectori* Mattioli explains that these are 'collected out of Dioscorides and our commentaries on the same'; and he adds hints on solution, trituration, etc., and certain things from Galen.

The text includes (p. 78 *seq.*) cuts of furnaces, stills, etc.

434. MATTIOLI, P. A. *De plantis epitome . . . novis plane et ad vivum expressis iconibus descriptionibusque longe & pluribus & accuratioribus nunc primum diligenter aucta & locupletata a D. Ioachimo Camerario . . . Accessit . . . liber singularis de itinere ab urbe Verona in Baldum montem plantarum ad rem medicam facientium feracissimum, auctore Francisco Calceolario, pharmacopoeo Veronensi.*

[No Name: device of P. Perna] Frankfurt, 1586

Ed. p. of the edn. revised by Camerarius; the original work (based on the *Commentarii*—No. 431) was published under the title *Compendium de plantis omnibus* by Valgrisius, Venice 1571.

Prov. Presented by the late Professor Trail, 1882. Glossed with English floral name equivalents etc. in 18th cent. hand.

Dedn. by Camerarius to Maurice, Landgrave of Hesse, 1586.

Dedn. by Mattioli of orig. edn. also included. The former contains the significant passage: '. . . to display not only complete [sc. plants], which has not often been the custom of others, but particularly those parts, such as roots, stems . . . as if by a kind of anatomy' (*quasi per anatomiam quandam*). This is accompanied by a ref. to Gesner.

Calzolari's opuscule (ded. to Mattioli 1571) is an early specimen of a local flora, with lists of plants found in various localities passed during the journey to Monte Baldo. The lists do not suggest a high altitude flora such as we know it, and seem to be mainly restricted to the valleys leading toward the summit.

435. [MEIETUS, PAULUS ed.—from dedn.] *Opuscula illustrium medicorum de dosibus, seu iusta quantitate et proportione medicamentorum. Nunc recens . . . edita & multis mendis vindicata.*
Io. Mareschallus, Lyon 1584
Only edn. BM, and not in Bibl. Nat., but title suggests earlier edn.
Dedn. Padua 1579.
Contents: Matthaeus Curtius, B. Faventinus, Barth. Montagnana, Gentilis Fulginas, Tho. de Garbo, 'Alchindi', Gul. Rondeletius, Petrus Gorraeus.

436. MELA, POMPONIUS. *De orbis situ libri iii,* and C. Iulius Solinus *Polyhistor . . . tabulis elegantibus illustratus. . . .*
Basel (Col. Off. Henricpetrana 1576)
[*Ed. p.* of *De orbis situ* Milan 1471—Hain 11,014.]
Prov. LIDDEL.
Lectori by Christianus Urstisius.
The *Universalis Cosmographia,* which with several additional maps follows the colophon, is dated Zürich 1546.

437. MELANCHTHON, PHILIP. *Initia doctrinae physicae dictata in Academia Vitebergensi. . . .*
Christ. Egenolph, Frankfurt (Col. 1550)
BM gives only edn. Wittenberg 1549.
The dedn. to Michael Meienburg defends the Aristotelian system as showing God's handiwork. He adds a warning against the errors of Democritus, the Epicureans, Stoics, Pyrrhonists, and 'horum similes'. (For some analysis of the work see Vol. I, p. 148.)

438. MENA, FERDINANDUS DE. . . . *Methodus febrium omnium et earum symptomatum curatoria; Hispaniae medicis potissimum ex usu. Accessit liber de septimestri partu & de purgantibus medicinis eodem & expositore & enarratore.*
Chr. Plantin, Antwerp 1568
Only edn. BM.
Prov. 'Hepburn'.
Secn. heading and sep. pagn. for *De septimestri partu;* cont. pagn. for *De medicamentis. . . .*
Refs. mainly to classical authors, but also to (e.g.) Fuchs.

439. MERCADO, LUIS DE (*syn.* Mercatus, Ludovicus). . . . *De essentia, causis, signis & curatione febris malignae in qua maculae rubentes similes morsibus pulicum per cutem erumpunt, cui accessit consilium continens summam totius praedictionis & curationis in eodem affectu.*

C. Waldkirch, Basel 1594

Only edn. BM (copy imperfect).

The *Praefatio* contains a general discussion of fevers in which spots like the bites of gnats or fleas appear without 'exanthemata' (*syn.* 'extuberationes') which 'among the barbarians are known as *variolae* and *morbilli;* or there may be *vibices* [weals], perhaps called *esserae* by the barbarians'.

The *Tractatus quartus* is much scored and glossed.

The disease (p. 114 gives a list of signs) was typhus (Castiglioni, p. 457).

440. MERCADO, L. DE. *Libri ii de communi et peculiari praesidiorum artis medicae indicatione. Accessit prooemii loco Methodus universalis in tres classes dissecta:* I *Affectas partes,* II *Affectus ipsos,* III *Medendi rationem monstrat.*

Io. Bapt. Ciotti, Cologne 1588

Not BM. Dedn. suggests *Ed. p.*

Prov. LIDDEL.

Dedn. by printer to Hiero. Mercurialis (No. 441) referring to Mercado's Chair in *Academia Vallesolatena* (Valladolid) dated Venice 1588.

Ad Lectorem by Mercado and his pupil, Michael Arindez; also by the latter's pupil, Petrus Sanctius.

See also Nos. 154, 738.

441. MERCURIALIS, HIERONYMUS. . . . *De arte gymnastica libri sex. In quibus exercitationum omnium vetustarum genera, loca, modi, facultates & quidquid denique ad corporis humani exercitationes pertinet diligenter explicatur. Secunda editione aucti & multis figuris ornati. Opus non modo medicis, verum etiam omnibus antiquarum rerum cognoscendarum & valetudinis conservandae studiosis admodum utile.* . . .

(*a*) Juntae, Venice 1573.
Prov. (i) Cargill (ii) William Robertson.
(*b*) Jacob de Puys, Paris 1577.
(*c*) Juntae, Venice 1601.
[*Ed. p.* Venice 1569.]

Dedn. to Maximilian II (dated Padua 1573) has ref. to Cardinal Farnese 'cui post Deum omnia debere me fateor'.
Many fine cuts of rope-work, discobuli, weight lifting, etc.

442. MERCURIALIS, H. (*a*) . . . *Variarum lectionum libri quatuor priores multo quam ante & auctiores & emendatiores, quintus plane novus. In quibus complurium, maximeque medicinae scriptorum, infinita paene loca vel corrupta restituuntur vel obscura declarantur.*
Petrus Perna, Basel 1576

Prov. PATRICK DUN. Auto and numerous glosses in pencil— Dun's use of this, here and elsewhere, is the earliest I have noted.
The dedn. by Andreas Sylvius to Cardinal Perenottus, giving biographical details about Mercuriali, is dated Venice 1570.

(*b*) *Variarum lectionum in medicinae scriptoribus & aliis libri sex. Ab auctore aucti & recogniti. Quibus adiecta sunt capita sex nunquam edita.*
Juntae, Venice 1588

Prov. 'Ex bibliotheca T. Moresini D.M.' (No. 466).
The dedn. by Bonaventura Grangerius is dated Paris 1585.
[*Ed. p.* Venice 1570].

443. MERCURIALIS, H. *De puerorum morbis tractatus locupletissimi varia doctrina referti . . . Ex ore . . . H . . . M . . . atque in libros tres digesti, opera Iohannis Groscesii. Addita Alexandri Tralliani . . . de Lumbricis epistola cum eiusdem Mercurialis versione. Eiusdem De venenis & morbis venenosis libri ii seorsim editi. . . .*
Andr. Wechel, Frankfurt 1584

[*Ed. p.* Venice 1583—different editor.]
Prov. LIDDEL.
Mercuriali's dedn. of the *Epistola* is dated Padua 1570. Sep. t.p., pagn. and dedn. in *De venenis* by Albertus Scheligius.

444. MERCURIALIS, H. . . . *De venenis . . . opera Alberti Scheligii.*
Juntae, Venice 1601

Earliest edn. as sep. work BM.

445. MERCURIALIS, H. *De morbis muliebribus praelectiones. Iam dudum a Gaspare Bauhino exceptae, ac paulo antea inscio autore editae. Postremo vero per Michaelem Columbum ex collatione plurium exemplarium consensu auctoris locupletiores & emendatiores factae. Tertia vero hac editione & auctiores & castigatiores adhuc redditae.*
Juntae, Venice 1591

[*Ed. p.* 1586].

The dedn. by Columbus (Padua 1591) refers to the delivery of these lectures, which (in his *Medicinae Studiosis*, dated Padua 1586), he states ('ni fallor') was fourteen years previously. An interesting example of the gradual production of an authentic text for publication of lectures for which the author himself had no such intention.

446. MERCURIALIS, H. *De decoratione liber non solum medicis et philosophis verum etiam omnium disciplinarum studiosis . . . utilis. Ex H— M— explicationibus. A Julio Mancino exceptus primum. . . . Additi nunc primum duo tractatus, alter de Varicibus, alter de Reficiendo Naso, nunquam antea editi. . . .*
(*a*) Io. Wechel, Frankfurt 1587.
Ed. p.
Prov. LIDDEL.
(*b*) Juntae, Venice 1601.

It is not quite obvious to the modern student why 'philosophers and students of all disciplines' should be concerned with such questions as 'Quid sit ars cosmetica—sub qua Medicinae parte collocetur?' and the treatment of 'spots on the face'!

447. MERCURIALIS, H. *Hippocrates Coi opera quae extant Graece et Latine veterum codicum collatione restituta . . . & scholiis illustrata.*
Juntae, Venice 1588
Only edn. BM.
Fine engraved t.p. *Lectori* by M. Columbus. Historical and bibliographical intro. Latin and Greek texts in parallel, with *scholia* at the end of each secn. Followed by *Linguarum explanatio, Erotiani onomasticum*, etc.

448. MERCURIALIS, H. *Liber responsorum et consultationum medicinalium. Nunc primum a Michaele Columbo collectus & . . . editus.*
C. Waldkirch, Basel 1588
There were numerous edns. in various forms. This has dedn. to Cardinal Cajetan from Padua 1587.
Prov. LIDDEL, and a second copy scored & glossed in 17th cent. hand.
Full index of consultations grouped according to organs and complaints. Names of patients given in many cases.

449. MERCURIALIS, H. *. . . de compositione medicamentorum tractatus tres libros complectens. I. De compositione medicamentorum. II. De medicamentorum dosi. III. Medicamentorum componendorum rationem & methodum tradit. Eiusdem de oculorum et aurium affectibus*

168

praelectiones seorsim editae. Omnia primum a Michaele Columbo . . .
edita: nunc vero a mendis . . . vindicata.

Io. Wechel, Frankfurt 1591

[*Ed. p.* 1590.]

Prov. READ, the *De oculorum* . . . glossed in his hand.

The *Lectori*, by Petrus de Wittendell, refers to a previous edn. by Columbus, but states that he himself hopes to publish before long the lectures which were applauded to the echo when they were delivered at the University of Bologna, and which he had taken down from Mercuriali's mouth over a period of three years.

De oculorum . . . has sep. t.p. and pagn.

450. MERCURIALIS, H. [A collection of works previously published.]

Juntae, Venice 1601

Includes *De Pestilentia . . . lectiones habitae Patavii MDLXXVII mense Januarii. In quibus de peste in universum, praesertim vero de Veneta & Patavina singulari quadam eruditione tractatur. Eiusdem tractatus de Maculis pestiferis & de Hydrophobia.*

Dedn. by H. Zacchus, who says that he took down the lectures and gave them to the printer so that they might be made public— dated Padua 1577. The *Prooemium* is by Mercuriali, so there seems to be no question of piracy. Sep. t.p. and pagn. for the *Tractatus*.

451. MERCURIALIS, H. *De peste in universum . . . Patavina. Item de Morbis cutaneis & omnibus humani corporis excrementis. . . .* [No name or date. Device of Perna.] Basel.

The *De peste* is the same work as *De pestilentia* . . . included in No. 450. The *De morbis cutaneis* . . . is said by Garrison (p. 348) to have been printed in Venice 1572 and to be the first systematic textbook on diseases of the skin. The earliest BM is 1577 and the dedn. by Paulus Aicardius is dated Padua 1576; but in his *Medicinae studiosis* the latter says that many corrupt versions were being circulated.

452. MINADOUS, AURELIUS. *Tractatus de Virulentia Venerea in quo omnium aliorum hac de re sententiae considerantur, mali natura explicatur, caussae & differentiae, aliaque cum Dogmatica curatione proponuntur.*

Rob. Meiettus, Venice 1596

Only edn. BM.

Prov. BUTE.

In his *Praefatio* the author says that he had composed the work three years previously. The text opens with an account of the name, origin, and opinions on, the disease (originally the 'morbus gallicus'). See also the last ch. p. 279.

169

453. MIZAUD, ANTOINE (*syn.* Mizaldus, Antonius). . . . *Memorabilium sive arcanorum omnis generis per aphorismos digestorum centuriae IX et Democritus Abderita de rebus naturalibus & mysticis. Cum Synesii et Pelagii commentariis. Interprete de Graeca lingua Dominico Pizimentio.* . . .

Io. Birckmannus, Cologne 1574

[*Ed. p.* 1573.]

The dedn., comprises a long historical and bibliographical introduction. Cont. pagn. with half-titles for the commentaries.

454. MIZAUD, A. *Historia hortensium quatuor opusculis methodicis contexta, quorum primum, Hortorum curam, ornatum & secreta quamplurima ostendit. Secundum, Insitionum artes proponit. Tertium, Auxiliares & medicas hortensium utilitates percurrit. Quartum, Iucunda et benefica medicandorum hortensium, olerum, radicum, fructuum, vuarum* [sic], *vinorum et carnium artificia explicat.* . . . *rerum variarum accessione nunc primum aucta.* . . .

Io. Gymnicus, Cologne 1577

This is a large collection—p. 16 gives a list of other works included—not cited in BM. A work with similar title occurs in Bibl. Nat. as 'Paris 1560'. The sep. works are variously divided by dif. pagn., sep. t.p.'s, etc. The *Quartum* of the title comes *after* the additional works.

455. MOLLERUS, IUSTUS. *Fasciculus remediorum ex Dioscoride et Mathiole collectorum.* . . . [from half-title,—t.p. and all sigs. before index wanting].

Dedn. 1579. Only edn. BM, whose copy likewise has no imprint.

456. MONARDES, NICOLAUS. (*a*) *De simplicibus medicamentis ex Occidentali India delatis, quorum in medicina usus est. Auctore D . . . M . . . interprete Carolo Clusio.* . . .

C. Plantin, Antwerp 1574

Ed. p. of trn. Original, Seville 1565. Bibl. Nat.

The dedn. by de l'Ecluse draws attention to his contraction of the original two parts into one book, and to his addition of notes and pictures: dated Malines 1573.

Prov. BUTE.

(*b*) *Simplicium medicamentorum ex novo orbe delatorum quorum in medicina usus est, historia.* . . . *Tertia editio, auctior et castigatior ex postrema auctoris recognitione.*

Off. Plant., Vidua & Moretus, Antwerp 1593

Though the title resembles that of (c) it is actually the 3rd edn. of (a). It is bound (cont. pagn.) with works of Garcia d'Orta (No. 487) and Christopher Acosta (No. 3).

Prov. GREGORY.

This work is full of interesting observations, e.g. 'Tabaco' (p. 20), 'Guayacan' (p. 29), 'China' root (p. 34), 'Carca-parilla' (p. 37—one of the numerous forms of what we call 'sarsaparilla'), 'Sassafras' (p. 44), and (p. 79) a letter written to Monardes '. . . ex Peruana usque regione, a nobile quodam viro', about the [in ?]famous Bezoar stone.

> (c) *Simplicium medicamentorum ex Novo Orbe delatorum quorum in medicina usus est, historiae liber tertius. Hispanico sermone nuper descriptus a D.N . . . M . . . . Nunc vero primum Latio [sic] donatus & notis illustratus a C . . . C . . . .*
>
> Off. C. Plantin, Antwerp 1582

Ed. p. of trn. Original, Seville 1574.

Prov. LIDDEL.

A supplement to (a). The dedn. by de l'Ecluse to Sir Philip Sidney and his friend, Edward Dier, refers to the two earlier parts (i.e. 'a'), translated by him eight years previously, and says that he translated this third part to while away the time while held up by adverse winds at Gravesend.

Dated Frankfurt 1581

On p. 14 occurs a description of the American Ground Nut (*Arache*) —'*Missus etiam ex Peru fructus sub terra natus nucleum intus continens . . . amygdalae similem. . . .*' There was also a note by de l'Ecluse concerning a gift from Sir Francis Drake.

> (d) . . . *Altera editio, auctior & castigatior.*
>
> Off. Plantiniana Vidua and Io. Moretus, Antwerp 1593
>
> Prov. Binding stamped 'C A D 1595'.

> (e) *The three bookes written in the Spanishe tonge by the famous physition D. Monardes resident in the citie of Sevill in Spaine and translated into Englishe by John Frampton Marchant.* [2nd t.p.:] *Joyfull neuues out of the newe founde worlde, wherein is declared the rare and singular vertues of diverse and sundrie hearbes. . . .*
>
> Wm. Norton, London 1577

Ed. p. S.T.C. 18005-6 (An alternative printing exists wanting 2nd t.p.).

Prov. READ—glossed in his hand.

'Dedicatorie happe unto Man, Woman, or Childe thei have fledde verie much from the olde order and manner of Phisicke . . .'. Also, 'To . . . Maister Edwarde Dier Esquire . . . returning . . . home into Englande out of Spaine . . .'.

457. MONTALTUS, HIERONYMUS. *De homine sano libri iii, in quorum primo agitur de natura & substantia hominis, in altero, de his quae ad ipsam substantiam labefactandam, eiusque functiones violandas, valent; in tertio denique, De facultate, qua haec propulsare & proinde illam tueri valemus.*

Io. Wechel & Petrus Fischer, Frankfurt 1591

Only edn. BM (copy lacks *Index rerum*).

Prov. LIDDEL.

Evidently based on the Aristotelian system, but the discussion of certain topics, e.g. 'concerning the right use of venesection' worth consulting.

MONDINUS, See Nos. 187, 359.

458. MONTANUS, IOANNES BAPTISTA. *In nonum librum Rhasis ad Almansorem regem Arabum expositio. A Valentino Lublino . . . communicata.*

Baltassares Constantinus, Venice 1554

Ed. p.

Prov. LIDDEL.

The *Lectori* is written by Nicolaus Rontana. The text of ar-Razi is not cited as such.

459. MONTANUS, I. B. . . . *Opuscula varia . . . Omnia post alios eruditos viros . . . nunc tandem Hieronymi Donzellini opera . . . in duo volumina digesta.*

P. Perna, Basel (Col. 1558)

Only edn. BM.

Prov. LIDDEL.

'Vol. I' dedn. by Donzellinus. Table of contents of both vols.

'Vol. II' has sep. t.p. and pagn. Table of contents similar to that given in Vol. I, but *De Excrementis, sive fecibus et urinis* omitted.

Dédn. by Gaspar Peucer to Chris. Leuschner from Wittenberg.

Contains lectures on fevers, *Morbus Gallicus, Methodus Medicinae Ioan. Caii Britanni.*

For the importance of the early chapters on *Method* see Vol. i, p. 222.

460. MONTANUS, I. B. *Consultationes medicae. Antea quidem Ioannis Cratonis Vratislavensis . . . opera atque studio correctae emendatae adauctae. Nunc vero . . . locupletatae.*

[Henricus Petrus & P. Perna?] 1572

Earliest BM 1572, Bibl. Nat. 1565—but not 1572. There is a col. on p. 1023 bearing the names of the above imprint followed by an appendix. The dedn. by Crato is dated Vienna 1565 and gives some details concerning the publication of the works of Montanus, who was held in high esteem. The printer's note on t.p. *v* states that some *consilia* remained after the rest had been printed, so these were added as an appendix; but whether this was seven years after the original printing is not stated.

PROV. LIDDEL. Binding stamped 1577.

In the appendix (columns 129-30) the printer states that a *consilium* of Vesalius was discovered among those of Montanus, which, since there were a few pages left over, he decided to print. It concerns a case of the loss of sight of one eye accompanied by impairment of that of the other, and is of interest in providing an exceptionally clear statement of signs and symptoms: signed, 'Patavii Calend. Maii MDXLII, Andreas Vuesalius'.

In the main text there is a good example of Montanus's attempt at a 'scientific approach': ' . . and since no disease can be understood or properly [*recte*] cured except as a result of knowledge of its cause, we shall for that reason see the nature and causes of the separate features . . . '. He illustrates this by reference to the gouty pains *praeter naturam* accompanying the *morbus Gallicus*. In this connection reference may also be made to Appendix, p. 104—*De Podagra* [i.e. gout] *cum suspicione morbi Gallici;* and to a 'cure' for *morbus Gallicus* translated from the Italian of Galeazzo Pico de Mirandola.

461. MONTANUS, I. B. *Medicina universa . . . ex lectionibus eius caeterisque opusculis, tum impressis tum scriptis collecta, & in tres tomos nunc primum . . . digesta, studio & opera Martini Windrichii Vratislavensis.*

Heirs of Andr. Wechel, Frankfurt 1587

Only edn. BM, also dedn. 1587.

Prov. FORBES.

The 2nd and 3rd 'tomi' have sep. t.p. and dedn. but the whole vol. is cont. paged.

Text opens with *De methodo docendi omnes artes & scientias*—an excellent exposition of the kind of 'induction' which had been originally outlined by Galen (as Montanus shows by quotation) and was specially cultivated and discussed at Padua in the 15th cent. (Cf. Vol. I, p. 220.) Note that this work is edited by another physician of the Breslau school—cf. No. 460.

462. MONTANUS, IOANNES FABRICIUS. *Differentiae animalium quadrupedum secundum locos communes, opus ad animalium cognitionem apprime conducibile. Similitudinum ab omni animalium genere*

desumptarum libri VI ex optimis quibusque authoribus sacris &
profanis Graecis & Latinis, per Othonem Vuerdmüllerum collecti &
. . . digesti. Omnia nunc primum edita.

Andr. and Jac. Gessner, Zürich 1555

Not BM. or Bibl. Nat.

Cont. pagn. Sep. Dedns.—that of the *Similitudinum* by Conrad
Gesner. The work is largely a compilation—in *Epistola* sig. H 2 *v*
Montanus refers to the fact that in writing it he made frequent use of
Gesner's works and advice.

In respect of frogs he says (p. 5) that recent writers numbered them
among 'worms', which was not regarded with favour by Gesner!
On p. 12 he states that they are generated out of the earth.

463. MONTUUS, HIERONYMUS. . . . *Anasceuas morborum tomus*
primus.

Io. Tornaesius, Lyon 1560

BM cites 4 vols.

Prov. LIDDEL?

Dedn. to . . . 'Galliarum Scotiaeque Regi Francisco eius nomine
secundo . . .' presumably on the grounds of his marriage to Mary
Stuart.

464. MONTUUS, SEBASTIAN. *Dialexeon medicinalium libri duo nunc*
recens . . . prolati. . . . Adiectus est de his quae ad rationalis medici
disciplinam, munus, laudes, consilia & praemia pertinent libellus. . . .
Mich. Parmenterius, Lyon 1537 (Col. excudebat Io. Barbarus)
Only edn. BM.

A polemical work, attacking in particular L. Fuch's *Paradoxorum . . .*
(No. 275).

465. MORESCOTTO, ALFONSO. *Compendium totius medicinae, quo de*
complexionum arcanis iudiciis, morborum praecipuorum causis,
prognosticis & signis, deque fabrica receptorum, breviter trac-
tatur . . . A— M— filii in utilitatem sui Ornedi Morescotti col-
lectum. Additis formulis remediorum Petri Gorraei. . . .

Christ. Corvinus, Herbornae 1588

(The title gives further details of the origin of the work.)

[*Ed. p.* Frankfurt 1583.]

Prov. LIDDEL.

The *Formulae* of Gorraeus follow with sep. t.p. but cont. pagn.

466. MORESINUS, THOMA [syn. Morison, Thomas] *Liber novus de*
metallorum causis et transmutatione, in quo chimicorum quorundam
inscitia & impostura philosophicis, medicis & chimicis rationibus

174

retegitur & demonstratur; & vera iis de rebus doctrina solide asseritur.

Io. Wechel, Frankfurt 1593

Only edn.

In his dedn. to James VI, Morison inveighs against Paracelsus who 'condemned medicine to the use of metallic remedies'—a doctrine which had been confuted by Thomas Erastus (No. 225).

Morison, who appears on t.p. as 'Aberdonanus Scotus', acted as intelligencer for the Earl of Essex in Scotland, also for Elizabeth, on whose behalf he tried unsuccessfully to enlist the interest of the Earl of Huntly. He corresponded with Anthony and Francis Bacon, on whose account he sought the ear of James VI (*D.N.B.*). His auto. appears on a few books in the Library (not to be confused with *James* Morison)—one of these being No. 84(6).

467. MÜLLER, PHILIP [Ed.]. *Miracula & mysteria Chymicomedica, libris quinque . . . enucleata . . . Ed. tertia . . . Accesserunt his* 1. *Tyrocinium Chymicum,* 2. *Novum Lumen Chymicum, etc.*

Clemens Bergerus Bibliop. 1616 [Wittenberg—Duv.]

Ferg. II, 115 lists no edn. earlier than 1611, but says later (116) 'the first Edition of Müller's book appeared in 1610'. Duv. (p. 416) states that the 2nd edn. of 1614 was the first to include the additions. For *Tyrocinium* see No. 69; for *Novum Lumen*, No. 639.

468. MUENSTER, SEBASTIAN. *Organicum Uranicum. . . .*

Henricus Petrus, Basel 1536

Only edn.

In the dedn. he refers to Jacobus Kobel, of pious memory, and also to Ioannes Stoefler (No. 658), 'meo praeceptore', in the *Praefatio*).

Mainly concerned with the motions of the planets, which are demonstrated by numerous fine volvelles.

469. MUENSTER, S. *Rudimenta mathematica, haec in duos digeruntur libros, quorum prior Geometriae tradit principia seu prima elementa una cum rerum & variarum figurarum dimensionibus. Posterior vero omnigenum horologiorum docet delineationes. . . .*

Basel (Col. Henricus Petrus 1551)

Only edn. BM, but dedn. and insert before t.p. dated 1531.

The dedn. refers to the dispersal of the 'Budensis Bibliotheca' of Mattheus Corvinus, King of Hungary (see Vol. I, p. 51 n).

Numerous cuts showing measurements of all kinds for ordnance, river crossings, etc. Front insert shows graduation for wall dial with signs, seasons, & hours. A 'star dial' (pp. 222-4) is described for marking the hours of darkness.

470. MUHAMMAD ibn AHMAD [*syn.* Ibn Roshd *syn.* Averroes). . . .
Colliget libri VII cum quibus etiam nunc primum in Quinto Libro
impressimus translationem trium illorum tam difficilium eiusdem libri
capitum LVII, LVIII, LIX *olim a Iacob* [sic] *Mantino* . . .
factam [further details of trans.] *Addidimus* . . . *tres illas sectiones*
collectaneorum Iohanne Bruyerino . . . *latinitate donatas. Eiusdem*
Averrois Commentaria in Avicennae Cantica una cum Avicennae textu
. . . *atque castigationibus Andreae Bellunensis exornato. Eiusdem*
Averrois Tractatus de Theriaca, nunquam antea impressus. Marci
Antonii Zimarae Solutiones contradictionum in dictis Averrois super
Colliget. Abimeron Abynzoahar. Omnia . . . *emendata.*

Juntae, Venice 1553

The work baldly referred to on t.p. as 'Abimeron Abynzoahar' is the
well-known 'theyzir' (*Liber Servitoris*) of Abd al-Malik ibn Zuhr
(Avenzohar) (No. 1).

[*Ed. p.* of *Colliget*, Ferrara or Venice 1482. GW III, 3017; of
Theyzir, Venice 1490/1.]

Prov. GORDON.

Incipits: (Intro.) 'Quando ventilata fuit super me voluntas' . . .
(ThK 546). (*Theriaca*) 'Inquit magnus medicus Amech Averrois'
(ThK 354—'Hamet' for 'Amech'). (*Liber Theizir:*) (Prologue). 'Dixit
servus regis, scilicet Abimeron Abinzoahar . . .' (ThK 216); (Text)
'Convenerunt omnes medici' (ThK 113). (*Antidotarium*) 'Haec est
particula Antidotarii Abumeronis Avenzoar libri . . . sui theyzir.'
(*Cantica:*) (Commentary) 'Postquam [theyzir] prius gratias egero Deo,
largienti vitam perpetuam animarum' (Avicenna's text) 'Medicina
est conservatio sanitatis. . . .' (Cf. ThK 502, 403.)

It was as a translator of the 'theyzir' that 'Paravicus' would have
attained immortality had not Thorndike (Isis **26,** 33) shown that 'he'
was a misprint for '[medicus] Patavinus'!

MUHAMMAD ibn ZAKARYA ar-RAZI (*syn.* Rhazes). See
Nos. 286, 306, 458, 708.

471. MYLIUS, IOANNES DANIEL. *Opus Medico-Chymicum.*

Luca Iennis, Frankfurt 1620, etc.

This vol. (rebound in 1936) is water-stained and linen-reinforced
throughout, but contains the *Liber Septimus* which Duv. (p. 420) says
'did never appear'. Apart from this slip the bibliographical description
of this involved work is most useful. Liber 7 is listed in Bibl. Nat.
Make up of No. 471:

(1) *Io . . . D . . . M . . . Medicinae Candidati. Tractatus secundi, seu*
Basilicae Chymicae liber septimus: de Animalibus (1620).

176

The printer explains that owing to pressure of business he had failed to make clear in the previously issued part of the work that the Seventh Book *On Animals* was then outstanding.

(2) *Io . . . D . . . M . . . Vettarani Hassi M.C. Tractatus* III *seu Basilica Philosophica continens lib. iii.* 1. *Horum prior continet philosophorum ac sapientum antiquorum consilia super Lapidem Philosophorum seu Medicinam Universalem.* 2. *Liber describit chymicorum vasa & fornaces.* 3. *Liber explicat quaedam philosophorum obscura.*

End papers, following *Liber* 2, with engravings of apparatus and 'philosophers'. *Liber* 3 has sep. pagn.

Followed by (3) *Tractatus* II *Basilicae Chymicae Liber* I with secn. heading only. The tail-piece (p. 184) suggests that *Liber* I is complete.

472. MYLIUS, IO. D. . . . *Philosophia Reformata. Continens libros binos: I Liber in septem partes divisus est . . . II Liber continet authoritates philosophorum.*

> Frid. Weiss for Luca Jennis, Frankfurt 1622

Ed. p. Duv. p. 420.

The contents of Liber I are: *De generatione metallorum in visceribus terrae; Principia artis philosophicae; Scientia Divina abbreviata; De duodecim gradibus philosophorum; De ambiguis in hac divina scientia; De recapitulatione artis divinae theorica & practica. Epilogus.*

The dedn. repeats the well-known tradition that 'Aloredus, Anglorum Rex, scholas publicas variarum artium apud Oxonium instituit, & tertiam bonorum suorum partem in Scholasticos contulit'.

The magnificent engravings of this work have been often reproduced (cf. Read, J. *Prelude to Chemistry*, 2nd edn., London 1939, opposite p. 262, etc.). A fine copy of this very rare work.

473. NABOD, VALENTINUS. *Ennaratio elementorum Astrologiae, in qua praeter Alcabicii qui Arabum doctrinam compendio prodidit expositionem atque cum Ptolemaei principiis collationem reiectis sortilegiis & absurdis vulgoque receptis opinionibus de verae artis praeceptorum origine & usu satis disseritur: in . . . Coloniensi Academia . . . proposita. . . .*

> Heirs of A. Birckmannus, Cologne 1560

Only edn.
Prov. JOHNSTON.
Allen p. 62.

474. NAPIER, JOHN. *Mirific[i] logarithmorum canonis descriptio eiusque usus in utraque trigonometria, ut etiam in omni logistica mathematica amplissimi, facillimi, expeditissimi explicatio liber i . . . Authore ac inventore, Ioanne Nepero Barone Merchistonii & Scoto.*

[A. Hart] Edinburgh 1614. [t.p. defective—but identical title-heading occurs on sig. B i *r*.]

Ed. p.

Prov. CARGILL.

Dedn. to Charles, son of James I, 'filio unico Walliae Principi Duci Eboraci & Rothesaiae . . .'. An early example of the union of the Scottish royal ducal title of Rothesay with those of Prince of Wales and Duke of York. The heir-apparent, Henry, had died in 1612.

The *Canon* was republished in a modified form in 1619 (same printer) with 'accesserunt opera posthuma'; and in the same year Hart also published a completely revised version *Mirifici logarithmorum canonis constructio . . . cum annotationibus aliquot doctissimi D. Henrici Briggii.* . . . (Ed. Robert Napier, John's nephew.) (There was a further edn. of the *Constructio* by Vincentius of Lyons in 1620, reprinted at Paris 1895.)

475. NAPIER, J. *Rabdologia seu numerationis per virgulas libri duo cum appendice de expeditissisimo* [sic] *multiplicationis promptuario. Quibus accessit & Arithmeticae Localis liber unus.*

Andr. Hart, Edinburgh 1617

Ed. p.

Prov. (i) 'Liber DAVIDII ANDERSONI Abredonensis Scoti'.
(ii) GREGORY.

'David Anderson' can have been none other than 'Davie do a' thing'—so called on account of his great ingenuity in hanging the great bells in the church of St. Nicholas and removing a huge rock obstructing the harbour mouth. Of even greater importance is the fact that his daughter, Janet, became, so to say, the Mother of the 'Academic Gregories'—an interesting example of the necessity for examining the characteristics of the maternal side of families whose names have been famous in the republic of letters. Alexander Anderson, the algebraist and editor of Viète's posthumously published works, was a close relative—probably her uncle.

The dedn. to Alexander Seton, Chancellor to James I, gives a short history of Napier's concern to simplify calculation and states 'ipsam autem novi canonis supputationem ob infirmam corporis nostri valetudinem viris in hoc studii genere versatis relinquimus im primis vero doctissimo viro D. Henrico Briggio Londini publico Geometriae Professori & amico mihi longe charissimo' ('On account of indifferent health I am leaving the actual calculation of the new canon of logarithms to men engaged in that sort of work, in particular to that most learned man, Henry Briggs, [Gresham] Professor of Geometry in London and by far my dearest friend'). Briggs visited Napier at Merchiston Castle, Edinburgh.

The text concerns the construction and use of the *virgulae*—better known as 'Napier's bones'—one of the earliest forms of calculating machines (the cylindrical slide-rule was invented by William Oughtred within twenty years). Another interesting topic in this remarkable book is the reference to 'doctissimus ille mathematicus, Simon Stevinus, in sua Decimali Arithmetica' (No. 657). (See Vol. I, p. 90.)

For a detailed bibliographical account of Napier and his works see (i) Macdonald, W. R. *The Construction of the Wonderful Canon of Logarithms*, Edinburgh 1889, (ii) *De Arte Logistica* (facsimile) Edinburgh 1839, with valuable historical introduction, (iii) Knott, C. G. *Napier Tercentary Volume* (pubd. for Royal Society of Edinburgh), London 1915.

476. NIKANDROS OF KOLOPHON. (*a*) *Nicandri Colophonii poetae et medici . . . Alexipharmaca. Io. Gorraeo . . . interprete. Eiusdem . . . in Alexipharmaca praefatio omnem de venenis disputationem summatim complectens & annotationes.*

Mich. Vascosanus, Paris 1549

Ed. p. (*Ed. p.* of Greek text accompanied the Aldine *Dioscorides* of 1499.)

Prov. READ.

The dedn. to Cardinal Bellaye contains the Aristotelian-Galenic theory of cures for poisons.

(*b*) *Nicandri Theriaca. Interprete Io. Gorraeo. . . .*

Gul. Morelius, Paris 1557

Greek and Latin texts on opposite pages, followed by annotations by Gorraeus. The whole followed by (sep. pagn.): *In Nicandri Theriaca scholia auctoris incerti & . . . in eiusdem Alexipharmaca diversorum auctorum scholia.*

Greek text only. Breaks off at p. 32 with catch-word ' το καλον '[sic.].

For a recent text and English trans. of these (2nd cent. B.C.?) 'pharmacological' poems see Gow, A.S.F. & Scholfield, A.F. *Nicander: The Poems and Poetical Fragments*, Cambridge 1953.

477. NICOLAUS 'PRAEPOSITUS'. *N . . . P . . . Dispensarium ad aromatorios, nuper . . . recognitum. Item cum pluribus additionibus . . . & multiplicium confectionum sive antidotorum quae in secundo serie alphabetica tractantur libro descriptionibus adauctum. Item modus debite dispensandi tyriacam ex solennium medicae facultatibus [sic] autorum voluminibus excerptus. Item errores qui circa eiusdem dispensationem saepenumero committuntur, his annectuntur. Platearius, vulgo Circa instans nuncupatus, de simplici medicina recognitus, ac*

*novis exornatus additionibus per . . . M. Michaelem de Capella . . .
additis.*

Nicolaus Petit & Hector Penet, Lyon, 1536

[*Ed. p.* See Vol. I, p. 241.]

Prov. 'Georgius Brown . . . 1545'.

This work has given rise to a whole tissue of fictions which have only gradually been tracked down and eliminated. For a detailed study, see F. K. Held, *Nicolaus Salernitanus und Nikolaos Myrepsos*, Leipzig 1916.

The original *Antidotarium Nicolai* occurs in many MSS from (XIII-1) and several incunable edns. (*Ed. p.* Jenson, Venice 1471). The usual incipit is 'Ego Nicolaus rogatus a quibusdam in practica medicinae studere volentibus ut eos recto ordine modum conficiendi dispensandique docerem . . .' whereas the works entitled *Nicolai Pr(a)epositi Dispensarium ad Aromatarios* begin: 'Querebat ille Saladinus solennis doctor qui & quot sunt libri apothecario necessarii . . .'. Held (*op. cit.* p. 11) was unable to find any of the latter earlier than the beginning of the 16th cent. Apparently he overlooked the edn. cited in BM old catalogue as '1490(?)'. Mr. G. D. Painter, Assistant Keeper, in a personal communication kindly informed me that, on the basis of the types used in the book, it is believed to have been printed between 1495 and 1500. Of the authorship of No. 477 there is little doubt: he was Nicole Prévost living in Tours in (XV-2), (E. Wickersheimer, *Dictionnaire Biographique des Médecins en France au Moyen Age*, Paris 1936, p. 581). Of the author of the earlier work nothing can be said with certainty: indeed K-H. Lebede goes so far as to say that 'von einem einheitlichen *Antidotarium Nicolai* kann nicht gesprochen werden, da es ein solches Antidotarium nicht mehr gibt': he further questions whether 'Nikolaos Myrepsos' ever existed. (*Das Antidotarium des Nicolaus von Salerno und sein Einfluss auf die Entwicklung des deutschen Arzneiwesens*, Hamburg 1939). (See also Vol. I, p. 241.)

Bound with *Luminare majus, opus . . . medicis & aromatariis perquam necessarium . . . Lumen apothecariorum . . . Index alphabetica serie contextus tam simplicium quam compositarum medicinarum. . . .*

(Device of) Lois Martin, 1528 on t.p. Col. following *Luminare* 'Lugduni apud Antonium Blanchard . . .'.

Incipits correspond to those given by ThK for Paulus Suardus (*Thesaurus Aromatariorum*. Milan 1496) and Io. Iac. Manlius de Bosco (*Luminare majus* 1528, of which, however, BM cites an earlier, Venice 1496.)

Make up:

Dedn. by compiler, Manlius de Bosco, undated, but names Michael Savonarola (No. 613). Followed by text ending fo. LIX *r*, followed by *Tabula Electuarium* and 5 pp. index, Col., and blank fo. Thereafter fo. 1

Lumen Apothecariorum. Incipit libellus . . . edito . . . doctore . . . Querico de
Augustis de Thertona, concludes fo. xxiv *v.* Fos. *xxv-xxvi* r *Tabula;* fo.
xxvi *v* blank. Bound in front of the whole block (inc. t.p.) is *Thesaurus*
aromatariorum, medicis atque aromatariis . . . necessarius, recens . . . recognitus
atque . . . impressus. Ad clarissimos dominos Mediolanenses Collegii physicos
Paulus Suardus aromatarius.

See also No. 740.

478. NON(N)OS, THEOPHANES. *De omnium particularium morborum*
curatione, sic ut febres quoque & tumores praeter naturam complectatur,
liber, nunc primum . . . editus & . . . conversus per Hieremiam
Martium. . . .

Iosephus Rihelius, Strassburg 1568

Ed. *p.*

Prov. LIDDEL.

In his dedn. to the 'Senate' of Augsburg Martius claims to have been
the first to give a Latin version of Nonnos.

He speaks of other works made 'at the expense of that great and
worthy man, patron of all scholars, Antony Fugger'; he mentions also
Marcus Fugger, brother of Antony. He indulges in long encomia on
Montpellier and Florence, mentioning many names connected with
those great seats of learning.

Greek and Latin texts in parallel columns, addressed to the Emperor
Constantine, on whose instructions Nonnos made this compendium of
(mainly) later Greek medical works (X—1).

479. NORDEN, JOHN (from dedn.). *The Surveior's Dialogue . . . as also*
the true and right use for the manuring of grounds . . . Newly im-
printed. And by the same author inlarged, and a sixt booke newly
added. . . .

I. Busby. London 1610

Ed. *p.* London 1607. STC 18639. 3rd edn. STC; but the 1618
edn. has 'now the third time imprinted'.

Prov. 'London 27 July 1629 . . . Clephane . . .'.

The temper of the times may be indicated by the strongly 'defensive'
style in which the work opens. The text contains many refs. to local
industries, e.g. 'sider', 'perry'; and also to the destruction of the timber
of the 'Welds' (Wealds) for iron and 'glass houses', estimates of the
numbers being given. He gives a warning concerning the possible
exhaustion of water power to drive the bellows in the winter.

480. NORMAN, ROBERT. (*a*) *The New Attractive, containing a short*
discourse of the Magnes or Loadstone and amongst other his vertues, of
a new discovered secret and subtil propertie concerning the declining of

the needle, touched therewith under the plaine of the horizon. Now first found out by R . . . N . . ., Hydrographer. Heer unto are annexed . . . rules for the Art of Navigation by the same R.N. Newly corrected and amended by M.W.B. 1585.

Thos. East for Ricard Ballard, London 1585
[*Ed. p.* London 1581 STC 18647. No. 480 is earliest edn. with Borough's addns.]

With sep. t.p. and registration: *A discourse of the variation of the Cumpas or magneticall needle, wherein is mathematicallie shewed the manner of the observation, effects, and application thereof, made by W.B. And is to be annexed to the New Attractive of R.N.* 1585.

'M.W.B.' is 'Master William Borough'—(signature to the Preface of the *Variation*, which is dated 1581).

> (*b*) *The New Attractive, shewing the nature propertie, & manifold vertues of the loadstone with the declination of the needle touched therewith, under the plaine of the horizon. . . . With the application thereof for finding the true variation of the compass: as also divers profitable rules and instruments for the more perfection & exactness in the Art of Navigation. By Maister W. Burrowes.*

John Tappe, London 1614
Note that 'declination' is used as the equivalent of the modern 'dip' or 'inclination'.

481. NUÑEZ, PEDRO [*syn.* Nonus, Petrus]. *Opera, quae complectuntur. . . .*

Off. Henricpetrina, Basel (Col. 1566)
Ed. p. Bibl. Nat.—No Portuguese version cited.

Not BM. From the intro. it appears that it is the Latin version of the original Portuguese edn. written thirty years previously.

The t.p. is in the form of a publisher's 'blurb': the following list of contents is compiled from t.p.'s of the various *opera* in the text.

I. *Rerum astronomicarum problemata geometrica* [viz:] *De duobus problematis circa navigandi artem li. i.* II. *De regulis & instrumentis ad varias rerum tam maritimarum quam coelestium apparentias deprehendendas, ex mathematicis disciplinis li. ii.* [Sep. work, but with cont. pagn.] *In theoricas planetarum Georgii Peurbachii annotationes aliquot.*

From the intro. it appears that the work grew out of reports of a voyage made by Martinus Alphonsus to the Plate in 1530. The origin of the term 'rhumb lines' is alluded to: 'Hispani porro eas lineas [cf. nautica vero acus . . . huiusmodi rectas lineas virtute magnetis representat] communi nomine rumbos appellant'. There is a fine cut of a compass rose—'figura nautica instrumenti quod Hispani acum appellant'. For Nuñez see Vol. I, p. 139.

482. OBERNDOERFFER, IOANNES. *Apologia chymico-medica practica . . . adversus illiberales Martini Rulandi . . . calumnias. . . .*

Typ. Forsterianus [Amberg?] 1610

Only edn. BM—given as 'Erfurt [?]'. Ferg. II, 149.
It contains Ruland's reply of *1611*. See No. 601

OCYORUS, TARQUINIUS (*syn.* SCHNELLENBERG) See
No. 603

483. DE ODDIS, ODDI. *In primam & secundam Aphorismorum Hippocratis sectionem . . . expositio. Nunc primum . . . Marci Oddi filii opera . . . edita.*

Paulus & Antonius Meietus, Venice 1572

[*Ed. p.* 1564.]
Dedn. by the son dated 1572.
In his preface the author says he intends to follow Galen's elucidation, and thereafter add his own in the light of more recent commentaries, e.g. of Giacomo da Forlì, Ugo Benzi, etc.

ODO of MEUNG. See No. 417.

484. ORIBASIUS. *Collectorum medicinalium libri xvii qui ex magno septuaginta librorum volumine ad nostram aetatem soli pervenerunt, Ioanne Baptista Rosario . . . interprete.*

Paulus Manutius, Venice [1555].

Ed. p.
Prov. LIDDEL.
The original—a collection from Galen, Rufus, Antyllus, etc.—is ded. to the Emperor Julian.
Liber 25 consists of an account of bones and muscles with their names, also their articulations, symphyses and functions.

485. ORIGANUS, DAVID [as signature in dedn.]. [*Ephemerides*]
(*a*) *Annorum priorum* 30 *incipientium ab Anno Christi* 1595 *& definientium in annum* 1624. *Ephemerides Brandenburgicae coelestium motuum et temporum.*

D. Reichard, Frankfurt 1609 (Col. Eichorn)

Dedn. mentions Tycho Brahe and his heirs, also Kepler.
(*b*) *Novae motuum coelestium E . . . B . . . annorum* LX *incipientes ab anno* 1595 *et definientium in anno* 1655. [Same imprint].
(*c*) *Annorum posteriorum* 30. *Incipientium . . . 1625 & definientium in anno* 1654. [continues as in (*a*)] [Same imprint.]
There are two copies of each of the above. (*b*) contains a reply to the 'mad dog-like barking' of Maginus and another against his work of

eleven years previously—also refs. to Peregrinus and Gilbert. The existence of a work published eleven years earlier (not BM) is attested in in Bibl. Nat. and Th. *Hist.*

Prov. One copy of each of (*b*) & (*c*) 'ex dono Mgtri. Iacobi Fraseri'; (*c*) has much MS historical matter on front- and end-papers.

486. OROSCIUS, CHRISTOPHORUS. *Annotationes in interpretes Aetii. . . .*

Basel (Col. Robt. Winter 1540)

Only edn. BM.

A critical commentary without text on the sixteen books of Aëtios.

487. ORTA, GARÇIA DE (*syn.* Garcia ab Horto). *Coloquios dos simples, e drogas he cousas medicinais da India, e assi dalgunas frutas achadas nella onde se tratam alguna cousas tocantes amediçina, pratica, e outras cousas boas, pera saber compostos pello Doutor Garcia Dorta fisico del Rey nosso senhor, vistos pello muyto Reverendo senhor, ho licienciado Alexos Diaz falcam desenbargador da casa da supricaçao inquisidor nestas partes. Com privilegio do Conde viso Rey.*

Ioannes de Endem, Goa 1563

Ed. p.

Prefaced by an enconium to Conde de Redondo, Viceroy of India, by Luis de Camoens.

The text takes the form of a dialogue between Garcia and his friend Ruano. On fo. 227 *v* there is a ref. to the removal of 'wild herbs' from the wooded mountains into the garden of the University of Coimbra.

Some terminal fos. mutilated.

488. ORTA, G. DE. *Aromatum et simplicium aliquot medicamentorum apud Indos nascentium historia. Primum quidem Lusitanica lingua διαλογικως conscripta . . . Deinde Latino sermone in epitomen contracta & iconibus ad vivum expressis locupletioribusque annotatiunculis illustrata a Carolo Clusio. . . . Quarta editio castigatior & aliquot locis auctior.*

Off. Plantiniana—Vidua and Io. Moretus, Antwerp 1593

[*Ed. p.* in Latin, Antwerp 1567.]

Prov. GREGORY, and another copy (No. 3b).

Numerous cuts, including one of pepper (p. 87). Text includes a long discussion of the 'Bezoar Stone' (p. 165), and a ref. to 'anil', which nearly three centuries later gave its name to 'aniline'.

489. ORTUS SANITATIS [Jacobus Meydenbach, Mainz 1491].

The t.p. contains no printed matter except the title, but much MS relating to prov. etc., including 'Georgius Pavius pharmacopoeus . . .

1619 ex dono praepositi Abredonensis' and (earlier hand) 'liber Alexri. Ogilvy'.

Coloured frontispiece, and 'Incipit prohemium praesentis operis omnipotentis eternique dei tocius nature creatoris . . .'.

'Capitulum primum / Aarona Pandecta . . .' (coloured cut of Arum Lily). Debased rubrication.

There is no sep. col., device, or register: but the last fo. *v* contains the passage: 'Impressum est autem hoc ipsum in inclita civitate Moguntina que ab antiquis aurea Moguntia dicta . . . in qua . . . ars ac scientia hec subtilissima caracterisandi seu imprimendi fuit primum inventa', also a ref. to Io. Meydenbach.

Arber (p. 28) describes this work as a modified Latin trn. of the *German Herbarius* (*Herbarius zu Teutsch*—also known as J. von Cube's *Herbal* of 1485), but with additional treatises on 'animals' and 'stones', together with new blocks, some copied, and some prints therefrom duplicated. There is a fine cut of a jeweller's *atelier* with machinery. The cuts are all coloured—the col. states that the 'flowers are in natural colours'. (See Vol. I, p. 186.)

OTHO, VALENTINUS. See No. 568.

490. PALMARIUS, JULIANUS [*syn.* Le Paulmier, Jules]. . . . *de morbis contagiosis libri septem.*
    (*a*) Dionysius Du-Val, Paris 1578.
    Ed. p.
    Prov. 'J. Gregorie L. B. 1745'.
    (*b*) Cl. Marnius & heirs of Io. Aubrius, Frankfurt 1601. Not BM.
    Prov. LIDDEL. And a second copy.

Contents: *De lue venerea li. ii. De hydrargyro li. i. De elephantiasi li. i. De morsu canis rabidi li. De febre pestilenti.*

The *epistola* before the last part contains a long ref. to the teaching of Fernel and to the desirability of building a 'Xenodochium' outside the city. The dedn. to the Senate of Paris refers to his work 'in Xenodochio Lutetiano aegris decumbentibus'. The word 'Xenodochium' in classical times meant a 'hospice', i.e. where 'strangers' might be given 'hospitality'. Here it seems as if it meant an 'isolation hospital', though within the Christian Church it seems to have been loosely used for any hospital for the sick from about the Fourth century.

In the text (p. 188) is a ref. to 'Hydrargyri cinis qui & pulvis mercurii & praecipitatus hydrargyrus vulgo dici solet . . .' ('the ash of mercury which is commonly called both powder of mercury and mercury precipitate'), with a further discussion on the use of 'calx' and its preparation by means of *aqua fortis*. The whole work gives a valuable picture of knowledge of, and speculation on, 'contagious' diseases.

491. PALMARIUS, PETRUS. *Lapis Philosophicus dogmaticorum, quo paracelsista Libavius restituitur, Scholae Medicae Parisiensis iudicium de chymicis declaratur, censura inadulteria* [sic] *& fraudes Parachymicorum deffenditur* [sic] *asserto verae Alchemiae honore per P . . . P . . . Doctorem Parisiensem Galeno-chymicum. Adiecta est historia laeprosae mulieris persanatae.*

David Doulceur, Paris 1609

Ferg. II, 163 states that there had been an edn. in the previous year. He gives no authority for this statement, which appears at variance with the date of the Privilege in this copy—Dec. 17th 1608. Only Edn. BM.

The account of the circumstances of its publication is valuable; the author himself in his *Lectori* explains why, though calling himself 'Doctor Galeno-Chymicus', he writes *against* Libavius, the object of attacks by the Paris Medical School (Vol. I, p. 256). Petrus Palmarius (nephew of Jules Le Paulmier) appears therefore to be a right-wing deviationist in the Libavius–Du Chesne bloc!

The nature of Palmarius's treatment may be judged from his account of the case of a 'mulier laeprosa' who had been treated in vain with 'Hermetic' drugs. The description of the 'syndrome signorum', reaction, and resolution is a model of clarity. But the remedies—'aurum potabile exuberatum, supra sphaeram exaltatum nec reductile' and 'Alexandri Suctenii [von Suchten?] antimonium non vulgare' have a distinctly unorthodox flavour! [Sigs. lii-liv].

492. PANIZZA, LUDOVICUS. *De Venae Sectione in inflammationibus quibuscunque fluxione genitis, per sanguinis missionem curandis . . . disputatio ac decisio.*

Off. Farrea, Venice 1544

[At least one sig. wanting].
Ed. p. Earliest BM 1561.
Prov. FORBES.

A disputation (apparently carried on by correspondence) on the famous controversy concerning 'revulsion' and 'derivation' (*Intro.* Vol. I, p. 233). The discussion appears to us to be largely irrelevant to the science of medicine, but at the worst the text provides information about numerous contemporary Chairs of Medicine!

493. PANIZZA, L. *Apologia commentarii olim aediti de Parca Evacuatione in gravium morborum principiis . . . facienda. . . .*

Venturinus Ruffinelus, Mantua 1556

Ed. p. ? Earliest BM 1561.
Prov. FORBES.

Preceded (sep. pagn. and sigs.) by the same author's *De Minoratione facienda opusculum.* Half title only and no col. No imprint, but final secn.

signed by Panizza 'ex Mantua . . . Anno LVI'. Nevertheless the work is referred to in the license of Julius III (1552) for the *Apologia* as having been received, together with *De Sectione Venae* by 'our predecessor' Paul III.

494. PAPPUS, ALEXANDRINUS. . . . *Mathematicae Collectiones a Federico Commandino in Latinum conversae & commentariis illustratae* [Ed. Valerius Spacciolus—Commandino's son-in-law].
t.p.: Franciscus de Franciscis, Venice 1588.
Col. Hiero. Concordia, Pesaro 1588.
Ed. p. of Latin trn. Part of the Greek text was first edited by John Wallis, 1686.

In his *Lectori* the printer says that the publication has been undertaken by the Duke of Urbino since the heirs of Commandino could never agree about the printing. He points out that Books I and II are still wanting despite all Commandino's efforts to find them.

Book VIII contains numerous cuts of machines (see esp. fo. 314*v* for a 'gear box') with problems on the design of toothed wheels, screws, etc. which will interlock.

495. PARÉ, AMBROISE. . . . *Oeuvres . . . divisee en vingt sept livres. Avec les figures & portraicts tant de l'anatomie que des instruments de Chirurgie & de plusieurs monstres. Reveuz & augmentez par l'autheur pour la seconde edition.*

Gabriol Buon, Paris 1579

Not BM. Monsr. P. Josserand, Conservateur en chef du Département des Imprimés, kindly informed me that the *Ed. p.* of 1575 is to be found at the Libraries of St. Geneviève and the Faculty of Medicine, but not at the Bibliothèque Nationale, where, however, there is a copy of No. 495, but lacking t.p. and sig. A.6.

Prov. (i) 'Albertus le Feure, Parisiensis Doctor Medicus'.
(ii) Forbes.

Dedn. to Henry III. Sig. a iii *v*—portrait of Paré, aged 65. Sig. b 1 *r*—sonnet on Paré by Ronsard. This is followed by sig. e; no sigs. c-d are mentioned in the collation in Janet Doe, *A Bibliography of the Works of Ambroise Paré*, Chicago 1937, p. 108.

Au Lecteur: 'Nous avons apris du bon pere Guido [Guy de Chauliac— No. 322] que nous sommes, comme l'enfant, qui est sur le col du Geant . . . . Les ancients n'ont peu tout veoir. . . .'

This is a magnificent copy, illustrated by numerous cuts; but nothing could better illustrate the confusion of thought and uncertain critical faculty of the age than the concluding plates depicting (LV) the 'Succarath' (whose existence is 'attested' by Thevet *Cosmographie*, II.

23.1), the famous 'plated' Rhinoceros (LX) of Dürer (see Vol. I, p. 26) and (LXII) 'Les dragons qui tuent les elephans'. For Paré's empirical treatment see Vol. I, p. 282.
[The Library also possesses Thos. Johnson's Engl. trans., London 1649.]

496. PARKINSON, JOHN. *Paradisi in Sole Paradisus terrestris or A garden of all sorts of pleasant flowers which our English ayre will permit to be noursed up with a Kitchen Garden of all maner of herbes, rootes, & fruites* . . . *and an Orchard of all sortes of fruit bearing trees* . . . *with the right orderinge planting & preserving of them and their uses & virtues.* . . .

(Col. Humfrey Lownes & Robt. Young, London 1629)

Ed. p. S.T.C. 19300.

Prov. READ.

Dedn. to Queen Henriette Marie. In the *Epistle to the Reader*, Parkinson explains that it is now possible to replace the errors of classical writers by Biblical truth! He refers to the works of de l'Ecluse, Dodoens, Turner, etc., and says that he will hasten the Fourth Part on Simples.

Although its publication lies outside our period, this work is significant as the basis of English domestic gardening for perhaps a couple of centuries: here for the first time is a scholarly work intended neither for the physician nor for the apothecary but for the lover of flowers for their beauty and of fruits for their savour.

Encomia include one by Turquet de Mayerne and a poem by Thomas Johnson (No. 495) 'utriusque Societatis consors' (cf. Alexander Read).

497. PAULUS AEGINETA. . . . *Opera a Ioanne Guinterio* . . . *conversa & illustrata commentariis. Adiectae sunt annotationes Iacobi Goupyli.* . . .

(*a*) Chris Wechel, Paris 1536 [*Liber tertius* only].

Prov. GREGORY.

(*b*) Andr. Cratander, Basel 1538.

Prov. FORDYCE. (Inscribed by David Fordyce, Regent at Marischal College from 1742, to his brother, John, physician at Uppingham, to whom the better known Dr. Geo. Fordyce, F.R.S., his nephew, was apprenticed).

(*c*) Gul. Rovillius, Lyon 1551.

Prov. LIDDEL.

Earliest BM 1554 *Ed. p.* of Guinter's edn. Paris 1532.

Greek text and brief notes. [*Ed. p.* of Greek text, Aldus, Venice 1525.]

498. PAULUS AEGINETA. . . . *de Chirurgia liber, inter caeteros eiusdem autoris medicae artis ordine sextus, a Ioanne Bernardo Feliciano . . . nunc primum Latinitate donatus . . . Castigationes praeterea Albani Torini in suam Aeginetae tralationem.*

Io. Bebel, Basel 1538

Neither BM nor Bibl. Nat.

499. PAULUS AEGINETA. *Pharmaca simplicia, Othone Brunfelsio interprete. Idem de Ratione Victus, Gulielmo Copo . . . interprete.*

Chris. Wechel, Paris 1532

[*Ed. p.* Strassburg 1532.]
Prov. 'Georgius Pavus' and Gregory.
Cont. pagn. with secn. heading only. The dedn. by Brunfels gives some bibliographical details.

500. PAULUS SILENTIARIUS. . . . *Hemiambia dimetra catalectica in Thermas Pythias Latine facta, epico carmine a Claudio Acanthero medico. Accesserunt . . . annotationes; brevis item . . . de Thermis dissertatio & . . . poemata eodem auctore. . . .*

Geo. Angelerius, Venice 1586

Ed. p.
Prov. BUTE.
Trans. from Greek.
Paul the 'Silentiary' lived VI-2. The above poem, originally attributed to him, is now believed to have been written by Leo Magister (IX-?).

501. PAYNELL, THOMAS. *Of the wood called Guaiacum that heale* [th] *the French Pockes and also helpeth the goute in the feete, the stoone, the palsey, lepree, dropsy, fallynge evyll, and other dyseases.*

Thos. Bertheletus, London 1539

[*Ed. p.* of trans. 1533 entitled *De Morbo Galico.*] This is 3rd edn. STC. 14026. *Ed. p.* of original work, Mainz 1519.
Prov. HENDERSON; formerly BM sale duplicate, 1787.
In his preface Thomas Paynell, 'Chanon of Marten Abbey', describes the circumstances of its printing and also of his trans. of *Regimen sanitatis Salerni* (cf. No. 502) both originally printed by Berthelet. He states that 'almost into every parte of this Realme this most foule and peynfulle dysease is crepte'.
The original work was *De guaiaci medicina & morbo Gallico* by Ulrich von Hutten. It contains the tradition of the origin of the disease, the nature of the wood, and the methods of treatment. (See Vol. I, p. 271.)

502. PAYNELL, T. *Regimen sanitatis Salerni. This booke teachyng all people to governe them in health is translated out of the Latine tongue into Englishe by T . . . P . . . whiche booke is amended, augmented.*
Wm. How, for Abraham Veale, London 1575
[*Ed. p.* of trans. 1528. STC. 21601]. This is the 6th edn.
Prov. HENDERSON.
Not strictly a trans. but a lengthy commentary, with documentation on the poem. Similar trans. were made by Sir John Harington (1607) and for the Edinburgh printer, A. Hart, 1613.

503. [PECKHAM, JOHN]. *Io. Archiepiscopi Cantuarensis Perspectiva communis.*
(Col. Diligentissime emendatum per L. Gauricum Neapolitanum,
Io. Baptista Sessa, Venice 1504)
[*Ed. p.* Leipzig 1504.]
T.p. has fine cut. Dedn. by Gauricus to Paulus Trivisanus. 'Est enim ut scis, optice media inter mathematicum & naturale, atque uti oculorum sensus caeteris est nobilior, ita & optice suam obtinet dignitatem . . .' ('Optics, as you know, is as it were midway between mathematics and natural philosophy; and just as vision is nobler than the other senses, so optics derives its peculiar importance . . .'.
Incipit: Inter philosophicae considerationis studia . . . . ThK 363 record a 14th cent. MS and printed version Milan *ca.* 1480 Hain, 9425, For the importance of this work see Vol. I, p. 160.

504. PEGELIUS, MAGNUS. *Universi seu mundi diatyposis.*
Stephanus Myliander, Rostock 1586
Not BM.
One of the pamphlets in a collected volume presumably bound to the orders of Duncan Liddel.

505. PELETIER DU MANS, JACQUES. (*syn.* PELETARIUS) . . . *de occulta parte numerorum quam Algebram vocant libri duo.*
Gul. Cavellat, Paris 1560
[*Ed. p.* French, Lyon 1554.] Earliest Latin version (Bibl. Nat. not BM).
Contains a useful historical introduction. See also No. 235(e).

PENA, IOANNES. See No. 234.

506. PENOTUS, BERNARDUS G. *Tractatus varii de vera praeparatione et usu medicamentorum chymicorum nunc primum editi.*
Io. Feyrabend, Frankfurt 1594

Ed. p. Duv. Ferg. also cites it but had not seen it. Earliest BM
1602.

Prov. LIDDEL—Also a second copy.

Following the t.p. is a table of the four tracts. Sep. half-titles and
dedns. An interesting 'period' preface.

507. PENOTUS, B. G. . . . *de Denario Medico quo decem medicaminibus*
omnibus morbis internis medendi via docetur.

Io. Le Preux, Berne 1618

[*Ed. p.* 1608. Duv. p. 464]. Ferg. II, 180 cites, but had not seen it.

On t.p. *v* are named *Apologetica adversus Guibertum* and Isaac of
Holland (No. 357) *Lapidis philosophorum theoria.* The main text consists of a number of tracts. Cont. pagn. throughout but several sep.
dedns.

508. PERSON, DAVID. *Varieties or a surveigh of rare & excellent matters*
necessary & delectable for all sorts of persons. Wherein the principall
heads of diverse sciences are illustrated, rare secrets of naturall things
infoulded. . . .

Richard Badger for Thos. Alchorn, London 1635

Only edn. S.T.C. 19781 (5th book wanting in this copy).

Contains epistles from the celebrated Scots poets Arthur Johnston,
William Drummond of Hawthornden, etc. Johnson (p.323) notes its
Aristotelian, anti-Copernican character even at this late date. See p. 37
for ref. to 'claick (i.e. Barnacle) geese'.

509. PETRONIUS, ALEXANDER (*syn.* Petronio Alessandro). *Del*
viver delli Romani, et di conservar la sanità . . . libri cinque. Dove
se tratta del sito di Roma, dell' aria, de venti . . . con dui libri appresso
dell istesso autore del mantenere il ventre molle, senza medicine.
Tradotti dalla lingua Latina nella volgare, dall . . . M. Basilio
Paravicino. . . .

Domenico Basa, Rome 1592

Only edn.

Privilege of Clement VIII 'omnibus & singulis utriusque sexus
Christi fidelibus'—indicating the common inheritance of a printing
establishment by a widow or daughter.

510. PETRUS DE ABANO. (*a*) *Conciliator differentiarum philosophorum*
& medicorum in primis doctoris in omni disciplinarum genere eminentis-
simi Petri Abani Patavini. Cum tabula Differentiarum . . . et tractatus
de Venenis . . . . Adiectis insuper Symphoriani Camperii . . . addition-
ibus eiusdam P . . . de A . . . hereses repellentibus . . . .

191

[Ed. Scipio Ferrarius.]
  Girardus de Zeis and Barth. de Morandis, Pavia 1523.
Prov. Scored and glossed in Liddel's and an earlier hand.
 (*b*) *Conciliator controversiarum quae inter philosophos et medicos versantur* . . . .
          Juntae, Venice 1565 (Col. 1564)
 Prov. HENDERSON.
 Ed. p. Mantua 1472.] Hain No. 1.
The contents of both edns. are the same, but the order differs.
This famous work aimed at showing that differences of teaching as between Aristotle and the medical faculties were not unresolvable. The aim may be considered ill-chosen, but the treatment reveals an exceptionally acute and critical mind—too critical for the liking of the Holy Office, which is said to have burned his exhumed body.
 In (*a*) the text is preceded by several fos. of corrections by Symphorien Champier.

511. PETTUS, SIR JOHN. *Fleta Minor—The laws of art and nature in knowing, judging, assaying, fining, refining and enlarging the bodies of confin'd metals. In two parts. The First contains assays of Lazarus Erckern, Chief Prover (or Assay Master General of the Empire of Germany) in V books originally written by him in the Teutonick language & now translated into English. The second contains essays on metallick words, as a dictionary to many pleasing discourses.*
        Stephen Bateman, London 1686
 [*Ed. p.* 1683, based on Ercker's original, Prague 1574 subsequently enlarged under another title. Ferg. I, 245] 2nd edn.—not BM.
 Prov. BUTE.
Ercker's work was one of the great classics of mining and metallurgy of the 16th cent. (See Vol. I, p. 176). Pettus gives the reason for his strange title in the preface. He uses the form 'Erckern' throughout.

512. PEUCER, CASPAR. *Hypotheses astronomicae seu theoriae planetarum ex Ptolemaei et aliorum veterum doctrina ad observaiones* [sic] *Nicolai Copernici & canones movum* [sic] *ab eo conditos accommodatae. Opera et studio Casparis Peuceri in Academia Witebergensi.*
        Io. Schwertel, Wittenberg 1571
 Not BM, Bibl. Nat., Bodleian nor Trinity College, Dublin; but recorded under same date in *Bibl. Univ.* No work under this title is indexed in Th. *Hist.*, though two others, including an *Elementa*, are. It had previously been published by Conrad Dasypodius as by an unknown author (No. 190).

Prov. 'G. Strachan 1611'. End papers covered with calculations.

In the dedn. to William Landgrave of Hesse, Peucer refers to the fact that from no existing theory could values be deduced conforming to those observed. He refers to Melanchthon, Rheticus, Reinhold, Dasypodius—'taking it very hard' that the last-named had published it as by an unknown author. And he claims that Bathodius, who, according to Dasypodius, had first brought the MS to his notice, could not have been ignorant of its true authorship.

513. PEURBACH, GEORG. [*syn.* (BM) Georgius, *Peurbachius*].
Theoricae Novae Planetarum.

(*a*) *T— N— P— Georgii Purbacchii Germani ab Erasmo Reinholdo Salveldensi pluribus figuris auctae & illustratae scholiis quibus studiosi praeparentur ac invitentur ad lectionem ipsius Ptolemaei. Inserta item methodica tractatio de Illuminatione Lunae.*

Carolus Perier, Paris 1553

Both BM and Bibl. Nat. assume this to be *Ed. p.*, the date of the dedn. (1542) being regarded as a mis-print for 1552; the former is, however, repeated in the edns. of 1558 and 1580.

Prov. JOHNSTON.

The dedn. has interesting comparisons of the relations between astronomical observations and contemporary events with the corresponding relations in the ancient world, e.g. the account by Thucydides of the disasters following a solar eclipse during the Peloponnesian War. There is also an *epistola* from Melanchthon to Simon Grynaeus, dated 1535—presumably 'lifted' from (Peter Apian's ?) edn. of Peurbach, Wittenberg 1535, to which Melanchthon wrote a preface.

[*Ed. p.* of orig. Venice 1495 (Hain 13596); but it was included with Sacrobosco's *Sphaera*, Bologna 1480 (Hain) 14109.]

(*b*) [Same title as (*a*) down to *Ptolemaei* but '*Purbachii*', then continues:] *recens editae & auctae novis scholiis in Theoria Solis ab ipso autore. Incerta item methodica tractatio de Illuminatione Lunae.*

Carolus Perier, Paris 1558

(*c*) [Same title as (*b*).] Same prelims. but between dedn. by Reinhold (No. 560) to Albert of Brandenburg and *Epistola* of Melanchthon are poems on the death of Reinhold.

Heirs of Io. Crato, Wittenberg 1580

Prov. LIDDEL.

(*d*) *T— N— P— G— P— Germani. Quibus accesserunt: Iohannis de Monteregio Disputationes super deliramenta Theoricarum Gerardi Cremonensis. Item. Ioannis Essler . . . Tractatus . . . ante LX annos conscriptus cui titulum fecit Speculum Astrologicorum in quo Astrolog*

o*   193

[sic] *errores ex neglecta temporis aequatione provenientes ostenduntur & multa quae ad Theoricarum, praesertim octavae sphaerae, intellectum faciunt explicantur. Quaestiones vero in Theoricas Planetarum Purbachii, authore Christiano Urstisio Basil. eadem forma damus. Omnia nunc iterum in gratiam candidatorum astronomiae edita.*

Seb. Henricpetrus, Basel (Col. 1596)

Not BM. Edn. of 1573 Bibl. Nat. with same imprint, but Urstisius not mentioned in title.

Epistola of Melanchthon reprinted but undated. No other prelims.

Text continuous, but differently paragraphed from (*a*), etc. Also different figs. The ' . . . contra Cremonensia' of Regiomontanus starts at p. 153. In its preface are refs. to the *Studia generalia*, giving an indication of current views on their content. The text is in dialogue form between 'Viennensis' and 'Cracoviensis'.

[*Ed. p.* 1494. Hain 13805 undated]. Regiomontanus was in error in ascribing the *Theoricae* to Gerard of Cremona: they are now attributed to Gerardus Sabtonetanus (BM).

On p. 196 starts the *Tractatus* of Io. Essler. Then with t.p., new pagn. but no date, there follows the work of Urstisius (which does not appear under his name—'Wurstisen'—in BM) ded. to 'clarissimis . . . viris scholae Tigurinae . . .' from Basel 1568. In the long *Praefatio Isagogica* there is (sig. b 5 *v*) a ref. to the astronomical model of Ioannulus Torianus Cremonensis—'instrument-maker (*artifex*) to the Emperor Charles V and of the most admirable ingenuity'. There follow (i) a classification of heavenly motions on Aristotelian lines, (ii) the *Quaestiones*. The pre-eminence of the sun is discussed on p. 11. See also Nos. 632, 669.

514. PEURBACH, G. *Tractatus . . . super propositiones Ptolemaei de sinubus et chordis. Item compositio tabularum sinuum per Ioannem de Regiomonte. Adiectae sunt & tabulae sinuum duplices per eundem Regiomontanum. Omnia nunc primum . . . impressa.*

Io. Petreius, Nürnberg 1541

Presumably *Ed. p.* Earliest BM 1561.
Edited by Io. Schonerus. See ThK. 693 and 262.
See also No. 669.

515. PHAYRE, THOMAS [*syn.* Phaer—BM]. *The Regiment of Life Whereunto is added a Treatise of the Pestilence; with the Book of Children. Latelye corrected and enlarged by Thomas Phayre.*

Edw. Alde, London 1596

[*Ed. p.* of the *Regiment* 1544. STC 11967; of the *Book of Children* 1530? STC 3294] Castiglioni refers to it as *The Boke on Children* 1545; Garrison, as *Book of Children* 1545.
Prov. 'Iohannes Forbes'.

The *Regiment* is alleged to be a trans. of No. 310, but has many additions (e.g. chapter on the stone) not in the orig. French. The 'brief treatise of the pestilence . . . composed and lately recognised by Thomas Phayre . . .' in fact follows closely the text of *'Regime et traicte singulier contre la peste fait et composé par Maistre Nycollas de Houssemaine docteur regent en Luniversite Dangiers'* [*sic*] which follows Goeurat's *Summaire* with cont. pagn. (No. 310).

A 'remedie for the gout' (sig. R s *v*) gives a good idea of 16th cent. medicaments.

516. PHILANDER, GULIELMUS. *In decem libros M. Vitruvii Pollionis de Architectura annotationes.*

Michalles Fezandat, Paris 1545

[*Ed. p.* Rome 1544. *Ed. p.* of Vitruvius, Rome 1487.]
Intro. by Ioannes Andreas Dossena, the printer of the *Ed. p.*

517. PICCOLOMINI, ALESSANDRO. . . . *de Sphaera libri quatuor ex Italico in Latinum sermonem conversi, eiusdem compendium de cognoscendis stellis fixis: & de Magnitudine terrae & aquae liber unus, itidem Latinus factus. Ioan. Nicol. Stupano Rheto interprete.*

P. Perna, Basel 1568

Ed. p. of Latin version. The 3rd Italian edn. is given as Venice 1552.
Sep. t.p. and pagn. for the *Compendium*, which itself has cont. pagn. but sep. t.p. for the part on the stars.

518. PICCOLOMINI, FRANCESCO. . . . *Librorum ad scientiam de natura attinentium: partes quinque.* . . .

Heirs of Andr. Wechel, Frankfurt 1597

[*Ed. p.* (Italian) 1st part 1551, Rome; two parts, 1565, Florence.]
No Latin BM.
Prov. 'ex don. And. Straquih . . .1635[?]'.

519. PICO, IOANNES FRANCISCUS DE MIRANDOLA . . . .
[*Opera Omnia*] [t.p. wanting.] Final col. Strassburg 1507.
A MS t.p. appears to have been composed by the owner since it corresponds precisely neither to that given for this edn. nor to any other in Bibl. Nat. The table of contents follows the actual order of the sep. works, which is not the case in the printed t.p.

Ed. p. of *Opera* Venice 1498 Bibl. Nat; but for *Commentationes* containing *Disputationes adversus Astrologos*—the most important work for our purpose—see Hain 12994 (Bologna 1495). The *Disputationes* in No. 519 has col. dated 1504.

Prov. 'Ex libris Pauli Emanuelis Werneri'.
The *Disputationes* is fully described in Allen, and is assessed in Th. *Hist.* passim.

520. PICO, IO. F. DE MIRANDOLA. . . . *Lib. iii de Auro. Opus sane novum ac aureum, in quo de auro tum aestimando, tum conficiendo, tum utendo . . . disseritur. Accessit Bernhardi Comitis antiqui Trevirensis . . . ΠΕΡΙ ΧΕΜΕΙΑΣ Opus historicum & dogmaticum ex Gallico in Latinum . . . versum & nunc primum editum.*

Cornelius Sutor, Urselli 1598

[*Ed. p.* of *Lib. de Auro* 1586. Duv. & BM; also Ferg. II, 202, who also cites this edn.]. The second work (half-title and cont. pagn.) has a preface by Gratarolus dated 1567 to which (Ferg. I, 102) 'Schenreuter' (Etschenreuter) replied in the same year. Th. *Hist.* thinks the first work supposititious (but see Ferg. *loc. cit.*). The second, which contains interesting historical material, raises numerous questions. First, who was 'Bernard'? Both 'Trevirensis' and 'Trevisanus' occur in the literature, but it is uncertain whether the two forms really referred to different people. The confusion may be merely orthographic since on p. 148 'Bernardus . . . comes de Tresne in Germania . . .' writes a preface to Thomas of Bologna, dated 1453; but the editor (presumably Gratarolus) refers to him (on the same p.) as 'Comes de Trevisio', and (on the next p.) as 'Comes Tarvisinus', and in a kind of colophon on p. 222 as 'Tarvisino aut Trevirensi'. Another question is that of the 'authentic' spelling of the word for 'chemistry' which appears here variously as χεμεια, *chaemice, chaemia, chimia.*

521. PICTORIUS, GEORGIUS. (*a*) *Tuendae sanitatis ratio, VII dialogis, per sex rerum (ut medici vocant) non naturalium ordinem, quae sunt, aer, cibus, potus, motus, quies, somnus, vigilia, . . . ex summorum medicorum sententia nunc denuo . . . conscripta & per autorem . . . locupletior reddita. Et haec omnia breviter summatim per epilogum seu epitomen ob oculos posita: id quod prior editio non habebat. Quibus accedunt antea non impressa Succisivarum lectionum IX dialogi . . . Praeterea Conviviorum libri iii. . . .*

(Col. Henricus Petrus, 1554) Basel

[Issued in various earlier forms BM].

(*b*) *Dialogi . . . del modo del conservare la sanita. Novamente dalla lingua latina nella volgar Italiano tradotto. Aggiuntovi un trattado di Arnoldo di Villa nuova, del modo di conservar la giouventu & ritardar la vecchiezza.*

Vinc. Valgrisius, Venice 1550

Earliest trn. BM but dedn. by Pittore 1548.
Prov. BUTE.

522. PINEAU, SEVERIN (*syn.* Pinaeus, Severinus). (*a*) . . . *Opusculum physiologum & anatomicum in duos libellos distinctum. In quibus* *primum de Integritatis & corruptionis virginum notis, deinde de Graviditate & partu naturali mulierum in quo ossa pubis & ilium distrahi dilucide tractatur.*

Steph. Prevosteau, Paris 1597

Dedn. to (i) Riverius, Royal physician, France, (ii) The Dean and members of the Royal College of Surgeons, Paris (many named), (iii) 'Utriusque medicinae studiosis'.

Numerous cuts of gravid uterus and foetus. On p. 56 is a reference to a phenomenon rejected as fictitious by Pineau, which we have come to accept as well authenticated—that there can be a change of sex—'I do not cease to marvel at the opinion of those who maintain that females can be changed into males, since there is no reason or argument sufficiently cogent by which it could possibly be proved, especially in human kind'. Note that it had to be 'proved by theory (*ratio*) or argument',—observation is not mentioned. But Pineau's attitude was at least more 'reasonable' than that of some of his contemporaries who could believe anything—by 'reasoning'.

(*b*) . . . *Tractans analytice notas primo integritatis & corruptionis virginum.* . . . remainder similar to, but not identical with, (*a*).

Zach. Palthenius, Frankfurt 1599

Prov. 'Georgius Pavus pharmacopoeus 1613 (?)'.

This work and its author appear to have been overlooked by Castiglioni.

523. PISANELLI, BALDASSARE. *Trattato della natura de cibi et del bere.* . . .

Gio. Battista Porta, Venice 1584

[*Ed. p.* Rome 1583—5 edns. in 3 yrs., BM].

Prov. BUTE.

Dedn. Rome 1583. There is a second copy (same prov.) with identical title, but slight typographical variations. No imprint; also sig. T (and perhaps sig. a) wanting.

524. PISO, NICOLAUS (*syn.* Le Pois). (*a*) *De cognoscendis et curandis praecipue internis humani corporis morbis libri tres: ex monumentis classicorum medicorum, tum veterum, tum vel recentium collecti* . . . *Accessit de Febribus liber unus.*

And. Wechel, Frankfurt 1580

Ed. p.

Prov. GORDON, and an earlier.

(*b*) [Slightly different title with 'ab eodem recogniti & aucti' added.]

Heirs of And. Wechel, Frankfurt 1585

Prov. One LIDDEL; a second copy 'Jo. Gregorie 1741'.

In his *Praefatio*, Piso (Le Pois) speaks of the lectures of Parisian physicians being passed round in MS. The following passage indicates the beginning of the 'ideological war' (see Vol. I, p. 256)—'chymisticam (quam hodie novam medicinam vocant nonnulli)' . . . sig. iii *v* in (*a*), fo. 4*r* in (*b*). *De Febribus* has sep. t.p. but cont. pagn.

525. PITATUS, PETRUS. *Almanach novum . . . ad annos undecim, incipiens ab anno Christi MDLII usque ad annum MDLXII. Isagogica in coelestem astronomicam disciplinam. Tractatus tres perbreves de electionibus, revolutionibus annorum & mutatione aeris. Omn. . . . recog. et emend* [sic].

Ulricus Morhardus, Tübingen 1553

[*Ed. p.* 1544.]
Prov. LIDDEL.

The *Isagogica* is a mixture of astronomy and astrology. The book is peculiar in having no *personal* dedn. but the *Prooemium* contains an encomium on Venice.

526. PITISCUS, BARTHOLOMAEUS [trans. by R. Handson].
Trigonometry or the doctrine of triangles first written in Latine by B . . . T . . . of Grunberg in Silesia and now translated into English by Ra. Handson. Wherunto is added (for the marriners use) certaine nauticall questions, together with the finding of the variation of the compasse. All performed arithmetically without mappe, sphere, globe or astrolabe by the said R.H.

(Col. Edw. Alde for Io. Tappe, London 1614

Ed. p. S.T.C. 19967.
[*Ed. p.* of Pitiscus *Trigonometriae . . . libri quinque . . .* Vienna 1600.
There was an earlier work, *Trigonometria . . .*, included in the *Sphaericorum libri tres* of A. Scultetus, 1595.]
The above work is followed by *A Canon of triangles or the tables of sines, tangents & secants the radius asumed* [sic] *to be* 100000, which has sep. registrn. In the *To the Reader* before the *Trigonometry* Handson states that these tables differ from those referred to by Pitiscus in his Second Book, which were 'of severall radiusses'.
Prov. 'William Jameson. London'.
In his dedn. Handson hails Sir Thomas Smith and John Wostenholme as 'the sole founders and erectors of the Lecture of Navigation in the honourable Cittie of London': Johnson (*J. Hist. Ids.* **1**, 425) gives reasons for believing that this lectureship was abandoned about 1592. Handson names Richard Hakluyt as among the friends who induced him to publish the translation.

The *Questions* by Handson have sep. pagn. These refer to the relations between the 'ordinarie sea-chart' (meridians parallel without compensation for change of latitude), the globe, and the 'true sea-chart made after Mercator's way or Mr. Edw. Wright's projection' (i.e. with compensation for change of latitude—see Vol. I, p. 138).

527. PLANERUS, ANDREAS. . . . *Orationes tres quarum prima continet definitionis explicationem Artis Medicae, quae est apud Platonem in dialogo qui inscribitur Symposium . . . Altera, utilitatem . . . Libelli Galeni qui inscribitur Ars parva recenset. Tertia est de Arte dialectica & Organo Aristotelis.*

G. Gruppenbachius, Tübingen 1579

Not BM nor Bibl. Nat.

Prov. READ.

Contains a list of professors at Tübingen, of whom Planer was one.

528. [PLANTIN, CHRISTOPHE Ed.] *Icones stirpium seu plantarum tam exoticarum quam indigenarum in gratiam herbariae studiosorum in duas partes digestae. Cum septem linguarum indicibus, ad diversarum nationum usum.*

Off. Plantiniana—Vidua et Io. Moretus, Antwerp 1591

There is no indication that this is not *Ed.p.* except that the dedn. is by Christophe Plantin himself (to Severinus Gobelius, Physician to the King of Denmark, showing Plantin's extensive personal acquaintance) dated 1581. C. Ruelens and A. de Backer (*Annales Plantiniennes*, Paris 1866, p. 230) give the *Ed. p.* as *Plantarum seu Stirpium Icones*, 1581: their collation corresponds to No. 528 except that the latter has 27 fos. of index instead of '(16 pages?)'. *Tomus* 2 has sep. pagn. but the text is otherwise continuous.

Prov. LIDDEL.

This is a work of great interest on many counts. Its shape is most unusual (22.5 × 17 cm.) and its binding almost limp: but these characteristics are in keeping with the fact that it is probably the earliest fore-runner of Hooker's *Illustrations of the British Flora*. It consists entirely of figures reproduced from the blocks prepared for Plantin's numerous botanical publications. Also, it is based on one of the earliest attempts at botanical classification—in his dedn. Plantin excuses himself for preferring de l'Obel's to that of Dodoens. (See Vol. I, p. 196.) Clair (*op. cit.* in No. 402), p. 118 states, that 'apart from the wood-blocks which the printer had purchased from Purfoot the illustrations in the Plantin herbals were the work of one man'—Pietet van der Borcht.

529. PLATERUS, FELIX. . . . *de Febribus liber . . . genera, causas &*
curationes febrium tribus capitibus proponens. . . .

Heirs of Andr. Wechel, Frankfurt, 1597

Ed. p.

Prov. LIDDEL.

A piece of t.p. has been excised. *Historiae viginti* follow p. 285 (wrongly printed 885—as also 884) continuously.

An original attempt at a rational classification, based largely, however, on humoral theory.

530. PLATERUS, F. *De corporis humani structura et usu libri iii.* . . . *Qui*
libri cum operi practico recens ab eodem autore edito plurimum
inserviant, denuo sunt publicati.

Ludwig König, Basel 1603

[*Ed. p.* 1583.]

In the dedn. to the printer, Egenolph, Plater says that 'for more than thirty years, first in France, later in Germany, but most of them in our University of Basel, at which I made anatomy a familiar and customary exercise, I have put my hands to the dissection of more than fifty human bodies'. The comprehensive tables of topographical anatomy which alone form books i and ii support his claim. These books are closed by Froben's device (fo. 199*v*); and book iii, consisting entirely of figures with detailed legends, follows with sep. pagn. The adaptation of the figures of Vesalius is described on fo. 1*r*. For Plat(t)er see Vol. I, p. 235.

PLINY (=C. Plinius Secundus) See Nos. 58, 748.

531. POBLACION, IOANNES MARTINUS. *De usu astrolabi*
compendium. . . .

Io. Barbaeus for Jac. Gazellus, Paris 1546

Only edn.

Prov. 'C. C. Sherrington, Paris 1928'.

In his *Lectori* the printer mentions that he has already published an elementary 'teach-yourself' guide to the astrolabe in French; but since not everyone understands French, or by some Latin is preferred, he has been asked to produce a similar work in the latter language.

The text is followed by (1) *Procli Diadochi . . . de fabrica usuque astrolabi*, and (2) *Gregorae Nicephori Astrolabus*, both trans. by Georgius Valla. Cont. pagn. throughout.

Illustrated with clear diagrams (from another source?).

532. PONS, JACOBUS. *De nimis licentiosa ac liberaliore intempestivaque sanguinis missione, qua hodie plerique abutuntur, brevis tractatio. . . .*

Paulus Frellon et Abraham Cloquemin, Lyon 1596

Ed. p.

Prov. LIDDEL (?).

The dedn. to Henry IV (of France & Navarre) draws an ingenious comparison between the theme of the book and the 'heart and blood' of the kingdom, the government of which Henry had recently assumed.

An interesting reminder of the age-long habit of physicians to indulge in excessive use of a fashionable remedy—in this case bloodletting. In the absence of any index it is difficult to assess the merits of the book, which names Galen on most pages and vilifies Botallus (see Castiglioni p. 444). Page 5 gives a useful hint on current practice.

533. PONTANUS, IOANNES IOVIANUS. *De rebus coelestibus libri xiiii . . . tomus tertius.*

[No imprint, date or col. except 'Basel']

[Only edn. BM is 1519, but there was a 4-vol. edn. by Henric-petrus, Basel 1556.]

Prov. LIDDEL.

Though this is entitled *Tomus tertius* and begins at p. 1963, it actually starts with *De rebus coelestibus liber primus*. The *De rebus* ends at p. 2591 and is followed by: (1) A fragment—*De luna liber*—addressed to Aegidius Eremita, to which is added a note by P. Summontius, (2) A letter to P. S. Valla and Io. Ferrarius concerning a letter on Herodotus written by Pontano and trans. into Latin by Lorenzo Valla (1450), (3) A letter to Syncerus, (4) A commentary on the *Centiloquium* at that time attributed to Ptolemy (No. 544). (1) and (4) are named in the main title.

The *De Rebus Coelestibus* was the 'Bible' of renaissance astrology.

534. PORTA, GIOVANNI BATTISTA DELLA. *Magiae naturalis libri viginti. Ab ipso . . . authore ante biennium adaucti, nunc vero ab . . . mendis . . . repurgati in quibus scientiarum naturalium divitiae & deliciae demonstrantur.*

Andr. Wechel, Frankfurt 1591

[*Ed. p.* Naples 1558 *li.* iv; *Ed. p.* of *li.* xx, Naples 1589.]

Two copies of 1591 edn; also one each of 1607 and 1619.

One of the most popular works of 'natural magic' based on the transactions of della Porta's circle at Naples (disbanded by order).

It is of very unequal value, but the impulse to the formation of 'societies' was of course of the highest importance. See p. 55 for 'Barnacle geese'.

535. PORTA, G. B. DELLA. *Phytognomonica . . . octo libris contenta; in quibus nova facillimaque affertur methodus, qua plantarum, animalium, metallorum, rerum denique omnium ex prima extimae faciei inspectione quivis abditas vires assequatur. Accedunt ad haec confirmanda infinita . . . secreta. . . . Nunc primum ab . . . mendis . . . vindicata. . . .*

    (*a*) Io. Wechel and Petrus Fischerus, Frankfurt 1591.

    (*b*) N. Hoffman for I. Rhodius, Frankfurt 1608.

    [*Ed. p.* Naples 1588.]

Numerous ingenious illustrations of forms common to the three 'kingdoms'.

536. PORTA, G. B. DELLA. *Villae Io. . . . B . . . P . . . libri xii, in quibus maiori ex parte cum verus plantarum cultus certaque insitionis ars & prioribus seculis non visos producendi fructus via monstrantur; tum ad frugum vini ac fructuum multiplicationem experimenta propemodum infinita exhibentur.*

                             Heirs of Andr. Wechel, Frankfurt 1592

Earliest BM; but Imperial privilege dated Vienna 1582. *Prooemium* by the author on his own villa.

537. PORTUS, ANTONIUS. (*a*) *De peste li. iii. Quibus accedit quartus de Variolis & morbillis.*

                        Petrus Dehuchinus, Venice 1580

    Only edn.

    Prov. BUTE.

In his *Lectori* the printer bewails 'the calamity which has been visited on the fairest cities of Italy in recent years'.

    (*b*) [A greatly enlarged version of (*a*) by Dominicus Basa, Rome 1589.]

    Prov. 'Sum Guil. Charis' [?]

The *Lectori* contains a statement about a visit of Bernardo Giunta concerning the publication of the book.

The text discusses such questions as 'Contagium, quid sit; de contagio per contactum; de contagio ad distans' (See Vol. I, p. 264.)

POSTEL, GUILLAUME. See No. 644.

538. PRISCIANUS [*syn.* Octavius Horatianus]. (*a*) *Octavii Horatiani Rerum Medicarum lib. quatuor:—I. Logicus, de curationibus omnium ferme morborum corporis humani, ad Euporistem. II. De acutis & chronicis passionibus, ad eundem. III. Gynecia, de mulierum accidentibus & curis eorundem, ad Victoriam. IIII. De physica scientia,* *experimentorum liber, ad Eusebium filium. Per Heremannum Comitem a Neuenahr, integro candori nuper restitutus autor.*

Io. Schott, Strassburg 1532

Ed. p.

Prov. 1. 'Cracovie 1542 And. Pelzerii . . .'.
     2. Gordon.

(*b*) *Theodori Prisciani Archiatri ad Timotheum fratrem.*
[Contents: '*Phaenomenon Euporiston'; Logicus; Gynaecea ad Salvinam*] *Opus nunc primum aeditum.*

Off. Frobeniana, Basel 1532. (Col. H. Frobenius and Nic. Episcopius.)
Ed. p.

Prov. READ; also a second copy.

The editor, Sigismundus Gelenius, states that 'Theodorus hic Priscianus sane autor & quem ex stili puritate ac elegantia ante linguam latinam collapsam apparet hoc universae medicae praxeos compendium prodidisse . . .'. I, II, III of (*a*) correspond to the '*li.* iii' of (*b*), IIII is additional. Both (*a*) and (*b*) contain the dedn. of the original work, but whereas (*a*) begins '*Amice charissime . . .*', (*b*) has '*Frater carissime . . .*'. The text of (*a*) begins: 'Si medicina minus eruditi . . .'. of (*b*): 'Si medicinam minus eruditi & rustici homines naturae tantum conscientia . . .'.

The author was Theodorus Priscianus (c. 400 A.D.). The false name, Octavius Horatianus, apparently derives from a MS of XV-2.

539. PROCLUS DIADOCHUS. . . . *In primum Euclidis Elementorum librum commentariorum ad universam mathematicam disciplinam principium eruditionis tradentium li. iiii.*

Gratiosus Perchacinus, Padua 1560

First Latin edn. BM.
Prov. (i) 'Andreas Leochaeus' [deleted]?, (ii) 'Gulielmus Robertsone'.

Ed. by Franciscus Barocius, who writes a long account of how, dissatisfied with the Basel edn., he had been given an 'exemplar vetustissimum' in Crete.

540. PROCLUS DIADOCHUS ΣΦΑΙΡΑ. *Thoma Linacro Britanno interprete.*

Io. Hervagius, Basel 1549

[*Ed. p.* of Linacre's trn. 1500, BM. Hain 13387 undated.]
See also Nos. 531, 542(c), 720.

541. PSELLUS, MICHAELES. . . . τῶν περι ἀριθμητικῆς συνοψις
. . . *Arithmetices compendium recens e Graeco conversum.*

Io. Lodovicus Tiletanus, Paris 1545

[*Ed. p.* Greek text, Paris 1538. No. 541 seems to be the arithmetical secn. of *Ed. p.* 1532. *Sapientissimi Pselli opus dilucidum in quattuor Mathematicas Disciplinas, arithmeticam geometriam musicam & astronomiam* . . .].
Smith p. 168.

542. PTOLEMY, CLAUDIUS. *Almagest.*

(*a*) . . . *Almagestum seu magnae constructionis mathematicae opus plane divinum Latina donatum lingua ab Georgio Trapezuntio.* . . . *Per Lucam Gauricum* . . . *recognitum anno salutis* MDXXVIII *labente.*

Juntae, Venice 1528

Earliest Latin edn. BM.

Prov. Bound (modern) with works autographed by Liddel. The glosses in this work are, with one or two possible exceptions, not in his hand.

Ded. to Dominicus Palavicinus; also printed is the Preface of George of Trebizond's trn. ded. by George's son, Andreas, to Sixtus IV.

(*b*) . . . *ΜΕΓΑΛΗΣ ΣΥΝΤΑΞΕΩΣ ΒΙΒΛ. ιγ. ΘΕΩΝΟΣ ΑΛΕΞΑΝΔΡΕΩΣ ΕΙΣ ΤΑ ΑΥΤΑ ΥΠΟΜΝΗΜΑΤΩΝ ΒΙΒΛΙΑ ια—*

Magnae constructionis, id est perfectae coelestium motuum pertractationis, li. xiii. Theonis Alexandrini in eosdem commentariorum li. xi.

Io. Walderus, Basel 1538

Earliest Greek text [2 copies.]

The *Almagest* is ed. by Simon Grynaeus, Theon's *Commentary* by J. Camerarius. Ded. to Henry VIII.

(*c*) . . . *Omnia quae extant opera praeter Geographiam* . . . *castigata* . . . *ab E. O. Schreckenfuchsio & ab eodem isagogica in Almagestum praefatione.*

Henricus Petrus (Col. Basel 1551)

Only edn. but based on same printer's edn. of 1541. (BM copy imperfect) [2 copies.]

George of Trebizond's trans. and dedn. of *Almagest.* Io. Camerarius's trans. of *Quadripartitum.* Pontanus's trans. of *Centiloquium.* Nicolaus Leonicenus's trans. of *Inerrantium stellarum seu fixarum significationes.* Followed by Proclus, *Hypotyposes astronomicarum positionum* . . . *Georgio Valla interprete.* Annotations of first three books of *Almagest* by Schreckenfuchs; of remainder by Lucas Gauricus.

543. PTOLEMY, C. *Cosmographia* (*syn. Geographia*).

(For a tabular review of the edns. of this work, see Tooley, R. V. *Maps and Map-makers*, 2nd edn., London 1952, p. 6 f.)

(*a*) [No t.p. in orig.—BM] Leonardus Hol, Ulm 1482.

'*Incipit* [Beatissimo Patri Paulo Secundo Pontifici Maximo, Donis [*sic*] Nicolaus Germanus] Non me fugit beatissime pater cumque summo ingenio exquisitaque doctrina Ptolemeus cosmographus pinxisse . . .' [All wanting between brackets except 'Nic.'] The initial of 'Non' displays a tonsured man (presumably Iacobus Angelius, the first translator of the work into Latin) kneeling before the Pope with a book in his hands.

Col. (on *verso* of a fo. inserted *after* the 'duodecima et ultima Asie tabula') *Claudii Ptolemei viri Alexandrini Cosmographie octavus et ultimus liber explicit. Opus donni* [*sic*] *Nicolai Germani secundum Ptolomeum finit. Anno* MccccLxxxii *Augusti vero kalendas xvii imprssum* [*sic*] *Ulme per ingeniosum virum Leonardum Hol prefati oppidi civis* [*sic*].

The order of the maps is confused and there are only twenty-one of two folios each, and one of one folio, of the complete set of thirty-two. This edn. is of special importance, being the first (*a*) to be printed in Germany [*Ed. p.* Vicenza, 1475; earliest, with maps, Bologna (1477 ?)] (*b*) to have wood-cut maps, (*c*) to include some information not available to Ptolemy—five of the maps were 'modern', (*d*) to include the engraver's name—Io. Schnitzer de Armssheim—on a map. The trans. (1410) of Angelius was edited by Nicolaus Germanus (see *c* below).

The *Cosmographia*, unlike the *Almagest* and *Quadripartitum* was unknown to the *Latin* medieval scholars (though well known to the Arabs), until Angelius's trans. of 1409. The original work contained no maps, but only the astronomically determined latitudes and longitudes of important geographical features. Byzantine geographers almost certainly added to the text, and Maximos Planudes provided a set of maps. The next important addition before Nicolaus Germanus's complete revision was the map of Northern lands prepared by Claudius Clavus (a Dane) for Cardinal Fillastre, who added it to his MS.

Prov. HECTOR BOECE.

(*b*) *C— P— Alexandrini Liber Geographiae cum tabulis et universali figura et cum additione locorum quae a recentioribus reperta sunt diligenti cura emendatus et impressus.*

(Col. at end of *Liber Octavus* and before maps: Iac. Pentius de Leucho, Venice 1511).

Prov. WILSON.

Dedn. and annotations by Bernardus Sylvanus, explaining the corrections made in this edn. including that of the easterly 'bend' in the axis of Scotland. Its twenty-eight woodcut maps include the first (heart-shaped) world map to show part of the American continental mass.

    (*c*) *C— P— Viri A— Geographiae opus novissima traductione e Graecorum architypis castigatissime pressum.* . . .
    (Col. Io. Schottus, Strassburg 1513.) Followed by *Locorum admirabilium mundi descriptio* sigs. A-C5. No sep. t.p. or col.; mentioned on main t.p. as 'adnexo'.
    Prov. Gordon.

Dedn. by Io. Franciscus Picus de Mirandola (The Younger, since he refers to the *Contra Astrologos* of his 'patruus') to Jacobus Etzler (dated 1508), and a second by Etzler and Georgius Ubelin to Emp. Maximilian, dated Strassburg 1513.

This edn. contains besides a text 'sine ulla corruptione' a good deal of 'modernisation', e.g. the first map specifically devoted to *Nova Terra* (i.e. West Indies). The complete set included twenty 'modern' maps prepared under the supervision of Martin Waldseemüller.

    (*d*) *C— P— Geographicae enarrationis libri octo Bilibaldo Pirckeymhero interprete.*

        Io. Grieningerus and Io. Koberger, Strassburg 1525
    Prov. Johnston, etc.

Dedn. by Pirckheimer (see Vol. I, p. 25) to the Bishop of Brixen, in which he justifies the new edn. on the grounds that 'Germanus vero, tametsi in Mathesi admodum excelluerit, in Graecis tamen adeo aliquando hallucinatus est', dated Nürnberg 1524.

Fo. 82r closes Book VIII; thereafter follow 'certain annotations of Johannes de Monte Regio . . . on the errors which Iacobus Angelus made in his trans. of Ptolemy'. Above the Col. is an *interpres*, and above that a note on changes in the annotations as compared with other printings. Index and maps follow. After '*Asiae* XII' (Ceylon) of Ptolemy's original set, there follow others, including 'Oceani Occidentalis seu Terre Nove' with long description of Christopher Columbus and his undertaking, also a pictorial map. There is no further Col. The 'eastern bend' of Scotland is rectified in a 'tabula nova', which is preceded by an account of the inhabitants and of a tree 'sed non repertam, cuius caduca folia in subiectum amnem prolapsa in volucres verti dicerentur'. According to Tooley, there should be twenty 'modern' maps, corresponding to those in (*b*) above: several of these are wanting. Bagrow, *Gesch. der Kartographie*, gives the number as twenty-four.

    (*e*) *C— P— A— philosophi et mathematici praestantissimi libri VIII de Geographia e Graeco denuo traducti. Nominibus Graecis e regione appositis . . . Ioannis Noviomagi opera. Nunc primum . . . editi.*

                Io. Ruremundanus, Cologne 1540

The data for the maps is set out, but no maps are included. The special feature is the inclusion of all Greek proper names with their Latin equivalents.

(f) *Tabulae Geographicae Cl. Ptolemaei ad mentem autoris restitutae & emendatae per Gerardum Mercatorem illustriss. Ducis Cliviae etc. Cosmographium.*

(Col. Godefridius Kempensis, Cologne 1578.)

Dedn. by Mercator from Duisburg 1578 shows *Ed. p.*

Consists of a series of coloured maps with annotations thereon and also *in recto* of the double sheet. In some cases a note is added 'Historica haec, praeter ea quae tabulae inscripsi, annotat Ptolemaeus' ; e.g. 'In Maniolis insulis tanta est vis magnetum ut navigia quae clavos ferreos habent, detineantur, ideo ligneis compaginant'. ('In the Isles of Maniola [Gulf of Ganges, S 2°] the magnetic force is so great that ships with iron nails are held back, for which reason wooden ones are used to hold them together' (*Asiae Tabula* XI, sig. Bb).

544. PTOLEMY, C. [Quadripartitum & Centiloquium]. . . . *de Praedictionibus astronomicis, cui titulum fecerunt Quadripartitum Grece & Latine, libri iiii, Philippo Melanthone interprete. Eiusdem fructus librorum suorum, sive Centum dicta, ex conversione Ioviani Pontani.*

Io. Oporinus, Basel (Col. 1593)

Ed. p. of this version. The Greek text is signed 'Joachimus' (*sc.* 'Camerarius'), has sep. t.p. (Norico 1535) and pagn., and had been separately published by Petreius, Nürnberg 1535. Earliest printed Latin versions, Ratdolt, Venice 1484, Hain 13544.

Prov. LIDDEL (also 'ex libris Ioachimius Crollius'?)

Dedn. by Melanchthon to Erasmus Ebnerus, Senator of Nürnberg, is worthy of special note in revealing the former's attitude to judicial astrology, e.g. 'Et voluit Deus prospici inclinationes in temperamentis, ut maiore consilio regantur actiones' ('God has willed the 'inclinations in the temperaments' to be observed in order that actions may be regulated with greater circumspection'), also, '. . . in magna humiditate qualis fuit anno 1524, propter coniunctionem multorum planetarum in Piscibus . . .' (reference to 'the great wetness of the year 1524 on account of the conjunction of several planets in the [zodiacal sign of the] Fishes').

See also Nos. 752, 558, 85.

QUERCETANUS, J. *See* DU CHESNE, J.

545. RAMELLI, AGOSTINO. *Le diverse et artificose machine del Capitano A— R— dal ponte della Tresia, ingegniero del christianissimo Re di Francia et di Pollonia. Nellequali si contengono varii et industriosi movimenti degni diagrandissima speculatione, per caverne beneficio infinito in ogni sorte d'operatione.* . . .

Paris (in the author's house) 1588

Ed. *p.*

Preface: *De l'excellence des mathematiques, ou il est demontré combien elles sont necessaires pour acquerir tous les arts liberaux.*

Text alternately in Italian and French. Superb cuts.

Notable for displaying the use of levers, 'lantern wheels', sprockets, pulleys, screws, syphons, etc. The extended use of the screw for breaking bolts and gratings is made possible (?) by the replacement of wood by metal. The question is forced upon one whether friction and the weight of the moved parts would not have rendered many of the projected machines useless. It is not clear what motive power was to be applied to the ornamental syphon fountains. The applications are mainly to decorative trifles, pumps for water supply and drainage, and siege works. (See Vol, 1 Plate 1.)

546. RAMUS, PETRUS (*syn.* La Ramée, Pierre). *Scholarum mathematicarum libri unus et triginta.*

(*a*) Eusebius, Basel 1569.
Prov. LIDDEL.

(*b*) *A Lazaro Schonero recogniti & emendati.*
Heirs of Andr. Wechel, Frankfurt 1599.
Prov. 'Gulielmus Leslaeus'.
Earliest BM 1599.

The dedn. to Catharine de'Medici refers to the patronage displayed by the Medici in Florence. There is also a *Diploma* by Charles IX, dated 1566, *de Regiorum Professorum institutione* by his grandfather, Francis I.

547. RAMUS, P. *Arithmeticae libri duo, Geometriae libri septem et viginti.*

Eusebius Episcopius and heirs of his brother Nicholas, Basel 1580
Earliest BM of *Arithmetica*, Paris 1581. No Latin version of either cited. Mention is made of a *Prooemium Mathematicum*, Paris 1567.

Prov. LIDDEL.

Sep. pagn. and secn. headings for *Geometria*.

The long *Lectori* contains numerous refs. to individuals e.g. Oronce Fine, the Optics of al-Hazen and Witelo, Elizabeth of England and James VI 'Scotiae rectorem principem', and Nonus (i.e. Pedro Nuñez) who is called 'Archimedeus vir'.

Woodcuts in *Geometriae* show use of cross-staff for surveying.

548. RAMUS, P. [ed., with additions, by Lazarus Schonerus]. . . .
Arithmetices libri duo, et Algebrae totidem . . . Eiusdem Schoneri libri duo: alter, de Numeris Figuratis; alter, de Logistica Sexagenaria.
Heirs of Andr. Wechel, Frankfurt, 1592

Two copies, same imprint and date, but clearly of two different settings. Also the pagination of one differs after p. 177 owing to several *verso*'s being unnumbered; and the inset *Canon sexagesimorum* is differently placed.

[*Ed. p.* in this form Frankfurt 1586; of *Arithmetic*, Paris 1555—Smith pp. 331–3.]

'Algebra' is described as 'part of arithmetic', and is said to mean the art or doctrine of an outstanding 'Syrian' who, according to tradition, sent his book to Alexander the Great (!) calling it 'Almucabala' that is, the book of 'hidden things' (!). The form is syncopated, the exponent being represented by l(atus), q(uadratus), bq(biquadratus), etc.

549. RANGONI, TOMMASO (*syn.* Thoma Philologus Ravennas).
. . . de Vita hominis ultra cxx annos protrahenda.
Andr. Arrivabenus, Venice 1562

[*Ed. p.* Venice 1553. Also Venice 1560, but not 1562, BM.]
Prov. 'Sum. Guil. Charis' (?)
A compilation.

550. RANTZOVIUS, HENRICUS (*syn.* Rantzau). *De conservanda valetudine liber in privatum liberorum suorum usum ab ipso conscriptus ac editus a Dethleno Silvio. . . . In quo de diaeta, itinere, annis charactericis et antidotis praestantissimis . . . praecepta continentur. Altera editio auctior & emendatior.*

(*a*) Chris. Plantin, Antwerp 1580.
[*Ed. p.* 1576].
Prov. LIDDEL.

(*b*) *. . . Tertia editio . . . Seorsim accessit Guilielmi Gratarolae de literatorum & eorum qui magistratum gerunt conservanda valetudine, liber.*
. . . Io. Wechel and P. Vischer, Frankfurt 1591

Prov. FORDYCE (previously, 'Joan. Fordyce' and MS in French).

The dedn. by Silvius to the sons of Rantzovius describes his discovery in the latter's library of an unpublished book of collections from medical authors, dated 'Segebergae 1573'. Preface by Rantzovius to his sons.

551. RANTZOVIUS, H. *Editio duorum librorum Macri de Virtutibus Herbarum, de quibusdam animalium partibus, ac terrae speciebus,*

*itemque medicamentis totius corporis humani, iam recenter ex biblio-
theca sua Bredenbergensi depromptorum, quorum prior ante hac non
tam emendate extitit, posterior antea typis nunquam fuit expressus
aut . . . editus. Accessit incerti auctoris speculum medicorum, rudi ac
inculto quidem stylo conscriptum, sed propter res tamen ex eadem
bibliotheca . . . editum.*

(Col. Jacob Wolf, Hamburg 1590)

Only edn.

In his *Lectori* Rantzovius says that, having leisure to look round his
library, he decided to publish this work from the MS. Of the three
'Macers' then recognised he believed this was the one alleged to be a
contemporary of Pliny.

Incipit: 'Herbarum varias dicturus carmine vires. . . .' (cf. No. 417).
The text is followed by lists of Greek, Latin and German equivalents,
thereafter an alphabetic summary of virtues. Sigs A–L: – 2 books of
'Macer'; M–Z unpaginated, edges mutilated. *Speculum Medicorum* at
end, followed by verses by Rantzovius.

552. RANTZOVIUS, H. . . . *Diarium sive calendarium Romanum
Oeconomicum, Ecclesiasticum, Astronomicum, et fere Perpetuum.
Opus astronomicis, medicis, patribusfamilias, militibus, viatoribus
utilissimum.*

Christ. Axinus, Wittenberg, 1593

Ed. p.

Two copies, one lacking t.p. and some other fos.

553. RAYNALDE, THOMAS *The Byrth of Mankynde otherwyse named
the woman's booke. Newly set foorth and augmented. . . . The
simplest mydwyfe which can reade may both understande. . . .*

London 1560, No Col.

STC. 21153 gives *Ed. p.* 1540 'newly translated out of Laten
(by R. Jones)' and states that No. 553 is of the 4th Engl. edn. See
Castiglioni p. 482 and p. 1015. It is not clear what part Raynalde
played—probably none in *Ed. p.*, but as a reviser for later edns.
BM cites no edn. earlier than 1564.

A book for midwives, with numerous cuts (somewhat imaginative)
of the foetus *in utero*—see especially fo.: xliiiir and lxiiv. (See No. 597.)

554. RECORD(E), ROBERT. *The Castle of Knowledge.*

Reginalde Wolfe, London 1556

Ed. p. STC 20796.

Fine ornamental t.p. has description of contents too long to quote,
but including 'the explication of the sphere both celestial and materiall'.

Dedn. (1) 'Marye . . . Queen of England, Spain, both Siciles, Fraunce, Jerusalem and Irelande, etc.'

(2) Cardinal Pole.

Written in dialogue form, of which Recorde made frequent skilful use. See Johnson for detailed description of this highly important work also Vol. I, p. 117.

555. RECORD(E), R. *The Whetstone of Witte which is the seconde parte of Arithmetike: containyng the xtraction* [sic] *of rootes: the cossike practise with the rule of equation and the woorkes of surde numbers.*

Jhon Kyngstone, London 1557

Only edn. STC 20820.

Dedn. to 'The . . . companie of venturers into Moscovia. . . .'

The phrase 'second parte' refers to the fact that Recorde had already published the 'first part' under the title *The Grounde of Artes* (London, 1542. STC 20798), which ran through twenty-eight edns.— the last being late in the 17th cent. The absence of any copy at Aberdeen is noteworthy. He wrote also a geometry, *The Pathway to Knowledge*, (London 1551, STC 20812).

See Smith pp. 286 and 213 *seq.* also Vol. I, p. 91 and Plate II.

556. REGIOMONTANUS, IOANNES [*syn.* Io. de Monte Regio and (BM) Muller, Io.] *In laudem operis calendarii a Ioanne de Monteregio . . . editi Jacobi Sentini Ricinensis carmina.*

Off. Petrus Liechtenstein Venice 1514

Bound with *Almanach perpetuum* of Abraham Zacuto (No. 742) and apparently usually published therewith; the BM copy (530 f. 16) of Zacuto's work is however separately bound in modern cover.

557. REGIOMONTANUS, IO. . . . *Io.* . . . *de* . . . *R* . . . *M* . . . *de Triangulis omnimodis libri quinque. Quibus explicantur res necessariae cognitu, volentibus ad scientiarum astronomicarum perfectionem devenire: quae cum nusquam alibi hoc tempore expositae habeantur frustra sine harum instructione ad illam quisquam aspirarit. Accesserunt . . . D. Nicolai Cusani de Quadratura circuli deque recti ac curvi commensuratione: itemque Io. de M . . . R . . . eadem de re ἐλεγκυνα hactenus a nemine publicata. Omnia recens . . . edita.*

Io. Petreius, Nürnberg 1533

Ed. p.

Prov. LIDDEL.

Dedn. by Io. Schoener, 1533. The *Lectoribus* opens with a note— presumably by Schoener—followed by two pages by Regiomontanus concerning the parts played by 'Georgius' (*sc.* Peurbach), (Cardinal) 'Bessarionus' and 'Bilibaldus Pircamerus' (Pirckheimer).

This copy contains only the *De Triangulis*, which is effectively the first book devoted to trigonometry as a separate discipline. (See Vol. I, p. 96.)

558. REGIOMONTANUS, IO. *In Ptolemaei Magnam Compositionem quam Almagestum vocant libri tredecim conscripti a Io. . . . R . . . in quibus universa doctrina de coelestibus motibus, magnitudinibus, eclipsibus & in epitomen redacta, proponitur.*

Io. Montanus and Ulricus Neuberus, Nürnberg 1560

Ed. p. in this form.   Based on *Ed. p.* Venice 1496 (Hain 13806).

Dedn. by Erasmus Flock, who, having noted the astrological portents, praises Peurbach and Regiomontanus for having restored the *Almagest* from the 'depraved and mutilated state' into which it had fallen.

Dedn. of *original* to Cardinal Bessarion by Regiomontanus, 'pupil of Georg. Peurbach'.   Having complained of the degeneracy of the times, in which men seek wealth rather than the advantage 'bonarum artium' and the difficulty of study is increased by the 'scabrositatem librorum qui ex peregrinis linguis in Latinum conversi' he alludes in a touching manner to the dying 'George's' last request: 'Vale, inquit, mi Ioanne, vale, & si quid apud te pii praeceptoris memoria poterit, opus Ptolemaei, quod ego imperfectum relinquo, absolve; hoc tibi ex testamento lego, ut etiam vita defunctus, parte tamen mei meliori superstite, Bessarionis nostri optimi ac dignissimi Principis desiderio satisfaciam'. ('Farewell, my John'; he said, 'farewell; and if the memory of a worthy teacher carries any weight with you, finish the work of Ptolemy which I leave incomplete.   I leave this to you as a legacy in order that, having departed this life (the better part of me nevertheless remaining) I may satisfy the desire of our best and most worthy lord, Bessarion.'

This is all the more saddening in that the pupil followed his master to the grave within a few years, neither of them attaining the age of forty.

Each secn. is followed by a long commentary.   ThK give 15th cent. MS *incipit*: 'Recte profecto meo iuditio nobiliores. . . .' A MS of Gerard of Cremona's trn. of the *Almagest* runs: 'Bonum scire fuit quod sapientibus non. . . .'

559. REGIOMONTANUS, IO. . . . *Tabulae directionum profectionumque non tam astrologiae iudiciariae, quam tabulis instrumentisque innumeris fabricandis utiles . . . Denuo nunc editae & . . . emendatae. Eiusdem*

R . . . Tabula sinuum . . . universam sphaericorum triangulorum scientiam complectens. Accesserunt his Tabulae ascensionum obliquarum . . . per Erasmum Reinholdum . . . supputatae.
Matthaeus Welack, Wittenberg 1584
Not BM in this form. Hain 13802 n.d. ThK, 186 refer to Ratdolt, Venice 1490.
Prov. LIDDEL.
Dedn. by Melanchthon. *Lectori* by Andreas Schatus, claiming its importance for chronology. Orig. dedn. by Regiomontanus refers to the difficulties involved in connection with judicial astrology; also to the *Studium Generale* at Vienna.

The text consists of a number of astronomical problems, many worked out. Proportional parts are used in the tables—see sig. B *v exemplum.*

See also Nos. 513, 669, 747.

560. REINHOLD, ERASMUS (*syn.* Rheinholt, etc.) *Ptolemaei Mathematicae Constructionis liber primus Graece & Latine editus. Additae explicationes aliquot locorum ab E . . . R . . . .*
Io. Lufft, Wittenberg 1549
Ed. p.
From dedn. 'And since younger students are not yet accustomed to reading Greek I have added a Latin translation—and I have illustrated some of the more abstruse parts with scholia to help those who are learning them'. Dated 'Easter, 3058 years from the first Easter in which the Israelites celebrated their exodus from Egypt'. These points illustrate the appearance of the 'modern' type of textbook and the growing custom of showing ('off'?) a knowledge of Old Testament scholarship and history.

On p. 97*v* a fig. of a 'kind of the first instrument described by Proclus'.

561. REINHOLD, E. *Prutenicae tabulae coelestium motuum.*
(*a*) Ulrich Morhard, Tübingen 1551.
Ed. p.
Prov. LIDDEL.

(*b*) Vidua Ulrici Morhardi, Tübingen 1562.
So far as I have been able to compare the two edns. they differ only in typographical details.
Contents: 'Diploma Caesareium concessum Erasmo Rheinholt [*sic*] Salveldensi' (from Ferdinand I), containing a list of Reinhold's works on astronomy, mathematics, and optics: dated 1549.

The dedn. to Albert of Brandenburg recalls (somewhat inaccurately) how Alphonso 'had restored to light these almost extinct arts'. Author's Preface. Half-title: *Logistice scrupulorum astronomicorum* with its own preface and text to fo. 68r. Followed by half-title, *Initium canonum Prutenicorum*, and new pagn. for tables.

562. REINHOLD, E. *Gründlicher und warer Bericht von Feldmessen sampt allem was dem anhengig . . . Desgleichen vom Marscheiden kurtzer und gründlicher Unterricht. . . .*

(Col. Georgius Bawman, Erfurt) 1574

Ed. p. Bibl. Nat. Not BM.

Dedn. wrongly dated 1547 (it is signed by Reinhold with a ref. to the *Prutenic Tables*, whose *Ed. p.* was 1552).

See also No. 513.

563. REISCH, GREGORIUS. *Margarita philosophica nova.*

Grüninger, Strassburg 1512

T.p. wanting. No col. but a 'half-col.' after the index is consistent with the above imprint and date.

Prov. Lib. II, Tr. viii, last fo. *v* signed 'Robertus Elphinstone' (brother of Bishop William Elphinstone, first Chancellor of the University of Aberdeen).

The first printed 'omnibus volume' of texts for the *Trivium* and *Quadrivium*. For a concise but representative account see A. R. Hall *The Scientific Revolution*, London 1953, Ch. I.

Condensed list of contents (not actual titles): Rhetoric, Logic, *Epistolandi modus*; Arithmetic, Geometry and Music—all both 'speculative' and 'practical'; *de principiis Astronomiae; de Astrologia; de principiis (& origine) rerum naturalium; de potentiis anime, vegetative, sensitive et intellective; de principiis Philosophiae Moralis. Tractatus de compositione astrolabe Messehalath.*

Miscellaneous data and fine cuts of Astrolabe, torquetum, and rules of perspective, etc.

564. RENODAEUS, IOANNES. (*a*) *Dispensatorium medicum continens institutionum pharmaceuticarum lib. v. De materia medica lib. iii . . . .*

Impensis Theo. Schönwetter prelo Richteriano, Frankfurt 1609

(*b*) As above but . . . *accessit Iosephi Quercetani . . . Pharmacopoea Dogmaticorum restituta. Item Nicolai Epiphanii . . . Empirica nunc e manuscripto luci data.*

Paulus Jacobus, for Schönwetter, Frankfurt 1615

Earliest BM, Hanover 1631.

Prov: (*a*) Thomas Read, brother of Alexander.
(*b*) Alexander Read—scored, glossed and indexed in MS.
Dedn. 'Vester Renodaeus' to '. . . Parisiensium medicorum ordinem contra Paracelsum'—a useful though wordy document. The work of Quercetanus (No. 212) has sep. t.p., pagn., and dedn. Secn. heading introduces 'Epiphanii Empyrica 1561, 21 Aprilis' with the added note that he was an empiric physician whom Gesner, in his youth, found wandering in Greece, and of whose book, which is here presented, he described certain parts in his own so-called *Euonymus* (No. 299).

In the margin of p. 536 in (*a*) on which ref. is made to the mountains where the gemstone, Smaragda, is found 'quos tamen frequentes habitant Gryphae aliter ferocissimae', Thomas Read has written 'fabula'.

565. REUCHLIN, IOANNES. *De Arte Cabalistica libri tres iam denuo adcurate revisi*

(Col. Io. Socerius, The Hague 1530.)

Ed. p. Speyer 1497 (not in Hain). The edn. of 1517 was the first to have the dedn. given below.

Prov. 'Alexander Arbuthnot' on t.p. inscribed over another auto. written in red ink.

The dedn. to Leo. X gives a good example of the contemporary appreciation of the 'Renaissance' by one of the foremost scholars of the day. (See Vol. I, p. 3.)

The work is a *locus classicus* for the attempt to establish a correspondence between Greek (especially Neo-platonist and Pythagorean) and Cabalist philosophy and number-mysticism.

566. REUSSNER, HIERONYMUS. . . . *Liber de Scorbuto. In quo praeter genuinam variorum novorum ac insolentium morborum prosapiam, solertemque lienis e priscorum mystarum pandectis erutam liturgiam exaugurata Paracelsitarum hieroglyphia, nec non exauctorata Parachymicorum chrysurgia enucleatur. Quomodo iuxta Hippocratis Galenique therapeusin selecta Sparigicorum magisteria tuto administrari possint.*

Off. Paltheniana, Frankfurt 1600

Only edn. BM.

A wordy diffuse outpouring burdened with innumerable classical and mythological allusions; but contains points of interest, e.g. (etymological) 'Dani, Hollandi, Saxones & Germani maritimi vocant Scorbuc, Scorbuk, Scwerbyck . . .', etc. (p. 120). Also a ref. (p. 436) to the outstanding value of *Cochlearia* (still called 'Scurvy Grass'—it is of course a cruciferous herb).

215

567. REYMERS, NICOLAUS. *Geodesia Ranzoviana. Landt rechnen und Feldmessen sampt Messen allerhand grösse. Alles auff eine lichte behende und vormals unbekandte newe Art künstlich gründlich und deutlich beschrieben zu . . . Herrn Heinrichen Rantzoven Herrn Iohans seligen Sohne.*

(Col. Georg. Defner, Leipzig 1583.)

Contains an arithmetical and geometrical introduction, but consists mainly of a text of practical surveying. Last section on '*Irrmessen*' (errors of measurement). See also Appendix B.

RHAZES See MUH AMMAD IBN ZAKARYAH

568. RHETICUS, GEORGIUS IOACHIMUS [under 'Joachimus' BM]. *Opus Palatinum de Triangulis.* . . .

(Matthaeus Harnisius, Neustadt 1596—in col. following Otho's text but before his *Meteoroscopium*).

Ed. p. in this form.

The book consists of several parts with sep. t.p.'s and (in part) sep. pagn., though there is no indication of this on t.p. or elsewhere. The works by Rheticus consist of a purely geometrical introduction followed by 15-figure tables of sines by $\frac{1}{4}°$, etc. The much larger work by Valentinus Otho, which follows, consists of a text on non-right angled spherical and plane triangles, followed by 10-figure tables of sines to 10″ intervals, etc.

The dedn. by Otho (who completed the work carried on for years by his teacher) gives a sketch of the history of mathematics in Germany —Frederic III's patronage of the University of Vienna, Frederic the Wise's patronage of Wittenberg, and the personal studies and instruments of Wilhelm, Landgrave of Hesse. His *Lectori* urges the importance of the 'doctrine of triangles' to astronomy, giving a brief history of the idea from the time of Hipparchos. (See Vol. I, p. 95.)

569. RICIUS, AUGUSTINUS. *De motu octavae sphaerae . . . ubi tam antiquorum quam iuniorum errores luce clarius demonstrantur. In quo & quam plurima Platonicorum & antiquae magiae (quam Cabalam Hebrei dicunt) dogmata videre licet intellectu suavissima. Eiusdem de astronomiae autoribus epistola.*

Simon de Collines, Paris 1521

Only edn.

Dedn. by Oronce Fine 'ex Regali Collegio Navarrae 1521'; a second dedn. and preface by the author. Note by printer at the end, explaining the omission of the *Epistola*.

A 'contribution' to the 'solution' of this three centuries old (imaginary) problem. (See Vol. I, p. 114.)

570. RIDLEY, MARKE. *A short treatise of magneticall bodies and motions by M— R— latly* [sic] *physition to the Emperour of Russia & one of ye eight principals or Elects of the Colledge of Physitions in London.*

Nich. Okes, London 1613

Only edn. STC 21075.

Finely engraved t.p. of relative sizes of planets, with instruments referred to in *To the Reader*. The *Preface Magneticall* contains useful historical refs. This work has been somewhat neglected, being overshadowed by its greater predecessor, Gilbert's *De Magnete* (No. 307).

RINCIO, CESARE. See No. 755.

571. RIOLAN, JEAN (The Elder). *Methodus medendi, tam generalis, quam particularis.*

Hadrianus Perier, Off. Plantiniana, Paris 1598

BM has only the 'special' section.

Prov. CARGILL, and later donated to King's College by James Morison. Also a second copy.

The 'special' (*particularis*) secn. has sep. t.p. ('to students of the true medicine') and pagn.

Dedn. to the Senate of the University of Paris. A document in the controversy concerning '. . . quidam chymista, qui suo stibio morbos deploratos suscipit sanandos' ('a certain chemist who undertakes the cure, with his antimony, of diseases given up as incurable'), and ending with the objurgation 'ut vestrum nomen sit terrori impostoribus qui . . . faciunt(que) per mortes experimenta' ('let your name be a terror to those impostors who . . . make experiments by sacrificing the lives of others'). (See Vol. I, p. 256.)

572. [RIOLAN, J. (the elder).] *Apologia pro Hippocratis et Galeni medicina adversus Quercetani librum de Priscorum Philosophorum verae medicinae materia, praeparationis modo atque in curandis morbis praestantia. Accessit censura Scholae Parisiensis.*

Imprint as in No. 571 (1603).

Not BM.

The work is anonymous, but concludes with the confirmation by the Faculty of the condemnation demanded by the 'censors' of whom the elder Riolan was one.

573. RIOLAN, J. (the elder). *Ad Libavimaniam . . . Responsio pro censura Scholae Parisiensis contra Alchymiam lata.*

Imprint as in No. 571, but 'Adrianus'. Paris 1606.

Only edn.

217

Prov. (i) 'Thomas Cantuaren', (ii) 'Lumley', (iii) Thomas Read. These names actually occur on t.p. of the first work in the volume (No. 708), but No. 573 is glossed almost certainly by Read. The 'autographs' of Archibishop Cranmer and Lord Lumley have been compared with those in BM 719 k 3 (1) and with the copies in J. Jayne and F. R. Johnson, *The Lumley Library* . . . BM, 1956; they correspond to those regarded by these authors as having been written by amanuenses.

The purpose of the tract is revealed by the opening words: 'In order that it may be understood how justly the alchemy of Quercitanus [No. 212] has been condemned by the decree of the Faculty of Paris, and so that the justice (*aequitas*) of our cause may be clearer (*evidentius*). . . .'

On pp. 25-6 occurs the record of the famous ban: 'Schola Parisiensis iam quadraginta plus minus annis damnavit usum Antimonii & Praecipitate ('The School [of Medicine] of Paris has now for forty years, more or less, condemned the use of antimony and [mercury] precipitate').

The *Conclusio* (last fo. *verso*) signed by du Port, Dean of the Faculty) is a reaffirmation in slightly different words of the ban first set out in No. 572.

574. [RIOLAN, J. (the younger).] *Brevis excursus in battologiam Quercetani quo Alchymiae principia funditus diruuntur & artis vanitas demonstratur. Accessit censura Scholae Parisiensis.*
Imprint as in No. 571 (1605).
Only edn.

575. RIOLAN, J. (the younger) *Comparatio veteris medicinae cum nova, Hippocraticae cum Hermetica, Dogmaticae cum Spagyrica.* . . .
Imprint as in No. 571 (1605).
Only edn.

576. RIOLAN, J. (the younger). *Incursionum Quercetani depulsio.* . . .
Imprint as in No. 571 (1605).
Only edn.
For an account of the circumstances of the publication of Nos. 571-6, see Vol. I, p. 256.

577. RIPLEY, GEORGE. *The Compound of Alchemy or the ancient hidden art of Archemie conteining the right & perfectest meanes to make the* PHILOSOPHERS STONE, *Aurum potabile, with other excellent experiments. Divided into Twelve Gates. First written by the learned and rare philosopher of our nation G . . . R . . . sometime*

Chanon of Bridlington in Yorkeshyre and dedicated to K. Edward the 4. Whereunto is adioyned his epistle to the King, his vision, his wheele, & other his workes, never before published: with certain briefe additions of other notable writers concerning the same. Set foorth by Raph Rabbards Gentleman, studious and expert in Archemicall Artes.

Thos. Orwin, London 1591

Ed. p. Duv. and STC 21057. The work is said to have been written about 1471.

Dedn. to Queen Elizabeth ('mauger the Divell, the Pope & the King of Spaine').

One of the greatest (and rarest) classics of alchemy. For an account of the 'Twelve Gates' see Kopp *Geschichte der Chemie,* II, 9 *seq.* and Vol. I, p. 290.)

578. ROBORETUS, OCTAVIANUS. *De particulari febre, Tridenti anno 1591 publice vagante. Deque vesicatorium in ea potissimum usu, tractatus. . . . In quibus de putredine, de contagio, de sanguinis missione. . . . multa . . . pertractantur . . .*

Io. Bapt. Gelminus, Trent 1592

Only edn.

Prov. BUTE.

The printer states that: 'while the disease in this region was approaching its height . . . it did not spare the printers.' On p. 5 is a ref. to Fracastoro's (No. 273) use of the term *lenticularis* instead of *peticularis.*

579. RÖSSLIN, EUCHARIUS. *Der Frawen Rosengarten*

Col. Hainrich Strayner, Augsburg 1532

T.p. wanting—The work is an earlier edn. of *Der schwangerenn Frawen und Hebamme Rosengarte* 1537, since the dedn. ('Der Fürstin Katherina Hertzogin zu Braunschweyck und Lunenburg . . .'), sigs. and col. agree in detail with ref. in BM. [*Ed. p.* Strassburg 1513—Klein].

Many cuts of presentation of foetus, birth chair, midwives, etc.

According to G. Klein (*Eucharius Rösslein's 'Rosengarten'*—München 1910) this was the first printed handbook for the guidance of midwives, the earlier *De Secretis Mulierum* (cf.No. 17) passing under the name of Albertus Magnus but more probably written by his pupil, Heinrich v. Sachsen, and the *Frauenbüchlein* (before 1500) said to be by Ortolff v. Bayerland, containing little in the way of instructions for the conduct of labour. All derive in greater or less degree from Soranos of Ephesos (II-1) partly translated by 'Moschion' and Caelius Aurelianus. The illustrations are based on medieval models whose provenance can be traced back to the time of Soranos.

580. ROMANUS, ADRIANUS. . . . *Ideae mathematicae pars prima,*
sive methodus polygonorum qua laterum, perimetrorum & arearum
cuiuscunque polygoni investigandorum ratio . . . una cum circuli
quadratura continentur.

W. Keebergius, Antwerp 1593

Only edn. Bibl. Nat. Not BM.

Privilege signed by 'Henricus Cuyckius Pontificus & Regius librorum censor'.

The dedn. (Louvain 1593) to Clavius states that only Oronce Fine
(No. 262) among the moderns had entered this field 'sed quam bene,
referre pudet'.

In his *Lectori* he tabulates the contemporary scholars to whom he is
indebted—and in alphabetical order to avoid criticism!

581. RONDELETIUS, GULIELMUS. . . . *Methodus curandorum*
omnium morborum corporis humani in tres libros distincta. Eiusdem
de Dignoscendis morbis, de Febribus, de Morbo Italico, de Internis
& externis, de Pharmacopolarum officina, de Fucis. Omnia nunc . . .
castigatius edita.

Gul. Rovillius, Lyon 1576

Earliest BM 1586.

Prov. LIDDEL.

In his *Lectori* the printer states that the revision had been carried out,
after Rondelet's death, by 'a certain physician who was a great friend
of the author'.

582. RONDELETIUS, G. (*a*) . . . *Libri de Piscibus marinis in quibus*
verae piscium effigies expressae sunt. . . .

Matthias Bonhomme, Lyon 1554

Ed. p.

(*b*) *La première partie de l'histoire entière des poissons composée première-*
ment en Latin par maistre G— R— Docteur regent en medecine en
l'université de Mompelier. Maintenant traduite en Francais sans avoir
rien omis estant necessaire a l'intelligence d'icelle. Avec leurs pourtraicts
au naif.

Same imprint 1558.

Introduction by 'le traducteur à l'autheur', also to whole work—
'Il ni [*sic*] a persone de bon sens qui ne confesse que l'univers ne soit
cree [*sic*] pour l'home, aussi que l'home ne soit fait pour la gloire de
Dieu'.

The qualification in the title is to be taken literally, since (*b*) is not a
direct transn. of (*a*). A *Seconde partie* with sep. t.p. and pagn. which is
not contained in (*a*) is bound with this.

The t.p. gives a long list of contents: members of the modern groups *Crustacea, Mollusca,* and even *Coelenterata* are included as 'Pisces'; also 'monsters', which however are given on hearsay of, e.g. 'Gilbertus Romanus qui Romae medicinam facit', as well as of classical authors. Ref. (p. 21) is made to the girl who lived for three years on air alone! For the ecological basis of his classification, and an attempt at a 'scientific' approach combined with respect for authority of Aristotle and Theophrastos, see Vol. I, p. 201.

See also No. 404.

583. RONSSEUS, BALDUINUS. *De magnis Hippocratis lienibus Pliniique stomacace ac sceletyrbe, seu vulgo dicto Scorbuto commentariolus. Accessere eiusdem epistolae quinque eiusdem argumenti Io. Echtii de Scorbuto epitome. Io. Viveri de Scorbuto observatio. Io. Langii epistolae duae de Scorbuto.*

Clemens Schleich, Wittenberg, 1585

Only edn.

Prov. GREGORY, and a second copy.

Contains cuts of, and refs. to, *Cruciferae,* including *Cochlearia* ('Scurvy Grass'), pp. 257-8.

584. RONSSEUS, B. *Venatio medica, continens remedia ad omnes . . . morbos. Cito, tuto, et iocunde.*

Off. Plantiniana, Leiden 1589

Only edn.

Contains an epistle to the author from Heurnius (No. 335) written 'ex Lycaeo Bataviensi Lugduno'—the university had recently been founded in 1574.

585. RONSSEUS, B. *De humanae vitae primordiis, hystericis affectibus, infantilibusque aliquot morbis, centones.*

Off. Plant. Franc. Raphelengius, Leiden 1594

Only edn. BM and Bibl. Nat. but in his dedn. (dated Gouda 1593) the author says that he issued a book under a similar title several years previously, but when he saw the first printed copy he found it so unsatisfactory that he would have gathered all copies together and committed them to the flames but for the protests of Cornelius Heyduis of Delft. (See Vol. I, p. 184.)

Prov. LIDDEL (?)

586. RORARIUS, NICOLAUS. *Contradictiones, dubia, et paradoxa in libros Hippocratis, Celsi, Galeni, Aetii, Aeginetae, Avicennae. Cum eorundum conciliationibus, nuper recognita. . . .*

Franc. and Gaspar Bindonus, Venice 1592

Only edn.

587. ROTHMANNUS, IOHANNES. *Chiromantiae theorica practica concordantia genethliaca, vetustis novitate addita.*

Io. Pistorius, Erfurt 1595

Only edn.
Numerous cuts of hands and horoscopes. Terminal essay—*De anima mundi et spiritu universi.*

588. ROUSSAT, RICHARD. *Les elements et principes d'astronomie avec les universelz jugements d'icelle. Item un traité tresexquis & recreatif, des elections de chose à faire ou desirée à faire. . . . Le tout de nouveau mis en lumiere par R . . . R . . ., Chanoine & Medicin de Langres.*

Nic. Chrestien, Paris 1552 [Col. 5552!]

Only edn. BM which has '*Des*' for '*Les*' elements. . . .
A popular guide, clear and well written.

589. ROUSSET, FRANCOIS. *Foetus vivi ex matre viva sine alterutrius vitae periculo caesura . . . Casparo Bauhino . . . latio [sic] reddita. Variis historiis aucta & confirmata. Adiecta est Iohannis Albosii . . . foetus per annos xxiix in utero contenti & lapidefacti historia elegantiss.*

C. Waldkirch, Basel 1591

Ed. p. in this form. BM entry has 'Latine' for 'latio'.
[French original, Paris 1580 (BM).]
The dedn. by Caspar Bauhin reviews the possible cases of resort to section, quoting instances and refs. to Hippocrates and Paré. In his *Ad Lectorem* he reviews the history of the subject, quoting authorities. According to Bauhin (and this view is still held) Scipio Africanus was the first *attested birth* by 'Caesarian section': the supposed association with (Julius) Caesar is based on a translation of 'caesus' = 'cut'.
The 'elegantissima historia' (!) of Albosius is wanting (as also in BM copy).
For contemporary views on Rousset and his publications see K. Quecke, *François Rousset unddie erste Monographie über den Kaiserschnitt. Centaurus* **2,** 349.

590. RUBEUS, HIERONYMUS. *De destillatione . . . Liber in quo stillatitiorum liquorum, qui ad medicinam faciunt, methodus ac vires explicantur. Et chemicae artis veritas, ratione & experimento comprobatur. Iampridem ab innumeris mendis repurgatus. . . .*

Seb. Henricpetrus, Basel (Col. 1585)

Duv. p. 519 refers to edns. of 1599, 1585 and Ravenna 1582 (also BM); also to an alleged Basel edn. of 1550 in the library of Falconet at Paris, 1763.

Prov. LIDDEL.
Contains a full historical treatment of the subject, with numerous refs. to Cardano, *De Subtilitate*; also (p. 99) to Cardano's account of the making of Scotch whiskey (*De Varietate* cap. 50).

591. RUDIUS, EUSTACHIUS. . . . *de humani corporis affectibus dignos- cendis, praedicendis, curandis & conservandis, liber primus. In quo singularum humani capitis ac thoracis partium affectus artificiosa methodo, incredibilique facilitate . . . explicantur.*
Rubertus Meiettus, Venice 1595
Bound with (sep. t.p. and pagn.) *Liber secundus* (same date) and *Liber tertius* ('apud Robertum Meietum 1592'). The privilege of Lib. 1 is 1592. Only edn.
Prov. LIDDEL.

592. RUEFF, JACOB. *De conceptu et generatione hominis, de matrice et eius partibus, necnon de conditione infantis in utero & gravidarum cura & officio. De partu & parturientium infantiumque cura omni- faria. De differentiis non naturalis partus & earundem curis. De mola aliisque falsis uteri humoribus, simulque de abortibus & monstris diversis necnon de conceptus signis variis. De sterilitatis causis diversis, a praecipuis matricis aegritudinibus, omniumque horum curis variis libri sex. . . . Nunc denuo recogniti . . . picturis insuper . . . illustrati.*
(Col. Petrus Fabricius for Sigismund Feyrabend (Frankfurt 1587)
Ed. p. 'Ruff. J.', Zürich 1554 *Bibl. Nat.*
Prov. GREGORY.
Dedn. by Feierabend to Leonhard Thurneysser zum Thurm 'Biblio- pola Francofordiensis' (No. 691). Intro. by C. Gesner.
Striking cut of Adam and Eve as heading. Numerous cuts of delivery room, stool, instruments (including speculum), astrologers casting horoscope, monsters, etc.
This is a plagiarised version of Roesslin, *q.v.* (No. 579).

593. RUELLIUS, IOHANNES. (*a*) *Veterinariae medicinae libri ii Io . . . R . . . interprete.*
Simon de Colines, Paris 1530
Col. '. . . ex chalcographia Ludovici Blaublomii impensis S . . . C . . .'.)
This beautiful folio edn. (with cut of horse and rider on t.p.) is *Ed. p.* of the Latin trans. by Ruel.
Prov. 'Io. Gregorie 1743', and another copy.

In his dedn. to Francis I (dated Paris 1528) Ruel says 'when you imposed upon me the task of turning into Latin the *Veterinary Medicine*, already diligently set forth by many Greek authors . . .'. The numerous short pieces by Greek authors are set out in a table.

> (*b*) *ΤΩΝ 'ΙΠΠΙΑΤΡΙΚΩΝ ΒΙΒΛΙΑ ΔΥΩ. Veterinariae medicinae libri duo a Io . . . R . . . olim quidem latinitate donati, nunc vero iidem sua, hoc est Graeca, lingua primum . . . aediti.*

> Io. Valderus, Basel 1538

BM 1537. (Presumably *Ed. p.* of Greek originals.)
Prov. 'Julius Borgarutius'.
Dedn. by Simon Grynaeus to Io. Zobelius.

594. [RUFFUS, GIRARDUS]. *Divi Severini Boetii* Arithmetica *duobus discretis libris adiecto commentario, mysticam numerorum applicationem perstringente, declarata.*

> Simon de Colines, Paris 1521

Only edn.
Text of Boethius: *Praefatio*, 'In dandis accipiendisque . . .' *Prooemium* ['In quo divisio mathematicae'] 'Inter omnes priscae autoritatis viros . . .'. Similar in 9th cent. MS, ThK 362. Each *caput* is followed by a lengthy commentary by Ruffus. (See No. 381.)

595. RUFFUS, IORDANUS (*syn.* Ruffo, Rusto, Rufus, Russo).

> (*a*) *Libro dell'arte de marascalchi per conoscere la natura delli cavalli & medicarli nello loro infirmità & l'arte di domarli . . . Novamente stampato* (Col. Francesco de Lano, 1563) Venice.

Prov. BUTE.

> (*b*) *Il dottismo libro non piu stampato delle malscalzie del cavallo . . . Di piy* [sic] *vi s'è aggionto un trattato di Alberto Magno dell'istessa materia, tradolto dal latino in questa nostra volgar lingua. Et alcuni altri belli secreti di diverso autori non piu stampati per l'adietro.*

> Gio. de'Rossi, Bologna 1561

Prov. BUTE.
The bibliography of these two volumes is obscure. There seems little doubt that (*a*) and the earlier pages of (*b*) are trns. of the same original though differing markedly in phraseology and later in the order of the topics dealt with. The following citations from (*a*) indicate what this original was:

'Queste opere compose Misser Iordano Ruffo de Calabria, cavalliere, & famigliare della sacro maiestede de Federico Barba rossa secondo Imperatore Romani' (fo. 41*v.*, also p. 2. of intro.)

'Questo e el prologo de la detta opera translatata de latino in volgare per Frate Gabrielo Bruno maestro in Theologia delli Frati Menori') dated—fo. 43r—Venice 1492).

Earliest Italian version 'Venice 1492?' BM, whereas Hain 14034 gives title in Latin (as is usual) followed by '*Italice* 1487', and Klebs (868.1) '*Arte di conoscere la natura dei cavalli* 1493'. Copinger 5189 gives neither place nor date. Besides the incunable edn. BM gives one under the *title* of (a) (but 'per gli Heredi di Giovanne Padovano, Venice 1554'), (b), and one other edn. (Venice 1561) in a different trans. before 1600. No Latin edn. appears before Padua 1818. The matter is further complicated by the printer's assertion—in the *Lettori* of (b)— that he 'obtained the original at a great price'. The 'non più stampato' at the beginning of the title seems also to be a dubious claim. The original seems to have been available to Pier Crescenzi for his famous *Opus Ruralium Commodorum*.

596. RULAND, MARTIN. *Balnearum restauratum in quo curantur morbi tam externi quam interni per balneas naturales, artificiales, insessus, lixivia, sudationes itemque scarificationes ac phlebotomias, distinctum in lib. iii. Opus antea non . . . editum.*

Off. Henricpetrina, Basel 1579

Only edn.

Prov. FORBES.

The dedn. to Philip Ludwig, Count Palatine of the Rhine, refers to the use of baths and to the institution by Philip's father of the *Schola Langingae*. The index provides a list of baths, biblio. of Ruland's works, and many *chemical* refs.

597. RULAND, M. (a) (i) *Curationum empiricarum et historicarum in certis locis et notis personis optime expertarum & rite probatarum Centuria Prima emendatius impressa.* . . .

(Col. Off. Henricpetrina, Basel 1581)

The dedn. to Frederic, Count Palatine, dated 1578, contains interesting remarks on method in medicine (esp. sig. B r and v).

(*ii*) Same title, imprint and dedn. but dated 1593.

(*b*) . . . *Centuria secunda.* . . . Same imprint, dated (col.) 1580.

The dedn. contains observations on the difficulty and danger of experiment in medicine.

Three copies, one of which is bound with (a) (i), and one with (a) (ii) and (c).

(*c*) . . . *Centuria tertia* . . . (Col. Seb. Henricpetrus, Basel 1591).

(*d*) . . . *Centuria quarta* . . . (Col. 1593).

(*e*) . . . *Centuria quinta* . . . (No imprint).

Dedn. to Otho Hainricus.

Followed (with half-title) by *Martini Rulandi Med. Problemata de ortu animae recitata.*

Two copies.

(*f*) . . . *Centuria sexta* . . . (No imprint).

Contains also *Prodromos* and *Epistle* by A. Libavius (No. 393). 'Edita studio et opera Martini Rulandi M. filii', whose *Praefatio* is dated 1593.

Two copies.

(*g*) . . . *Centuria septima* . . . (Col. S. Henricpetrus, Basel 1595). Dedn. by his son 1593; terminal letter by Caspar Bauhin to Ruland *senr.* (1592), and final col.

The above *Centuriae* occur in five volumes, all bound in contemporary vellum. BM cites no earlier edns. than 1595.

598. RULAND, M. *Progymnasmata Alchemiae, sive problemata chymica nonaginta et una quaestionibus dilucidata, cum Lapidis Philosophici vera conficiendi ratione.* . . .

E Collegio Musarum Paltheniano, Frankfurt 1607

Only edn. BM, but with col. 1606.

The *Lapidis* . . . has sep. t.p. (dated 1606) and pagn. *Quaestiones* LXI-XCI incl. are placed in an appendix with sep. pagn. after an *Elenchus Remediorum Spagyricorum* with h.t. The question-headings provide a useful index of contemporary thought.

599. RULAND, M. *Lexicon Alchemiae sive dictionarium alchemisticum cum obscuriorum verborum & rerum hermeticarum, tum Theophrast-Paracelsicarum phrasium, planam explicationem continens.*

Zach. Palthenius, Frankfurt 1612

Ed. p. Duv. Ferg. II, 302 BM.

Prov. READ; also another copy.

The dedn. to Henricus Postulatus, Bishop of Halberstadt, is dated 1611, but the Imperial Privilege, Prague 1607. Ruland was physician to Rudolph II.

Mostly Latin or neo-Latin words and phrases with vernacular equivalents or Latin definitions.

600. RULAND, MARTIN (the younger). *De perniciosae Luis Ungaricae Tecmarsi et Curatione tractatus. Historicis curis atque observationibus nec non quaestionibus aliquot homogeneis locupletatus.*

R. Beatus for Nic. Basseus, Frankfurt 1600

Ed. p.

601. RULAND, M. (the younger). *Alexicacus Chymiatricus, puris putis mendaciis atque calumniis atrocissimis Ioannis Oberndorferi, quibus larvatus ille medicus apologiam suam, chymicomedicam practicam nequissimo ausu iniuriosissime consarcinavit, oppositus asserendae veritatis & famae integritatis suae jure. . . .*

Missus ab auctore Francofurtum, prostat apud Palthenium 1611
Only edn. Ferg. II, 304. (Duv. does not distinguish, as does Ferg. between Ruland father and son.) Oberndorffer's offending work is bound with it—see No. 482.

602. RUMBAUM, CHRISTOPHER. *De partibus corporis humani exercitationes quaedam. Quibus generatio, substantia, usus, sanitas, morbus, & curatio illarum exponitur. . . .*
Seb. Henricpetrus, Basel (Col. 1586)

Only edn.
Prov. LIDDEL.
The author is here concerned with the problem of harmonising Aristotle's teaching on generation with theology.

603. [SACHSE, MELCHIOR]. *Arzneybuch vast wunderkostlich für allerley von ime selbst zufelligen inner oder eusserlichen offen oder heymlichen des gantzen Leybs Gebrechligkeyt. Wie nur die mügen Namen haben für Mannss und Frawen Personen, iungen unnd alten, sehr nutz und dienstlich zu gebrauchen. Vor allen newen und alten erfarnen berhümbtisten Ertzten zusammen getragen gleich als für ein. Hauss Apoteck oder Hausschatz.*
(Col. Iohann Daubmann, Nürnberg 1549)

Catalogued BM under 'Schnellenberg' (*alias* 'Tarquinius Ocyorus'), but the preface by Sachse suggests that the latter made the collection (see t.p. for the interesting table of contents). Sachse's dedn. 'Rathsmeistern' of Erfurt (dated there 1546) suggests an earlier edn., but no earlier BM. Moreover, the first work—*Experimenta von XX Pestilentz wurtzeln und kreuttern*, with dedn. by 'Tarquinius Ocyorus alias Schnellenberg Med. D.' is dated 1549.

Prov. LIDDEL.
The collection includes a trans. by O. Brunfels, a work by Hieronymus Bo(c)k, and an '*Albertus Magnus Frauenbuch*'. Numerous cuts—some copied from Ryff (No. 592).

604. SACROBOSCO, IO. DE. *Sphaera.*
Io. Richardus, Antwerp 1559

[*Ed. p.* Ferrara, 1472 Hain 14100].
Prov. 'Sum ex libris Ulrici Holzeri Sti . . .' (early 17th cent?)

Text, followed by extracts: Prop. xxii of lib. III of Regiomontanus's *Epitome* and the *De ortu & occasu planetarum* . . . of Alfraganus.

Numerous cuts. Followed by extracts from the *Epitome* of Regiomontanus.

605. SACROBOSCO, IO. *Libellus de anni ratione, seu, ut vocatur vulgo, Computus Ecclesiasticus.*

Io. Richardus, Antwerp 1559

[*Ed. p.* Antwerp, 1547].

Incipit: 'Prooemium. Computus est scientia considerans tempora ex solis & lunae motibus. . . .'

See also Nos. 324. 633, 720,.

606. SALIGNACIUS, BERNARDUS. *Mesolabii expositio.*

Artusius Calvinus, Geneva 1574

Only edn.

The 'mesolabium' was a device (different forms are here described) used in 'the art of finding a mean proportional to two unequal straight lines'.

607. SALIGNACUS, BERNARDUS. . . . *Arithmeticae libri duo et algebrae totidem cum demonstrationibus.*

Heirs of Andr. Wechel, Frankfurt 15[93] [t.p. damaged]

[*Ed. p.* Frankfurt 1580—Smith p. 359.]

The *Algebra* has sep. t.p. (1593), and *Lectori* (in which the author says that without the aid of Balth. Gerlach of Eschwegen he would not have dared to write it) but cont. pagn.

The dedn. (dated Heidelberg 1579) refers to 'Scholam in octo classes distributam non modo linguarum Latinae, Graecae, & Hebraicae . . .'.

The *Arithmetic* is fairly practical but lays great stress on proportion.

Algebra is called the 'arithmetic of figurate numbers' and 'the algebraic symbols (*notae*), with their *indices* corresponding to a geometrical progression from unity, are:

$$1 \quad 2 \quad 4 \quad 8 \quad 16$$
$$u \quad 1 \quad q \quad c \quad bq$$
$$0 \quad 1 \quad 2 \quad 3 \quad 4' \text{ (cf. No. 555.)}$$

The second book contains a few examples, such as the one in which 'Two travellers start at different times . . .', still to be found in textbooks.

608. SALIUS-DIVERSUS, PETRUS. *De febri pestilenti tractatus et curationes quorundam particularium morborum quorum tractatio ab ordinariis practicis non habetur. Atque annotationes in artem medicam de medendis humani corporis malis a Donato Altomari . . . conditam.*

Heirs of Andr. Wechel, Frankfurt 1586

228

[*Ed. p.* Bologna, 1584.]

Prov. LIDDEL; also another copy.

Cont. pagn. throughout incl. the *Annotationes* of Donatus ab Altomari (No. 28).

A useful 'source book' of contemporary views on 'pestilential fevers' (Vol. I, p. 264). See esp. pp. 16-18 on the necessity of air for life and the origin of plague by its 'corruption', for instance by 'light' or 'the motion of the stars'.

609. [SANSOVINO, FRANCESCO.] *Della Materia Medicinale libri quattro. Nel primo & secondo de quali si contengono i semplici medicamenti con le figure delle herbe & con le lor virtù ritratte dal naturale & la maneira di conoscerle & di conservarle. Nel terzo s'insegna il modo di preparare & comporre i medicamenti secondo l'uso de i medici approvati, cosi antichi come moderni. Nel quarto et ultimo son poste le malattie che vengono al corpo humano, con i loro rimedi contenuti nel presente volume.*

Gio. Andrea Valvassori detto Guadagnino, Venice 1562

Only edn.

Prov. Several names cancelled. 'Di Alessandro Cherubini'.

Dedn. by Sansovino.

Tables in Italian, Latin, Greek, Arabic (transliterated), German, Spanish, French. Numerous fine cuts of plants. Not in Arber.

In *Lettori* the differences between the reports of Pietro Crescentio (see Vol. I, p. 189) and Leonhard Fuchs are attributed to the fact that the former wrote 'according to the temperature of Italy' (with regard to the cultivation of plants), the latter according to that of Germany.

610. SARACENUS, IANUS ANTONIUS. *Pedacii Dioscoridis Anazarbaei Opera quae extant omnia. Ex nova interpretatione. . . .*

Heirs of Andr. Wechel, 1598

Ed. p.

Long scholia with ref. to codices, ancient and modern commentaries, corruptions, obscurities. See No. 198.

611. SARACENUS, I. A. *De Peste commentarius. Editio altera: cui recens accessit selectorum tam externorum quam internorum remediorum particularis descriptio. . . .*

Io. Tornaesius, Geneva 1589

[*Ed. p.* Lyon 1572.]

Prov. (i) du Poirier . . . Med. Doct. 1665, (ii) Bute.

Ded. to his father, Philibert, in 1571, to which the latter replies referring to the plague at Geneva many years previously.

Numerous citations of classical authors, including Thucydides, Virgil, Livy, Hippokrates, Galen, and the Bible. 'Contagium, quid'— p. 22.

612. SARZOSUS, FRANCISCUS. . . . *In aequatorem planetarum libri duo. Prior fabricam aequatoris complectitur. Posterior usum atque utilitatem, hoc est veros motus ac passiones in zodiaci decursu contingentes, aequatoris ministerio investigare docet.*

Simon de Collines, Paris 1526

Ed. p.

Prov. Bound with No. 246 bearing David Rait's autograph.

An 'equatorie' was an instrument for calculating planetary positions. See Derek Price *The Equatorie of the Planetis*, Cambridge 1955.

613. SAVONAROLA, GIOVANNE MICHAELIS. *Libreto delo excellentissimo physico maistro M— S— de tute le cose che se manzano comunamente e piu che comune: e di quelle se beveno per Italia: e de sei cose non naturale: e le regule per conservare la sanita deli corpi humani con dubii notabilissimi novamente stampato.*

Simone de Luere, Venice 1508

Ed. p.

Prov. BUTE.

Articles on diet and hygiene.

614. SAVONAROLA, G. M. *Practica maior . . . in qua morbos omnes quibus singulae humani corporis partes afficiuntur . . . curare docet. . . .*

Iuntae, Venice 1559

[*Ed. p.* Florence 1479. Hain 14480.]

Prov. LIDDEL.

There is a gynaecological secn. See Castiglioni, p. 371. See also No. 313.

615. SAXONIA, HERCULES. *De Phoenigmis libri tres. In quibus agitur de universa rubificantium natura; deque differentiis omnibus atque usu psilotris, smegmatibus, dropacibus, sinapsimis simplicibus compositis vulgo vesicantibus; de quorum usu in febribus pestilentibus multa disputantur . . . Nunc primum . . . editi.*

Paulus Meiettus, Padua 1593

Ed. p.

Prov. BUTE; and a second copy with which is bound the same author's *De Plica*. . . . L. Pasquatus, Padua 1600—sep. pagn.

Typographus ad lectorem, 'You'll wonder perhaps why these books, having been completed by the author at the end of September of last year have only just appeared' (owing to the illness of the author). After three and a half centuries of 'mechanisation and progress' the reader has come to regard such an interval as unexpectedly short!

616. SAXONIA, H. *Luis Venereae perfectissimus tractatus, ex ore H— S— exceptus.* . . . [Ed. by Andreghetti Andreghetti.]

Laurentius Pasquatus, Padua 1597

Only edn.

617. SCALIGER, JULIUS CAESAR. *Exotericarum exercitationum liber quintus decimus* de Subtilitate *ad Hieronymum Cardanum.*

(*a*) Michaeles Vascosanus, Paris 1557.
Ed. p.
Prov. 'Liber Gulielmi [?]'—scored and glossed in several hands.
Encyc. Britt. (art. 'Scaliger, J. C.') regards this as the elder Scaliger's greatest work, displaying enormous learning; yet in correcting Cardano he commits as many errors in his turn. In his *Lectori* Scaliger refers in not ungenerous terms to the recent 'death' of Cardan, which like that of Mark Twain had been 'greatly exaggerated' as a hoax against Scaliger!

(*b*) Andr. Wechel, Frankfurt 1582.
Not BM.
Prov. LIDDEL.
An *epistola* from Crato à Kraftheim to Joseph Scaliger urges him to persuade Wechel to reprint his father's work, as it had been unobtainable in Germany since 1576.

618. SCALIGER, J. C. . . . *Commentarii et animadversiones in sex libros* de Causis Plantarum *Theophrasti.*

Io. Crispinus, Geneva 1566

Only edn.
Ed. p of the original (in the Latin trans. of Theodore of Gaza) was Treviso 1483 (Hain 15491, and PW Supp. VII, 1437). The Greek text was included in the Aldine *Aristotle* of 1498.
Prov. 'Jo. Gregorie 1767'.
The printer refers to the editorial assistance of R. Constantinus.

619. SCALIGER, J. C. *Animadversiones in* Historias *Theophrasti.*

Io. Iacobus Junta, Lyon 1584

Only edn.
Ed. p. of the original Latin and Greek texts was the same as that of *De Causis.*

620. SCALIGER, JOSEPH JUSTE. *Hippolyti episcopi Canon Paschales cum J . . . S . . . commentariis. Excerpta ex computo Graeco Isaaci Argyri de Correctione Paschalis. . . . J . . . S . . . Elenchus & castigatio Anni Gregoriani.*

F. Raphelengius, Leiden 1595

Only edn. in this form: an *Elenchus* . . . . ed. by Chris. Clavius and published by Aloysius Zannettus, Rome 1595 is bound with No. 374.

Prov. LIDDEL.

In his dedn. to Iohan ab Oldenbarneveldt ('Advocate' of the province of Holland, whose diplomatic skill gave him almost dictatorial powers over the remaining provinces) Scaliger refers to his *De Emendatione Temporum* published eleven years previously.

The text refers to the discovery of the text of Hippolytos on an exhumed marble statue, the establishment of the date of Argyros (1372), and the subsequent history of the reform movement, including the parts played by Regiomontanus, Gregory XIII, and Christopher Clavius.

621. SCHEGKIUS, JACOBUS. *Disputationum physicarum et medicarum libri viii.*

Io. Wechel, Frankfurt 1590

Not BM.

Dedn., dated 'Tybingae 1585', probably of *Tractationum physicarum et medicarum tomus unus*, Io. Wechel Frankfurt 1585 (BM).

Prov. LIDDEL, and another copy.

A dogmatically Aristotelian work, as shown by (e.g.) the questions 'Can disease be said to be something substantial or according to (*secundum*) substance?', the theses that 'the spirit which is contained in the ventricles of the brain is not vital but animal', and the Parthian shot 'quales ἐναντιολογίας etiam iners philosophaster P. Ramus contra Aristotelis philosophiam scripsit'.

622. SCHENCKIUS, IOANNES. *Observationes medicae de capite humano . . . ex clariss. medicorum, veterum simul & recentiorum, scriptis.*

Off. Frobeniana, Basel 1584

Two copies.

623. SCHENCKIUS, IO. *Observationum medicarum rararum, novarum, admirabilium et monstrosarum libri tertii de partibus naturalibus sectio prior. De ventriculo, intestinis & mesenterio. In qua quae medici doctissimi . . . abdita, vulgo incognita . . . accidere compererunt, exemplis ut plurimum & historiis proposita, exhibentur.*

232

Studio atque opera Ioan. Schenckii a Grafenberg . . . Continentur praeterea illustrium aetatis nostrae medicorum ἀνεκδοτα complura & nunquam publicata. . . .

Martinus Beckler, Freiburg-in-Breisgau 1595

Prov. 'Liber Coll. D. Ioannis Bapt. Oxon. ex dono Gulielmi Paddei Medicina Doct. eiusdem Coll. convictorii 1602.'

Another vol. contains also *Observationum . . . monstrosarum liber secundus de partibus vitalibus thorace contentis.* (Same imprint) 1594.

Earliest BM Lyon 1644.

'Thomas Moresinus' (No. 466) is named among those from whose works the collection was made. On p. 149 occurs Gesner's story of serpents discovered in a man's and a girl's stomach.

624. SCHEUBELIUS, IOANNES. *De numeris et diversis rationibus seu regulis computationum . . . Non solum ad usum quendam vulgarem sed etiam cognitionem & scientiam exquisitiorem arithmeticae accomodatum* [sic] (Col. Michaeles Blum, Leipzig 1545).

Only edn. (Smith p. 233). This may be due to the fact that it fell between the 'usum vulgarem' and the 'cognitionem exquisitiorem'.

Dedn. to the Doctors and Masters of Tübingen, where he taught.

An interesting note on nomenclature occurs on sig. R 3*r*: 'This rule which is commonly called "of three" . . . also the "merchants' rule", for whose use it is particularly well adapted. It has also been called the "golden rule . . .". We shall just call it the rule of proportion'.

Sig. b 2*r* has a table of binomial coefficients—'these numbers or parts of quantities are of use for the extraction of roots'.

625. SCHEUBELIUS, IO. *Algebrae compendiosa facilisque descriptio qua depromuntur magna arithmetices miracula. . . .* Gul. Cavellat, Paris 1552 (Col. 1551).

Only edn.

Prov. going back to 1566 incl. David Rait.

For Scheubel see also No. 235(d).

626. SCHOENER, ANDREA. *Gnomonice . . . Hoc est de descriptionibus horologiorum sciotericorum omnis generis proiectionibus circulorum sphaericorum ad superficies cum planas tum convexas concavasque sphaericas, cylindricas, ac conicas. Item delineationibus quadrantium annulorum etc. libri tres. His addita sunt eiusdem autoris Gnomonices Mechanices, seu de designandis per instrumenta horologiis, libellus. De Inventione lineae meridianae ac instrumenti ad hoc necessarii compositione. De Compositione Astrolabii plani & columnaris directorii astrologici plani. Omnia recens nata et edita.*

Io. Montanus & Ulricus Neuberus, Nürnberg 1562

Only edn. BM.
Fine cuts throughout.
Fo. xciii r gives an account of the method of observing the altitude of the pole, practised by William Landgrave of Hesse.
For William, Landgrave of Hesse, see Vol. I, p. 120.

627. SCHOENER, IOHANN (*syn.* Schonerus). *Algorithmus demonstratus. Habes in hoc libello . . . mathematicas demonstrationes, in eam calculandi artem, quam vulgus Algorithmum vocat . . . quare eme, lege & iuvaberis.*

Io. Petreius, Nürnberg 1534

Ed. p. Smith, p. 178.

628. SCHOENER, IO. *Tabulae resolutae astronomicae . . . ex quibus omnium siderum motus facillime calculari possunt secundum praecepta in planetarum theoricis tradita.*

Matthaeus Welack, Wittenberg, 1588

Ed. p. 1536 (Dedn. by Schoener to Senate of Nürnberg).
Prov. LIDDEL.
Dedn. by Io. Hagius from Wittenberg 1587.
Tables preceded by half-title, 1587.

629. SCHOENFELDT, VICTORINUS. *Consilium oder Rathschlag vor die beschwerliche jetz regierende Plage der Roten Ruhr und andere schädliche Bauchflüsse. Daneben ein kurtzer Bericht wie ein jeder sich vor der Pestilentz bewahren und von derselben sich erledigen soll. . . .*

Heirs of Chris. Egenolph, Frankfurt 1584

Only edn.
Ded. to the Bürgermeister, etc., of Marburg at whose 'Fürstlichen und loblichlen Hohen Schul' Schoenfeldt was 'Medicus und Mathematicus'.
A tract on dysentry ('Rote Ruhr' = 'bloody flux'). The text includes refs. to the heavenly portents (fo. 16*v*); 'cholera, das ist von der gelben Galle herkompt' (fo. 18*r*); and (last fo.) prayer for daily use in time of pestilence.

630. SCHOLZIUS, LAURENTIUS. *Consiliorum medicinalium conscriptorum a praestantiss. atque exercitatiss. nostrorum temporum medicis liber singularis. Nunc primum studio & opera L . . . S . . . editus.*

Heirs of Andr. Wechel, Frankfurt 1598

Ed. p.
List of *consilia* and of those physicians who helped him.

631. SCHOLZIUS, L. *Epistolarum philosophicarum medicinalium ac chymicarum a summis nostrae aetatis philosophis ac medicis exaratarum volumen . . . Nunc primum labore . . . L . . . S . . . datum.*

Heirs of Andr. Wechel, Frankfurt 1598

Bound with No. 360, but has sep. t.p. and pagn. The table of correspondents, of biographical interest, contains Thos. Moufet, who writes to Monavius (p. 532) about *mumia* (see Vol. I, p. 297 n.) and to Crato (p. 533) in 1582 referring explicitly to 'primary and secondary qualities'. See also No. 182

632. SCHRECKENFUCHS, ERASMUS OSWALD. *Commentaria in Novas Theoricas Planetarum Georgii Purbachii quas etiam brevibus tabulis pro eliciendis tum mediis tum veris motibus omnium planetarum, item tabulis coniunctionum & oppositionum ac eclipsium luminarium ad summum illustravit lucemque maximam iis adiecit. His quoque accesserunt varia exempla & demonstrationes, quibus astronomiae studiosus suo marte omnis generis tabulas secundorum mobilium facile conficiet & qua ratione veteres confectae sint videbit . . . Philippi insuper Imsseri . . . in eiusdem Purbachii Theoricas tabulis utilissimis adiectis.*

Henricus Petrus, Basel [Col. (before tables) 1546]

Earliest BM [1556.]

Text and very full commentary, followed by Imsserus's tabular summary and outstandingly clear figs. of the planetary orbs.

Dedn. to Michael, Archbishop of Salzburg, dated Freiburg-in-Breisgau 1546.

Bound with it (same imprint, but 1547) is Glareanus, *Dodekachordon*.

633. SCHRECKENFUCHS, E. O. . . . *Commentaria in Sphaeram Ioannis de Sacrobusto . . . quibus non solum ea quae in autoris contextu sunt sed alia etiam ad sphaericam doctrinam necessaria explicantur, tabularumque constructio ex suis principiis per demonstrationum seriem . . . docetur. His adiecti sunt eiusdem autoris Canones quibus usus tabularum quae operi ex libro Directionum Ioannis Regiomontani passim inseruntur ad . . . inquisitiones astronomicas continetur. Reliqua ad consummatam doctrinam hanc pertinentia ex illius PRIM[O] MOBILI eadem forma edito petes.*

(Col. Off. Henricpetrina, 1569) Basel

[*Ed. p.* Basel 1556 (?)]

The dedication stresses the importance of Sacrobosco's work during the Middle Ages, but does not refer to his errors.

The text opens with the usual *incipit*: 'Tractatum de Sphaero quatuor capitulis distinguimus. . . .' The commentary is very full, contains many allusions of historical importance, and is illustrated by figs.

634. SCHYLANDER, CORNELIUS. *Medicina astrologica omnibus medicine studiosis . . . necessaria, iam denuo . . . aucta, una cum Practica Chirurgiae brevi. . . .*

Ant. Tilenius, Antwerp 1577

Earliest BM but dedn. dated Antwerp 1569.

Prov. Bound with No. 572.

Practica Chirurgiae has sep. t.p. as above with the addition of . . . *singularem & facilem modum extrahendi olea ex floribus herbis vulnerariis, Ligno Guaiaco, & cera continens.*

Only edn.

Dedn. contains a dialogue between 'chirurgus' and 'Doctor' on the relation of medicine to surgery.

The *Medicina Astrologica* is a complete guide to the 'irrelevances' of renaissance medicine: critical days; relation of affected part to con-figuration of the moon and other planets with respect to the Signs; inspection of urine, including 'non viso aegroto aut aliquo eius ex-cremento'; and times expedient for phlebotomy and purgation. Tables and horoscopes.

635. SCRIBONUS LARGUS. . . . *de Compositione Medicamentorum liber iam pridem Io. Ruellii opera e tenebris erutus & a situ vindicatus. Antonii Benivenii Libellus de abditis nonnullis ac mirandis morborum & sanatationum causis. Polybus de Salubri victus ratione privatorum Gunterio Ioanne Andernaco interprete.*

(Col. Andr. Cratander, Basel 1529)

[*Ed. p.* Paris, the same year or possibly 1528.]

Prov. 'John Gregorie 1742'.

Printer's note on author and edn. Cont. pagn. but secn. headings for second and third works, with sep. dedn. for second, which opens with Benivieni's well known account of the origin and nature of the *Morbus Gallicus.*

636. SCUTELLARIUS, JACOBUS. . . . *in Librum Hippocratis de Natura Humana commentarius . . . Nunc primum in lucem editus.*

Seth Viottus, Parma 1568

Only edn.

Dedn. to Ottavio Farnese.

Lectori—the author intends to keep as close as possible to Galen's commentary.

The text of Hippokrates is followed *seriatim* by long commentary.

637. SEIDELIUS, BRUNO. *Liber morborum incurabilium causas . . . explicans.*

Io. Wechel, Frankfurt 1593

Ed. p.

Prov. GREGORY.

The dedn. (signed by Zacharias Palthenius, 'Ioannis Wecheli Typographi nomine'), says that the heir of Seidelius brought the work to him.

638. SEIDELIUS, B. . . . *de Ebrietate libri iii nunc primum . . . editi . . . .*
    Addita est . . . Basilii Magni Homilia contra ebrios, Simone Stenio . . .
    interprete.

                Gul. Antonius for Petrus Fischer, Hanover 1594

Only edn.

Prov. (i) 'Sum Guil. Charei', (ii) Patrick Dun (in pencil).

639. [SENDIVOGIUS, MICHAEL] (*syn.* Sandivogius). (*a*) *Novum*
    Lumen Chymicum. E naturae fonte & manuali experientia depromp-
    tum & in duodecim tractatus divisum. Accessit dialogus Mercurii,
    Alchymistae & Naturae.

                Renatus Ruellius, Paris 1608

Ferg. I, 94 states that this was Jean Beguin's first published work (no copy in Young or Duveen Collections), being a transcription of the work of Sendivogius and acknowledged by the editor in his *Epistola*, as 'hos ab amico tractatus acceptos' but giving no name.

    (*b*) Another edn. with title as above as far as '*depromptum*';
    continues: *in duas partes divisum, quarum prior xii tractatibus de*
    Mercurio agit, posterior de Sulphure, altero Naturae principio.

                de Tournes, Geneva 1628

    (*c*) *A New Light of Alchemy* . . . . Trans. J. F. M. D.

            Rd. Coles for Thos. Williams, London 1650

Wing III, 227. Two copies.

    (*d*) Ditto. A. Clark for Thos. Williams, London 1674.

Wing III, 227, Two copies.

    (*e*) The *Dialogus Mercurii* . . . (Servatus Erftens, Cologne 1607)
    in a sep. edn. is bound with No. 600.

Several sigs uncut.

See Ferg II, 365; Duv. p. 542; and Patterson, *Annals of Science*, **2,** 246. See also No. 476.

640. SENNERTUS, DANIEL. *Physica hypomnemata.*

                Petrus Ravaud, Lyon 1637

[*Ed. p.* Frankfurt 1636 as *Hypomnemata Physica*, BM.]

Prov. READ.

This work belongs rather to the background of Boyle's *Sceptical Chymist* than to the 'renaissance'; it contains, however, a very full description and discussion of a typical renaissance problem—the 'Barnacle Goose' (p. 441 *f.*).

641. (QUINTUS) SERENUS SAMMONICUS. (*a*) . . . *de Re Medica sive morborum curationibus liber . . . emendatus. Gabrielis Humelbergii Ravenspurgensis . . . commentarii.*

[Device of:] Froschouer, Zürich 1540

Prov. HENDERSON.

A word-for-word commentary on the poem of Serenus with much historical and traditional allusion. For Humelberg see also No. 276.

(*b*) (Similar title followed by:)—*His recens accesserunt emendationes novae ex vetere manuscripto codice collectis.* [sic.]

No imprint or col. except (t.p.) 'MDLXXXI' and device of Froschouer.

Not BM.

Emendations by Caspar Wolph are set out *seriatim* on pages preceding the text and commentary of Humelberg.

See also No. 130.

642. SERLIO, SEBASTIANO. (*a*) *De Architectura libri quinque . . . a Ioanne Carolo Saraceno ex Italica in Latinam linguam nunc primum translati atque conversi.*

Franciscus de Franciscis, Venice 1959.

Ed. p. of Latin edn.

Prov. Tarradale House, 1960, previously purchased from the Sunderland Library, Blenheim Palace by B. Quaritch 1883.

(*b*) *The first (-fift* [sic]) *booke of Architecture made by Sebastian Serly, entreating of geometrie. Translated out of Italian into Dutch, and out of Dutch into English.*

Robt. Peake, London 1611

Il libro primo (—*quinto*) *d'Architettura* was first published Venice, 1551, trans into 'nederlandts duer P. Coeke', Antwerp 1553, and 'aus dem Italiänischen und Niderländischen . . . in die gemeine Hochteutsche Sprache übergesetzt', Basel 1608 (BM).

The five 'books' are separately foliated, with sep. t.p.'s. all of same date. Fine cuts throughout. Book II on Perspective is especially valuable.

In his dedn. to Prince Henry, Peake refers to the ignorance of geometry in England which 'hath left many lame workers, with shame of many workemen . . .'.

643. SEVERINUS, PETRUS (*syn.* Sôrensen). *Idea Medicinae Philo-sophicae, fundamenta continens totius doctrinae Paracelsicae Hippo-craticae & Galenicae.* . . .

Sixtus Henricpetrus, Basel 1571

Ed. p.

Prov. Three copies: (1) Liddel; (2) 'J. Gregorie L.B.1745';
(3) 'Wm. Moir', etc.

In his dedn. to Frederick II of Denmark the author stresses the necessity for observation and correlation of conjunctions with plagues etc; also for the search for 'hidden seeds' ('ad Naturae penetralia propius accedentes, invisibilium seminum artificiosas lithurgias . . .'). He admits that he 'found himself drowning in the obscure diction' of Paracelsus. See Th. *Hist.* V, 627-30. This work seems to have been little studied in relation to the 'Paracelsus problem'—Liddel's 'paradoxa Paracelsi' (sig. B 4r among prelims) and long glosses might be illuminating.

644. SEVERTUS, IACOBUS DE. *Orbis Catoptrici seu Mapparum Mundi principiis, descriptione ac usu libri tres. Opus Cosmographicorum . . . recens . . . editum . . . cuius usus faciliori opera ex Gulielmi Postelii Mappa quam ex reliquis elicitur. Editio secunda.*

Laur. Sonnius, Paris 1598

BM and Bibl. Nat. lack *Ed. p.*

'Gul. Postellus' was Guillaume Postel, one of the most remarkable 'wandering scholars' of (XVI-2).

The 'cosmography' refers to the whole visible universe and is based on ancient and modern authors.

645. SFONDRATUS, PANDULPHUS. *Causa Aestus Maris.*

Benedictus Mammarettus, Ferrara 1590

Only edn.
Prov. Liddel (?)
Dedn. to Gregory XIV.

In his *Lectori* the author says that he is writing about 'that science to which, in experience and observation on the seas, I have devoted all my life'.

Ch. III, in which the observations of Columbus and Magellan are recorded, is of special interest.

646. SILVATICUS, IOANNES BAPTISTA. *De compositione et usu theriacae libri duo . . . Recogniti denuo.*

Beatus for Nicolaus Basaeus, Frankfurt 1600

Ed. p. Heidelberg 1597. This edn. not BM.

Dedn. 'prudentissimis viris . . . Mediolanensis Collegii medicis . . .'
includes the prayer that 'no one in our city be allowed to compound
theriac without your permission'. (See Vol. I, p. 238.)

647. SILVATICUS, MATTHEUS (*a*) [Opus Pandectarum] t.p.
wanting.
                    B. Locatellus for Octavianus Scottus, Venice 1498
    (*b*) *Opus Pandectarum Medicine M . . . S . . . nuper impressum cum*
quottationibus omnium auctorum in locis propriis et cum Simone
Januensi additis etiam nonnullis capitulis simplicium medicinarum in
aliis non repertis necnon et tractatu declarante quantum ex solutivis
laboriosis ingrediatur pro singula dragma pillarum et ellectuariorum
solutivorum nec non et tabula addita, compillatis per . . . Baptismam
Sardum qui pro amovendis erroribus non paucis in opere Pandectarum
comptis maxima cum diligentia opus hoc castigavit.
                Simon Bivilaqua for Nich. de Girardenghis 1512.
   Ed. p.

Prov.: Two copies: (1) 'Stephanus Everardus es [*sic*] verus possessor
huius libri ex dono Mgri Gulihelmi [*sic*] Barwell 1591 de Ravenston'.
(2) Several names unidentified, one of 1564. *Incipit:* (added by editor)
'Opus Pandectarum quod aggregavit . . . M . . . S . . . ad serenissimum
Sicilie regem Robertum qui fuerunt anno mundi 6516 anno vero
Christi 1317 coevi Petro de Abano, Dino de Garbo, Gentili, Bona-
venture, Francisco Mayroni et Nicholao de Lira additur Simon
Ianuensis ubique per alphabetum'.

Dedicatory note by Baptista Sardus to 'Universitas Ticinensis'
(i.e. Pavia). Medical dictionary with numerous entries which would
hardly find a place in such a work today. Many entries are followed by
abstracts from authorities easily identified; but 'posse' (presumably
equivalent to 'δυναμις', that is 'power',—hence the medieval 'dyna-
midia') might puzzle some readers.

648. SIMONIUS, SIMON. *Artificiosa curandae pestis methodus. Libellis*
    duobus comprehensa. . . .
                  (Col. Io. Steinmann, 1576) Leipzig
   Only edn.
   Prov. LIDDEL.

The dedn. refers to two books written while the plague raged in the
whole region but rather more fiercely in Leipzig itself. He seems especially concerned to expose the pretensions of quacks and the avarice
of physicians in general.

The text refers to Fracastoro (No. 273) and Fernel's *De abditis rerum*
causis, as well as to the classical authors.

649. S)CIO, NOBILE. *De Temporibus et modis recte purgandi in morbis, tractatus . . . ad Hippocratis, Galeni, Avicennae & ceterorum medicorum scripta intelligenda. . . Nunc primum . . . editus.*

Seb. B. Honoratus, Lyon 1555

)nly edn.

rov. (i) 'Albertus le feure Parisius Doct. Med.'
(ii) Arbuthnot.

650. SC LENANDER, REINERUS. *De caloris fontium medicatorum causa eorumque temperatione, li. ii. . . .*

Franc. de Gabiano, Lyon 1558

C ily edn.

P ɔv. BUTE.

In the dedn. to William, Duke of Cleves, the author refers to G. Agricola (No. 8) as the first to account for the heat of subterranean springs.

The te t refers to the two 'kinds' of 'sulphur' (p. 98).

651. SOI ENANDER, R. *Consiliorum medicinalium . . . sectiones quinque, uarum prima ante annos triginta octo a Ioanne Francisco de Gabiano ugduni edita, & cum consiliis celeberrimi medici Ioannis Montani 16 excusa. Reliquae quatuor ab auctore iam recens additae.*

Heirs of Andr. Wechel, Frankfurt 1596

On y edn.

Pro . LIDDEL (?); also another copy.

Text (p 505)—'Interea abste D. D. Godefride (*sc.* Achtius] amice peto ut isti s Tabaci Nicotianae seminum quaedam granula . . .' ('Anno 88').

SOLI IUS, C. JULIUS See No. 436.

652. SOVI ɹOLIUS, GULIELMUS. . . . *de Peste disputatio.*

Mathurinus Prevotius, Paris 1571

Onl edn.

Prov BUTE.

The dedr refers to the difficulty of the times in Paris.

The text Ch. 6) refers to the importance of observations of the stars as well as of he 'inferior' bodies. At the end there is a note ('medicinae studiosis') b Arnaldus Ferratus, pharmacist of Toulouse, dated 1571. Ch. 2 atten ts a correlation between the incidence of well attested past plagues nd the distribution of the heavenly bodies preceding them. The whole w ʼk is a valuable 'text book' on the phenomena and theories of the plagu

R*

653. [ED. SPACHIUS, ISRAEL.] *Gynaeciorum sive de mulierum tum communibus tum gravidarum, parientium, et puerperarum affectibus & morbis. Libri Graecorum, Arabum, Latinorum, veterum et recentium quotquot extant, partim nunc primum editi, partim vero denuo recogniti . . . Opere & studio Israeli Spachii.*

Lazarus Zetzner, Strassburg 1597

Only edn.

Numerous cuts.

654. STADIUS, IOANNES. *Ephemerides novae et auctae . . . Ab anno 1554 ad annum 1576. . . .*

Heirs of Arnold Birckmann, Cologne 1560

[*Ed. p.* 1556.]

Prov. 'Io. Rowe 1647' (Principal of Marischal College).

Dedn. to 'Philip of Spain, England, etc.' Includes a string of predictions based on eclipses and conjunctions. Copernicus and Erasmus Reinhold (No. 560) are referred to approvingly for their improved tables.

Contents include a letter from Gemma Frisius (No. 293) and a ref. to 'Hermetis Trismegistae Iatromathematica (hoc est medicinae cum mathematica conjunctio)'.

655. STADIUS, IO. *Tabulae Bergenses aequabilis et adparentis motus orbium coelestium . . . Opus astronomis, astrologis, medicis, politicis, oeconomicis, poetis, theologis, historiographis, grammaticis necessarium.*

Heirs of A. Birckmann, Cologne 1560

Only edn.

In the dedn. to Robert de Bergis, Bishop of Liége, he says: 'Solis absidem magna devexitate ad terrae centrum descendere, ut dubites, orbisne qui solem vehit constringatur, an tellus ipsa sursum contendat.' ('The marked descent of the sun's apse towards the centre of the earth makes one doubt whether the orb which bears the sun is drawn down or whether it is the earth itself which strains upwards?') (See Th. *Hist.* VI, 28.)

There is a secn. with half-title on The Motion of the Eighth Sphere (see Vol. I, p. 106).

656. STEPHANUS ATHENIENSIS (*syn.* Stephen of Athens). . . . *Explanationes in Galeni priorem librum Therapeuticum ad Glauconem. . . .*

(Col. Michaeles Silvii. typis) Anton. Vincentius, Lyon 1555.

Not BM.

Text of Galen with long *seriatim* commentary by S . . . A . . . and terminal scholia by Augustinus Gadaldinus.

657. STEVIN, SIMON. *Les Oeuvres Mathematiques de S . . . S . . . de Bruge. Ou sont inserées les memoires mathematiques, esquelles s'est exerce le Prince Maurice de Nassau, Prince d'Aurenge . . . Le tout reveu, corrigé et augmenté par Albert Girard. . . .*

Bonaventure and Abraham Elsevir, Leiden 1634

Ed. p. in this form.

Contents: *L'Arithmetique . . . aussi L'Algebre . . .*
Les six livres d'Algebre de Diophante d'Alexandrie [first four trd. Stevin, last two trd. Girard].
La Pratique d'Arithmetique contenant les Tables d'Interest. La Disme. . . .
[All the above had appeared in *L'Arithmetique* (Leiden, 1585); the *Tafelen van Interest* (1582) and *De Thiende* (nd.—1585?) had been published first in Flemish.]
Suivent les memoires mathematiques du Prince Maurice de Nassau. Cosmographie . . . La Practique de Geometrie. L'Art Ponderaire, ou la Statique. L'Optique. Et après les susdites memoires. La Castremetation, La Fortification par escluses. La Fortification.

These 'mathematical memoirs' had been composed in MS for the use of Prince Maurice. They were based on *De Havenvinding* (1599— Englished by Edw. Wright as *The Haven-Finding Art*); *De Beghinselen der Weeghconst, Der Weeghdaet,* and *De Beghinselen der Waterwichts*— published together in 1586 and constituting *La Statique*; and other works; and were printed on Stevin's instructions in 1608, both as *Wisconstighe Ghedachtenissen* and in a French trans.

There is a touching dedn. by 'les tres-humbles & tres obeissans subjets serviteurs & servantes, la vefue & les enfans orphelins de feu Albert Girard' to the 'Estats Generaux et . . . le Prince d'Aurenge gouverneur desdites Provinces' (brother of Maurice, then deceased).

Sep. dedn. of *La Disme* to 'astrologues, arpenteurs, mesureurs de tapisserie, gavieurs, stereometriens en general, maistres de monnoye & à tous marchans'.

For further details see Smith pp. 386-9 and especially E. J. Dijksterhuis *The Principal Works of Simon Stevin*, Vol. I, Amsterdam 1955, *Intro.*

658. STOEFFLER, IOANNES (*syn.* Stoflerinus). (*a*) *Elucidatio fabricae ususque astrolabii . . . atque totius sphericae doctissimo autore, iam denuo ab eodem vix aestimandis sudoribus recognita diligenter locupletataque & tandem non minore diligentia Cobelianis typis excusa. . . .*

(Col. Jac. Cobelius) Oppenheim 1524. (Col. includes the fine device of Cobelius.)

[*Ed. p.* Oppenheim 1513 (t.p.) 1512 (col.) BM. But in the opening para. of the text of No. 658 the author concludes: 'Valete sydere foelici lectores candidi. Tubingae. Anno Salutis sesquimillesimo decimo'.]
Prov. JOHNSTON.
Numerous cuts of manufacture and use of the astrolabe (see Vol. I, p. 118), including one showing a twenty-four hour day.

> (*b*) Same title *Io. S . . . authore, cui perbrevis eiusdem astrolabii declaratio, a Iac. Koebellio adiecta est.* . . .

Cologne 1594

Not BM.
Prov. LIDDEL.

659. STRUPPE, IOACHIM (*syn.* Strüppius).
ΣΙΤΟΠΟΤΙΑΜΑΤΕΧΝΙΑ. Antidotarii Antitrimastigi, id est medelaetrium extremorum Dei flagellorum libri i adumbratio, qui est de corporali nec non spirituali anchora famis, sitis, valetudinisque mortalium. Durch Gottes segen newe speiszkammer und speisz-keller. . . .

(Col. Martin Lechler—at the author's expense—Frankfurt 1573), t.p. and dedn. 1574.
Only edn.
Prov. 'Georgius Pavus pharmacopaeus', and an earlier.
Dietetics—kinds of foods and modes of preparation.

660. STRUTHIUS, IOSEPHUS. *Artis Sphygmicae iam mille ducentos annos perditae et desideratae, libri quinque . . . Nunc denuo . . . emendati.*

(t.p.: 'ad instantiam Jacobi Anelli de Maria, bibliopolae Nea[po]-litani', Venice 1573. Col. Petrus Dehuchinus]
[*Ed. p.* Basel [1555] ['Sphygmicae Artes . . .']
Prov. BUTE.
The dedn. 'Scholae philosophorum & medicorum Patavinae' contains the characteristic passage 'Vos anatomicen, vos Botanicen, artes iam diu sepultas, in lucem nunc [! 1555] revocastis: iam vero & Sphygmicen quam ex Latinis scriptoribus hucusque quod ego sciam adtigit nemo, vobis redivivam adfero . . .' (Posznan 1555). On a superficial view it does not appear that the 'Latin writers' (i.e. the medieval 'barbarians') missed much in ignoring this highly formalised study of the 'pulse' in relation to the 'viscidity', etc., of the humours; but a careful study of 'motions' of the arteries here described might be rewarding.

661. SUBERVILLE, HENRY. *L'Henry-metre instrument royal et universel, avec sa theorique, usage, et pratique demonstree par les propositions elementaires d'Euclide et regles familieres d'arithmetique & aussi sans arithmetique. Lequel prend toutes mesures geometriques & astronomiques qui luy sont circulairement opposées tant au ciel qu'en la terre,* SUR UNE SEULE STATION, *par un seul triangle orthogone, sans le bouger de sa place nyaller mesurer aucune distance de station . . . Item un petit traicté sur la theorique & pratique d'extraction des racines quarrées pour dresser les scadrons & bataillons quarrées.*

Adrien Perier, Paris 1598

Only edn.
Prov. 'George Andersone'.
Dedn. to Henry IV (with portrait) contains a list of contemporary instruments and a history of the astrolabe.

662. SUCHTEN, ALEXANDER VON. *Liber unus de Secretis Antimonii, Das is von der grossen heymligkeit des antimonii die artznei belangent. . . .*

Heirs of Chris. Müller, Strassburg 1570

Ed. p. BM. This edn. not in Ferg. or Duv.
Glossed throughout in Latin and German.
The long *Vorrede* is by Mich. Toxites who refers to having 'recently prepared the *Archidoxa* of Theophrastus [*sc.* Paracelsus] for the press'. The text, in German interspersed with Latin phrases, is provided with chapter, and marginal section, headings in Latin.

663. SUMMERUS, FABIANUS [*syn.* Sommerus). *De inventione, descriptione, temperie, viribus et in primis usu thermarum D. Caroli IIII Imperatoris. . . .*

(Col. Io. Steinmann, Leipzig 1571).

(t.p. mutilated).
Ed. p.
Prov. LIDDEL. On ft. cover *verso*: 'Emp. M.C. 1618'. The significance of this is not clear, since Liddel died in 1613.

664. SYLVIUS, JACOBUS (*syn.* Du Bois, Jacques). *De medicamentorum simplicium delectu, praeparationibus, mistionis modo, libri tres.*

Io. Tornaesius and Gul. Gazeius, Lyon 1548

[*Ed. p.* Paris, 1542. This edn. not BM.]
Prov. 'Georgius pacole pharmacop'.
The *Ad Lectorem* draws attention to the contemporary ignorance of the nature and uses of simples.

245

There is a preface: *De pharmacopoei officio.*

There is a second copy, which is of another edn. T.p. and dedn. wanting, but a page stuck to back cover reads: 'Lugduni exc[u]debat Philibertus Rolletus'. Several autographs undeciphered.

Refs. *passim* to 'Nicolaus', and on p. 400 to 'Nicolaus Alexandrinus' (see No. 477). The question of metallic remedies is taken up on p. 444, where 'lithargyrus' is several times mentioned.

See also Nos. 174, 291, 400.

665. SYNGRAPHEUS, POLYONIMUS. *Schola Apiciana. Ex optimis quibusdam authoribus diligenter constructa, qua continentur officia conviviatoris cultus & habitus boni convivii.*

(Col. Chris. Egenolph, Frankfurt 1534)

Ed. p.

Prov. FORDYCE.

M. Gavius Apicius was an epicure of the time of the Emperor Tiberius; but the extant book purporting to be by him is of much later date.

SWINESHEAD, RICHARD(?) of. (*Syn.* Suiseth.) See No. 680.

TABERNAEMONTANUS. See THEODORUS.

666. TAGAULTIUS, IOANNES. *Commentariorum . . . de purgantibus medicamentis simplicibus li. ii. In gratiam Pharmacopoeiae candidatorum nuper . . . editi.*

Gul. Rovillius, Lyon 1549

Not BM.

Prov. 'Liber est Thomas Tayler' (17th cent?)

The dedn. to Cardinal Bellaye, Bishop of Paris, refers to his service to the state, contempt for danger, shipwreck in 'Oceano Britannico', likening him and his brother to Nestor and Ulysses, Francis I to Agamemnon. He refers also to the reform of pharmacy. Dated Paris 1537, so there was presumably an earlier edn. (Bibl. Nat. Cat. for 'T' not available).

667. TAGAULTIUS, IO. . . . *de Chirurgica institutione libri quinque. His accessit sextus liber de materia chirurgica authore Iacobo Hollerio. . . .*

Gul. Rovillius, Lyon 1549

[*Ed. p.* Paris 1543; another edn. of Venice 1544 is mentioned but not No. 667]. The privilege has: 'Secunda editio—reveu corrigé & augmenté . . . par ledict Tagaut.'

PLATE V

One of the many illustrations of skin-grafting technique
from Tagliacozzi's *De Curtorum Chirurgia* (No. 668a).

The *Praefatio* explains that, disgusted by the errors and barbarisms of Guy de Chauliac's surgery ('apud medicos iuxta ac chirurgos in pretio habitam'), Tagault decided to write a work of the same kind purged of the errors and of 'vocabulis illis sordidioribus, atque ex intima fere plebeii sermonis faece acceptis . . .'.

Cuts of 'wound man', instruments & skeleton. Some of these were used in 'Vesalius's' *Chirurgia Magna* (No. 711).

668. TALIACOTIUS, GASPAR (*syn.* Tagliacozzi). (*a*) . . . *De Curtorum Chirurgia per insitionem libri duo.*

Gaspares Bindonus junr., Venice 1597

[Followed by expanded t.p. in red and black. As above, then] *In quibus ea omnia quae ad huius chirurgiae narium scilicet, aurium ac labiorum per insitionem restaurandorum cum theoricen tum practicen pertinere videbantur clarissima methodo . . . declarantur. Additis cutis traducis instrumentorum omnium atque deligationum iconibus & tabulis.*

Ed. *p.* (Folio).

Prov. READ. and a second copy.

(*b*) *Cheirurgia Nova . . . de narium . . . defectu per insitionem cutis ex humero, arte, hactenus omnibus ignota, sarciendo. . . .*

Io. Saurius for P. Kopff, Frankfurt 1598

Prov. 'Robrti Smith'

Contains the printer's dedn. to his patron, H. Rantzovius (No. 550) as well as the original (author's) to Vincenzio Gonzaga.

(*a*) has dedn. by Tagliacozzi to Vincenze Gonzaga of Mantua dated Bologna 1597. There are two t.p.'s., the second of which contains a papal privilege *in recto* and, *in verso*, a *nihil obstat* (in Italian) of the Council of Ten.

Both edns. have numerous cuts at end showing instruments, supporting jacket, site of graft, etc. of which an example is shown in Plate V. (For a detailed study of this work see M. T. Gnudi & J. P. Webster *The Life and Times of Gaspare Tagliacozzi*, New York n.d.)

669. [TANSTETTER, GEORG. Ed.] *Tabulae eclypsium Magistri Georgii Peurbachii, Tabula primi mobilis Ioannis de Monteregio. Indices praeterea monumentorum quae clarissimi viri Studii Viennensis alumni in astronomia & aliis mathematicis disciplinis scripta reliquerunt; quae si lector te oblectaverint, curabimus ut & alia in lucem bono auspicio aliquando progrediantur. Postremo . . . invenies . . . pene omnium instrumentorum puta astrolabii, sapheae, organi Ptolemaei, metheoroscopii, armillarum, torqueti rectanguli, quadrantium & id genus aliorum (quae recensere longum esset) usum & expeditam praxim. . . .*

247

(Col. Io. Winterburg for Leonardus and Luca Alantse, Vienna 1514. Elimatum denuo & recognitum . . . a Georgio Tanstetter . . . Imperante . . . Maximiliano Caes.)

An important record of the Vienna school of astronomers, which was the leading one in the 15th cent. (See Vol. I, p. 102.) There are short biographical notices of all those who took part in astronomical studies from the time when 'Henricus de Hassia, coming from Paris, first introduced the study of theology and astronomy into the *studium* of Vienna. He died 1397 and was buried in the cathedral.'

Ed. p. of Tanstetter's edn.

670. TANSTETTER, G. *Artificium de applicatione astrologiae ad medicinam deque convenientia earundem Georgii Tansteteri Canones aliquot & quaedam alia.*

Geo. Ulricherus, Strassburg 1531

Ed. p.

The dedn. by Otto Brunfels (No. 109) ends: 'I would prefer one astrologer-physician to ten high-and-mighty (*purpuratos*) followers of Avicenna ignorant of astrology.'

Allen, who states that the writing of the dedn. by Brunfels was due to his having received MS notes from one of Tanstetter's students, rightly commends the work as a valuable source of renaissance medical teaching.

671. TAP(P), JOHN. *The Pathway to Knowledge containing the whole art of arithmeticke both in whole numbers & fractions; with the extraction of roots; as also a brief introduction or entrance into the Art of Cossicke Numbers . . . Wherewith is also adioyned a briefe order for the keeping of marchants bookes of accompts by way of debitor and creditor.*

T. Purfoot for T. Pavier, London 1613

Ed. p. S.T.C. 23677.

Dedn. to Sir Thos. Smith, 'Governor of the Company of Marchants of London', etc., commending him for his patronage of navigation, lectures on which were then being read.

In *To the Reader* is explained the origin of the book as a revision of one written under the same title and 'as I take it translated out of Dutch'. The 'entrance' into 'Cossicke numbers' is 'only a literall translation collected out of M. Valentine Menher his arithmeticke' (see also p. 321 of the text).

Johnson (p. 319) calls his Seaman's Calendar (1602) 'an excellent book containing the astronomical tables necessary for the navigator'. See also Smith pps. 279, 281, 283.

PLATE VI

Armillary sphere (probably of the type used for teaching)
from Tanstetter's *Tabulae Eclypsium Magistri Georgii
Peurbachii* . . . (No. 669, fo. aa 8*v*.).

672. TARANTA, VALESCUS DE (*syn.* Vallastus de Tarenta).
[*Practica*—t.p. wanting.]
(Col. Jac. Myt, Lyon 1526)
[*Ed. p.* Lyon 1488 (printed '1478' in error—Hain 15249)].
Prologus: 'Incipit prologus in practicam usualem V . . . de Tharanta
. . . que alias philonium dicit. . . '.
Incipit of text: 'Deus ad laudem tui honorem . . .'; of Cap. 1. 'Primo
nota quod multa nomina sunt dolorem [*sic*] capitis . . .' ThK 190, 516.
From col: '. . . una cum introductorio . . . Ioannis de Tornamira'.
At the end of 424 double folios: '. . . et hec sufficiant pro facili intro-
ductione iuvenum practicam exercere cupientium quorum deus opera
dirigat atque felicitet'!
See also No. 192.

673. TARTAGLIA, NICOLO. (*a*) *Quesiti et inventioni diverse.* . . .
(Col. Venturino Ruffinelli for the author, Venice 1546
Ed. p.
Prov. 'Jacobi Moresini & amicorum' (?—partially obliterated).
Dedn. to Henry VIII ends '. . . alli piedi della quale, prostrato in
terra con le man gionto & capo chino himilmente me raccomando . . .'
—surely one of the most abject in a century in which such custom was
not unusual.
The *Lettori* commences 'Chi Brama di veder nove inventioni
'Non tolte da Platon, ne da Plotino
'Ne d'alcun altro greco, over Latino
'Ma sol da Larte, misura, e Ragioni
The mention of Plotinus and omission of Aristotle is significant.
Consists of nine 'books' of questions posed by various persons,
including Richard Wentworth, and discussed in dialogue between the
questioner and Tartaglia himself. The first is by Francesco Maria,
Duke of Urbino, who is made to refer to 'your book dedicated to me'
(presumably the *Nova Scientia inventa da N— T— Brisciano*, Venice
1537—examined at BM.) The purpose of the first six 'books' is wholly
the improvement of artillery, fortification, and the line of battle; but
the discussion raises fundamental questions of the theory of motion and
of statics, which are elaborated in the seventh and eighth 'books'. The
ninth consists of a variety of problems in 'Scientia Arithmetica,
Geometrica, & in la Practica Speculativa de Algebra . . .'. The work
contains many *loci classici* for the history of mechanics; the 'chemical'
function of saltpetre is also dealt with (fo. 41*v*).

(*b*) *Three Bookes of Colloquies* concerning the arte of shooting in
great and small peeces of artillerie. . . .
Thos. Dawson, for John Harrison the elder, London 1588

249

Only edn. in this form—STC 23689.

Prov. 'James Bagg'.

The full title—too long to quote—gives a vivid description of the contents, which consist of a trn. of the first three 'books' of (*a*), to many of the questions in which Cyprian Lucar has added *corollaries*. It is followed by a much longer *Treatise named Lucar Appendix* (cont. sigs.; sep. foliation, t.p. and date, but with col. to whole work at end).

For Tartaglia, see Vol. I, p. 32, also appendix B to this Vol.

674. TAURELLUS, NICOLAUS. *Medicae praedictionis methodus. Hoc est recta brevisque ratio coram aegris ... praedicendi morbos .... Quam ex Hippocratis & Galleni* [sic] *monumentis aliorumque ... doctorum ... condidit.*

Io. Feierabend for B. Iobinus, Frankfurt 1581

Only edn.

Prov. (i) Iohannes Adamsonus. (ii) 'Simion Henryson chirurgien'.

In his dedn. to the Senate of Nürnberg the author deplores the rift between medicine and metaphysics.

675. THEODORUS, JACOBUS (*syn.* Tabernaemontanus). *New Wasserschatz Das is von allen heylsamen metallischen minerischen bädern und wassern* [*many named*] *... Auch wie man dieselbigen unnd alle metallische wasser zu mancherley kranckheiten und leibs gebrechen wider den alten bösen gebrauch nützlich ... gebrauchen soll. Desgleichen sehen ... und am tag geben durch J ... T ... Tabernaemontanum.*

Nich. Bassaeus, Frankfurt 1593

[*Ed. p.* Frankfurt 1584.]

Prov. LIDDEL.

In the dedn. there is a ref. (fo. xiiii *r*) to 'der newen vermeinten ertyten und ketzerischen erstandenen secten der Paracelsisten'; also to a petrifying well (sig. x vi *r*), and to asphalt and oil wells.

The text covers the famous 'volcanic' area of the Rhine, Mosel, and Eifel.

676. THEODORUS, J. *Ein new Arzney Buch darim fast alle eusserliche unnd innerliche Glieder dess menschlichen Leibs sampt ihren Kranckheiten und Gebrechen von dem Haupt an biss zu den Füssen, und wie man dieselbigen durch Gotteshülf und seine dazu geschaffene Mittel auff mancherley weiss wenden und curieren soll. Durch ... Christophorum Wirsung aus den berhumbtesten Artzten ... ersten in Druck*

250

verfertigt. Folgends aber . . . und auffem newes mit einem lesslichern Buchstaben in Druck uebergehen durch J . . . Th . . . Tabernaemontanum. . . .

Matth. Harrnisch, Neustadt 1582

Only edn.

Prov. LIDDEL, etc.

Dedn. (i) by Wirsung, dated 'Heydelberg 1568'. (ii) Theodorus to Elizabeth, Herzogin of Saxony, dated 'Heydelberg, anno 77'.

This copy bears the marks of severe burning, probably in the 16th cent., since many pp. of text have been replaced by MS in Liddel's and possibly another hand.

677. THEODORUS, J. [*syn.* Tabernaemontanus]. *Regiment und kurtzer Bericht wie man sich in Sterbensläufften da die Pestilenz einreisset halten auch wie man sich durch Gottes dess Allmächtigen Hülff vor dieser vergifften Sucht mit guten erfahrnen Mitteln bewahren . . . Jetzund auff ein newes obersehen und gemehret . . zu Ehren und Wohlfahrt unserm geliebten Vatterland.*

Col. Nic. Basseus, Frankfurt 1586

Only edn. BM.

Fo. 69*v.* gives particulars of a 'Teutscher Theriack', and fo. 97*v* instructions for the cleansing of bedding, etc.

678. THEODOSIUS OF TRIPOLI. . . .*Sphaericorum libri iii a Christophoro Clavio Bambergensi Societatis Jesu . . . Demonstrationibus ac scholiis illustrati. Item eiusdem C . . . C . . . [list of works in geometry & trigonometry].*

Dominicus Basa, Rome 1586

[*Ed. p.* of Latin trans. 1518, of Greek original 1558.]

Pp. 237-40 inc. bound out of place and in wrong order.

The dedn. hints at the origin of trigonometry in Ptolemy's method of chords: see Vol. I, p. 96.

The *Praefatio* attempts to fix the time & place of Theodosius. According to Strabo he came not from Tripoli, but from Bithynia. Clavius says that he has followed Io. Pena's trans. from the Greek, also making some use of that of Francesco Maurolyco from the Arabic.

679. THEOPHILUS PROTOSPATHARIUS. . . . *in Galeni de Usu Partium libros epitome quam de corporis humani fabrica inscripsit. Iunio Paulo Crasso Patavino interprete.*

Conrad Neobarius, Paris 1540

Only edn.

In his dedn. Crassus refers to vivisection of human beings 'per eam quae a Graecis sapientibus anatome a nostris dissectio nominatur'. It would be interesting to know when 'dissection' replaced 'anatomy' in the sense of *process*.

For Theophilos see No. 47.

680. THOMA, ALVARUS. *Liber de triplici motu proportionibus annexis magistri Aluari Thome Ulixbonensis philosophicas Suiseth Calculatones* [sic] *ex parte declarans.*

(From t.p. col. and explicit: Guillermus Anabat for Ponset le Preux, Paris 1509.)

> Only edn. (BM French books).
> Prov. Bound in contemporary leather with work inscribed '. . . Steuart & suos'. Fine MS on end papers. Finely engraved t.p. *Incipit Prohemium*: 'Praeclara philonis in libro sapientie extat.' 'Incipiunt proportiones—Omnis numerus et similiter omnis quantitas . . .'

Explicit: 'liber de triplici motu compositus per Magistrum Alvarum Thomam ulixbonensem Regentem Parrhisius in Collegio Coquereti [Collège de Coqueret] Anno domini 1509 die Februarii 11.'

This is one of the several 'introductions' to the *Liber Calculationum* (*syn. Calculationes*) of Richard of Swineshead ('Suiseth' *syn.* 'Calculator') in which the germs of 17th cent. kinematics and fluxions are to be found; ThK 481 cite MSS without dates. The *Ed. p.* of the *Calculationes* is undated, but probably 1477 (Padua—Hain 15136).

'Ulixbonensis' is an uncommon variant of 'Ulyssibonensis', etc., i.e. 'of Lisbon'. My colleague, Mr. T. E. May, informs me that Rey Pastor claims that Alvaro Thomaz replaced the purely geometrical methods of Nicole Oresme by arithmetical, and demonstrated fully the theorem, attributed to Galileo, on the distance moved by a uniformly accelerated body. He seems to have been overlooked by other than Spanish authors. (See Vol. I p. 156.)

681. [*Ed.* TINGUS, PHILIPPUS.] *Epistolae medicinales diversorum authorum* . . . . [The following authors are included: *Io. Monardus, N. Massa, Aloisius Mundella, Io. Bapt. Theodosius, Io. Langius.*

Heirs of Iac. Iunta, Lyon 1556

Ed. p.
Prov. 'Gulielmus Baronsdale'.

TOMITANO, BERNARDINO. See No. 753.

682. [*Ed.* TOXITES, MICHAEL.] *Introductio in divinam chemiae artem integra magistri Boni Lombardi Ferrariensis physici: nunc primum integra . . . edita.*

P. Perna, Basel 1572

Only edn. BM: Ferg. I, 115 cites an edn. of 1602 (with different imprint) and discusses authorship. Not Duv.

Prov. LIDDEL (?).

The dedn. by Toxites to Albert, Count Palatine of the Rhine, praises chemistry ('. . . tam pauci ad secretioris physicae cognitionem, quam chemiam appellamus, pervenerint'), power of stars, the 'Stone'—all of which are necessary to effect cures. He also gives his views of the authorship of the work.

The main text is divided into a *Praefatio*; *Incipit numerus et ordo omnium pretiosae margaritae novellae in arte alchemiae; incipit preciosa margarita novella* (*prooemium* p. 27, cap. I, p. 31). The *numerus et ordo* sets out the dialectic method: 'Capitulum primum in quo disputat & probat artem Alchemiae non esse veram. Capitulum secundum in quo disputat artem Alchemiae esse veram.' (See Vol. I, p. 295.)

683. TRAFFICHETTI, BARTOLOMEO. *L'arte de conservare la sanità, tutta intiera trattata in sei libri per B . . . T . . . .*

(Col. Geronimo Concordia) Pesaro 1565

Only edn.

Prov. BUTE.

684. TRINCAVELLIUS, VICTOR. *Enchiridium medicum de cognoscendis curandisque tam externis quam internis humani corporis morbis. Ex V . . . T . . . praelectionibus de compositione medicamentorum atque morbis particularibus concinnatum, studio et opera Andreae Christiani. . . .*

Typis Oporinianis, Basel 1583

Ed. p. in this form (from dedn.). Earliest BM, Basel 1630.

Prov. 'Hafniae 1590 . . . Leslie'. Several pages cursive MS on end papers.

Dedn. by Christianus to Christian IV 'designato regi Daniae & Norvegii' and his brother Hulderich, sons of Frederick II, dated Basel 1583.

Praefatio unsigned—presumably by printer.

685. TRINCAVELLIUS, V. *Consilia medica post editionem Venetam & Lugdunensem accessione CXXVIII consiliorum locupletata & per locos communes digesta. Epistolae item philosophicis & medicis*

*quaestionibus insignitae expolitaeque. Accessere tractatus tres:—
De Ratione. De Venae Sectione in pleuriticis, & de Febre pestilenti
plane novus.*

      C. Waldkirch (Col: ad Lecythum Perneam) Basel 1587
Not BM.
    Ed. p. from dedn.
Dedn. by printer to Petrus Severinus (No. 643) Basel 1587.
Printed in two columns to the page, continuously numbered.
    The main text is prefaced by a brief life of Trincavelli for whose part
in the great controversy with Brissot about the site of phlebotomy
see Castiglioni pp. 441-2. The later *Quaestiones* deal (p. 942 *seq.*) with
the controversy between Aristotelians and Averroists (see Vol. I,
passim) and contains refs. to 'Suiseth Calculator' (No. 680), 'Hentisber'
[= William of Heytesbury) 'Marilianus' (= Marliani) and '*De motu
locali*'.

686. TRINCAVELLIUS, V. . . . *Omnia opera, partim ex diversis edition-
    ibus . . . partim nunc primum . . . emissa ac in duos tomos digesta . . .
    correcta & impressa.*

                    Off. Juntarum, Lyon 1592
    [*Ed. p.* 1586 (dedn.)]
    Prov. LIDDEL.
Tomus secundus has sep. prelims and pagn. with secn. headings and
sep. pagn. for parts.
    Dedn. by Io. Bapt. Regnauld to Alphonso Beccara, 1586.
    Short life and funeral oration for Trincavelli.
    See also No. 155.

687. TRISINUS, ALVISIUS. *Problematum medicinalium ex Galeni
    sententia libri sex.*

                Iac. Parcus, Basel 1547 (Col. 1546)
    Ed. p.
Prov. (i) 'Edwardus Lapworth Magdalenensis Oxon—Mors Christi
vita hominis 1602'. (ii) 'Ioan. Fordyce'. (iii) 'Ex don. Gul. Fordyce'.
The 'Problemata' appear to have more than the usual interest.

688. TRITHEIM, IOHANN (*syn.* Trithemius). *Polygraphiae libri sex
    Ioannis Trithemii, abbatis Peapolitani, quondam Spanheimensis,
    ad Maximilianum I Caesarem. Accessit clavis polygraphiae liber
    unus, eodem authore. . . . Additae sunt etiam aliquot locorum explica-
    tiones, eorum praesertim in quibus admirandi operis Stenographiae
    principia latent. . . . Per . . . Adolphum a Glauburg. . . .*

      Io. Birckmannus & Theodorus Baumius, Cologne 1571

PLATE VII

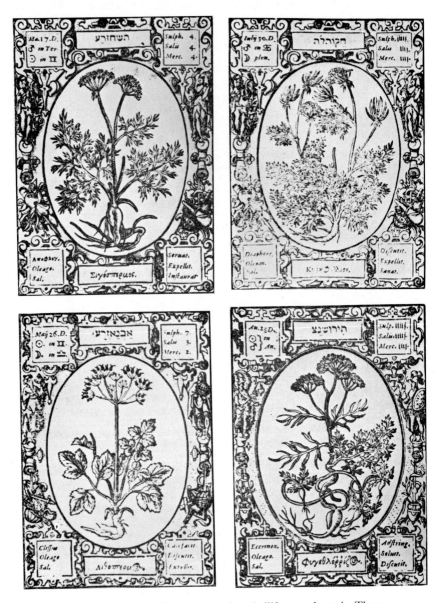

Four of the many illustrations of umbelliferous plants in Thurneisser's *De herbis quibusdam quae totum humanum corpus occulte signant*. The panels in each figure indicate astrological, 'spagyrist', and medical characteristics of each species (No. 691).

[*Ed. p.* (From dedn. to Maximilian I) 1508. Earliest BM, Oppenheim 1518.]

A life of Trithemius is extracted from a letter by Io. Duraclusius (his disciple) to Nicolaus Hamerius Emelanus, 1515.

A system of 'Cabalistic' writing and methods of 'coding'. See Biog. Univ., 'Trithemius'.

689. TRIVERIUS, HIEREMIAS (*syn.* Thriverus). *Universae medicinae . . . methodus . . . per Dionysium Thriverum Lovaniensem, ejus filium . . . edita. Cum praefatione ad . . . Zirizaeae Senatum qua non nihil detegitur nocentissima empiricorum impostura.*

Off. Plantiniana. Franc. Raphelengius, Leiden 1592

Only edn.

The *Praefatio* refers to the 'wretchedness of our most turbulent times, in which almost all learning and disciplines are treated with contempt . . . '. The acute attacks on the Paracelsians, and the refs. to Louvain (at which he was a professor), are of interest. Dated Zirikzee 1591.

Several fos. uncut.

690. TUNSTALL, CUTHBERT. *De arte supputandi libri quattuor.*

(Col. Richard Pynson, London 1522)

Only Engl. edn. (Seven further continental edns.) STC 24319.

Dedn. to Sir Thomas More on the eve of Tunstall's enthronement as Bishop of London—he was later translated to Durham.

This work is mentioned by Rabelais (Oeuvres II, 222) as 'required reading' for the young Gargantua in Paris. Johnson (p. 83) states that the Oxford Statutes of 1549 prescribed it as a text in arithmetic, together with that of Cardano. Its failure to reach a 2nd Engl. edn. may be contrasted with the twenty-eight edns. of Recorde's *Grounde of Artes* (see No. 554): it is indeed a rather long-winded book, wholly unsuited to the needs of merchants, surveyors, etc.

691. T(H)URNEISSER, LEONARD [t.p. wanting—begins sig. x. 2].
Opus per singula hominis membra digestum. De radicum, lignorum, herbarum, florum, fructuum et seminum nomenclaturis, generibus occulta habitudinis cum corpore humano (quae signatura est) & naturis coelestibus (quae constellatio) sympathia quadam seu consensu. . . . Ex doctrina partim nobilis illius & celebri fama noti Philippi Theophrasti Paracelsi, partim aliorum . . . artificum . . . partim denique ex ipsius autoris . . . scientia. De herbis quibusdam quae totum humanum corpus occulte signant. . . .introductio quaedam . . . Liber primus. [From heading of sig. A r].

255

[No imprint; but the privileges are dated (Imperial) Ratisbon, 1576, Florence 1575, Polish 1577. The dedn. to Stephen of Poland dated Berlin 1578.] Last p. has 'Finis libri primi', but there is no catchword at the foot.

The plants described (in relation to their 'signatures', celestial signs, and extractable essences) appear—from the accompanying cuts—to be all Umbelliferae. The indications used crudely foreshadow the 'isotype' system. See plate No. VII.

692. UBALDUS, GUIDUS, E MARCHIONIBUS MONTIS (*syn.* Monte, Guidubaldo del). *Mechanicorum liber.*

Hiero. Concordia, Pisa 1577

Ed. p.

The t.p. displays the Eastern Hemisphere placed on a lever with the motto 'Tolleret quis si consisteret', referring to Archimedes' alleged dictum.

Contents: *De libra* (scales); *vecte* (lever); *trochlea* (pulley); *axe in peritrochio* (wheel and axle); *cuneo* (wedge); *cochlea* (screw). The uses of machines for everyday purposes, trades, etc., are described, and the Greek and Roman writers on mechanics named. The development of all types of machines from the idea of the lever is demonstrated, and the *oblique* lever is implicit. (See Vol. I, p. 159.) Numerous clear figs.

693. UBALDUS, G. *Planisphaeriorum universalium theorica.*

Hiero. Concordia, Pisa 1579

Ed. p.

Theory of projection.

694. ULSTADT, PHILIP (*syn.* Ulstadius). *Coelum philosophorum seu de secretis naturae liber.* . . .

(Col. Io. Grienyger, Strassburg 1528)

[Earliest BM has same imprint but 'Argentoragi' instead of 'Argentoraci', as here. Dated 1526. Ferg. II, 482 states that *Ed. p.* was Freiburg 1525; but Duv. p. 591 believes this to be a mistake, printing not having been established at Freiburg till late 16th cent. He regards No. 694 as of the 3rd edn., the 1st (1525) and 2nd being by the same printer.]

Dedn. is dated 'Ex Friburgo Helvetico . . . 1525'.

Prov. 'Alexander Arbuthnot' written over another name.

The dedn. and prologue claim that the work is based on those of Lull, Arnald of Villanova, Albertus Magnus, and Io. de Rupescissa, whose works were capable of being understood by 'few if any of our contemporaries', or were condemned as 'old wives' tales' (*anicularum*).

The special interest of the book is its continuation of Io. de Rupescissa's ideas on the *Fifth Essence* (No. 313) and the parallel development of this fantasy with sound practical methods of extraction (See Vol. I, p. 173). The medical aspect is illustrated by ref. (fo. xxix) to 'Servitor' (No. 1) and Jean de St. Amand's commentary on Mesue (No. 740).

URSTISIUS, CHRISTIANUS. See No. 513(d).

695. VALENTINUS, PETRUS POMARIUS. *Enchiridion medicum containing an epitome of the whole course of physicke. With the examination of a chyrurgian by way of dialogue betweene the Doctor and the Student. . . . Published for the benefite of yong students in Physicke, Chyrurgians, and Apothecaries . . . The second impression enlarged. . . .*
　　　　　N. Okes for John Royston & Wm. Blaydon, London 1612
[*Ed. p.* London 1608 (STC 24577). BM regards the author's name as possibly a pseudonym.]
　　Latin dedn. to numerous London practitioners. Nominally 'surgical', it contains little reference to operative measures, but relies on herbal remedies (see Vol. I, p. 216).

696. VALLERIOLA, FRANCISCUS. *Loci medicinae communes tribus libris digesti. . . .*
　　　　　Heirs of Seb. Gryphius, Lyon 1562
　　Only edn.
A compendium of extracts from literature—both medical and general.

697. VALLERIOLA, F. *Observationum medicinalium libri sex nunc primum editi. . . . In quibus gravissimorum morborum historiae eorundem causae . . . describuntur.*
　　　　　Anton Gryphius, Lyon 1573
　　Earliest BM Lyon 1588.
　　Prov. LIDDEL, and another copy 'Ex lib...Ambrosii Langerii...'.

698. VALLERIOLA, F. *Commentarii in sex Galeni libros de Morbis et Symptomatibus, Francisco Valleriola autore.*
　　　　　Off. Erasmiana, Venice 1548

699. VAL(L)ESIUS, FRANCISCUS. *Controversiarum medicarum et philosophicarum libri decem.*
　　(a) Heirs of Andr. Wechel, Frankfurt 1582.
　　Ed. p.
　　Prov. LIDDEL.

s*　　　　　　　　　257

(*b*) . . . *Editio tertia ab autore denuo recognita & aucta. Accessit libellus de locis manifeste pugnantibus apud Galenum eodem . . . autore.* Same imprint, 1590.

The dedn. by Ioannes Crato refers to several contemporary physicians, e.g. Fernel, Cardano, Ferrier. Dated Prague 1581.

Interesting tables (p. 384) of degrees of heat and cold, including 'O' (zero).

700. VAL(L)ESIUS, F. . . . *in libros Hippocratis de Morbis Popularibus commentaria . . . Ioanni Petri Ayroldi . . . opera & industria nunc fidelius quam prius excusa.*

Io. Bapt. Ciotti; Cologne 1588

Prov. JAMES MORESON.

701. VALVERDE, IOANNES. *De animi et corporis sanitate tuenda libellus.*

C. Etienne, Paris 1552

Only edn.

Prov. (i) Wm. Gordon, Bishop of Aberdeen 1552. (ii) 'N (?) Stewart'. (iii) 'Walterus Stewart 1585'. (iv) Arbuthnot.

Diet and hygiene based on Aristotelian and Galenic physiology.

702. VARIGNANA, GULIELMUS. *Secreta medicinae . . . ad varios curandos morbos veriss. auctoritatibus illustrata: nonnullis flosculis in studiosorum gratiam additis nunc a Casparo Bauhino ad plurium exemplariorum* [sic] *collationem . . . castigata & obscurorum vocabulorum explicatione notisque marginalibus illustrati.*

Seb. Henricpetrus, Basel (Col. 1597)

Ed. p.

Prov. LIDDEL.

The editorial dedn. claims that the original work was finished in 1319. ThK 125 report no MS earlier than 15th cent. *Incipit:* 'Creavit altissimus cuius imperio. . . .' The author's dedn. is also printed.

703. VARRO, MICHAEL. . . . *De motu tractatus.*

Iacobus Stoer, Geneva 1584

Only edn.

In his dedn. the author claims to be able to 'demonstrate by reason and prove by experiment, those things which are attributed to Archimedes of Syracuse', and that 'these have a use not only in mechanics, in which it is certainly the greatest, but also in politics and economics; for in these there are motions, forces & resistances'. He anticipates the Royal Society in claiming with Horace 'me nullius unquam in verba magistri iurasse . . .'.

Neither the author nor the work is mentioned in Th. *Hist.* It is curious that the comparative rarity of works on *motion* during the 16th cent. does not seem to have been generally noted. (See Vol. I, p. 152.)

The use of sulphur and nitre in relation to motion is referred to at the end.

704. VASSEUS, IOANNES. *De iudiciis urinarum tractatus. Ex probatis collectus autoribus & in tabulae formam confectus, adiectis etiam causis quae hanc vel illam urinam reddant . . .*

Gul. Rovillius [Lyon] 1549

[*Ed. p.* Paris 1549.]
Prov. LIDDEL.

No dedn. but preface gives reasons for the inspection of urine.

705. VEGA, CHRISTOPHORUS à. *Liber Prognosticorum Hippocratis . . . additis annotationibus in Galeni commentarios . . . .*

Godefridus and Marcellus Beringi, Lyon 1551

Prov. LIDDEL.

706. VEGA, C. à. . . . *Liber de arte medendi.*

Gul. Rovillius, Lyon 1564

Not BM. No col. but on last page (673) before the index occurs the passage 'Ch . . . a V . . . librum de A . . . M . . . peregit suae aetatis anno quadragesimo septimo nostrae salutis 1557'. This suggests an earlier edn. (Bibl. Nat. Cat. 'V' not available).

Prov. LIDDEL. 'Lesleyus (?)' occurs on ft. paper.

A *Practica* conceived on broadly 'physiological' lines by one who was physician to the Emperor Charles V while he was King of Spain. It is instructive to compare it with Fernel's nearly contemporary works.

707. VEGETIUS, PUBLIUS. . . . *Mulomedicina. Ex trib. vetustiss. Codd. varietate adiecta unde infiniti loci addi & expurgari a quovis poterunt usu magno publico. Opera Ioan. Sambuci.*

. . . P. Perna, Basel 1574

[*Ed. p.* Basel 1528—*Artis veterinariae sive mulomedicinae libri quatuor*].
Prov. (1) Liddel (?)
(2) 'Io. Gregory 1769' with MS notes on ft. paper.

708. VESALIUS, ANDREAS. *Paraphrasis in Nonum Librum Rhazae medici Arabis clariss. ad Regem Almansorem de Affectuum singularum corporis partium curatione; Andrea Vesalio Bruxellense autore.*

259

(*a*) Robt. Winter, Basel 1537.
[*Ed. p.* Louvain 1537—poorly printed: very rare.]
Prov. (i) 'Thomas Cantuar—'. (ii) 'Lumley'. (iii) Read.
(See No. 573).

(*b*) Zacharia Lehmann, Wittenberg, 1592.
Prov. Scored and glossed in Liddel's hand.
The dedn. refers to 'Parisienses medici' and 'praeceptori meo D.
Iacobo Sylvio', to whom Vesalius was prosector in Paris together with
Servetus.

709. VESALIUS, A. . . . *de Humani corporis fabrica libri septem.*
(*a*) (Col. Ioannes Oporinus 1543) Basel.
Ed. p.
(*b*) Io. Oporinus, Basel (Col. 1555).
(*c*) F. Franciscius & Ioannes Greigher, Venice 1568.
(Prov. FORDYCE.)
(*b*) was re-set in larger type and with new blocks for t.p.
(*c*) has different t.p., has no portrait of Vesalius, and contains
dedn. by printer. See Harvey Cushing, *A Bibliography of Andreas
Vesalius.* N.Y. 1943, and C. Singer and C. Rabin *Prelude to Modern
Science,* Camb. 1946.

710. VESALIUS, A. *Anatomes totius aere insculpta delineatio, cui addita
est epitome innumeris mendis repurgata, quam de Corporis Humani
Fabrica conscripsit clariss. And. Vesalius. Eique accessit partium
corporis tum simplicium tum compositarum brevis elucidatio, per
Iacobum Grevinum . . . medicum Paris.*
Andr. Wechel, Paris 1564
The *Ed. p.* on which was based the vellum edn. of *Fabrica,* Basel
1543. This appears to be the first issue in this form—1565 BM.]
Prov. 'Liber Accademie Marschallane Ex Dono Mgtri Duncani
Liddelii Medicine Doctoris Neaberdonensis'. (Not Liddel's hand.)
Liddel's auto. t.p. This is the only volume in Liddel's library (so
far discovered) which bears any mark of 'donation' to the Library.

711. VESALIUS, A. . . . *Chirurgia Magna in septem libros digesta. In
qua nihil desiderari potest, quod ad perfectam atque integram de
curandis humani corporis malis methodum pertineat. Ab . . . Prospero
Borgarutio recognita, emendata, ac . . . edita. Formae etiam instru-
mentorum, quibus chirurgi utuntur, . . . descriptae sunt.*
Off. Valgrisiana, Venice 1569
Only edn.
Prov. Glossed throughout in 16th cent. hand—'1577'?

The dedn. by Borgarutius refers to Vesalius's 'untimely death' (as a result of exposure after shipwreck—for latest assessment see G. Sarton, also C. O'Malley *ISIS*, **45,** 131–44). List of famous contemporary physicians. Dated Padua 1568. The work was a compilation (made by the editor) in which Vesalius had no part (Cushing, *op. cit.* p. 216).

Fo. 210*r seq.* 'There follow delineations (*formulae*) of some instruments suitable for removing missiles, taken from Tagaultius' (No. 667). On fo. 215*r* is shown a method whereby a 'plumbeus globus immissus' may be extracted.

712. VICARY, THOMAS. *The Englishman's Treasure with the true anatomie of man's body compiled by that excellent chyrurgion Mr. T . . . V . . . Esquire Sergeant Chyrurgion to King Henry the 8. To K. Edward the 6. To Queene Mary. And to our late Soveraigne Qu. Elizabeth. And also chiefe chyrurgion to St. Bartholomewes Hospitall. Whereunto are annexed remedies for all captaines and souldiers . . . and . . . for all diseases . . . in man or woman . . . Also the rare treasure of the English Bathes written by William Turner . . . Gathered and set forth for the benefit and cure of the poorer sort of people, who are not able to goe to the physitians, by William Breemer, practitioner in physicke and chyrurgerie. And now eighthly augmented . . . by W[illiam] B[oraston].*

Bart. Alsop and Thos. Fawcet, London 1633

[*Ed. p.* 1585? STC 24706].

Dedn. to Sir Rowland Hayward, President of Little St. Bartholomew's in West Smithfield, and other Governors, by a number of surgeons, including William Clowes (No. 164).

Both dedn. and text emphasise the importance of anatomy, on the authority of Galen, 'Guido' (Guy de Chauliac—No. 322), and others. The 'diet for wounded' (p. 64) is worth noting, also Wm. Boraston's theory of the origin of pestilence from the firmament of heaven (p. 246). It is noteworthy that the anatomy is purely descriptive, without figures.

713. VICTORIIS, LEONELLUS DE. *Practica medicinalis . . . cum scholiis Ioannis Kufneri. De Aegritudinibus Infantium, eodem authore, tractatus. Item, appendix ad eundem per Georgium Kufnerum iuniorem.*

(*a*) Io. Frellonius, Lyon 1562.
Prov. (i) 'Albertus Lefeure Parisens, Doct. Medic.'
　　　(ii) Arbuthnot.

(*b*) Barth. Vincentius, Lyon 1574.
Prov. 'Lumisdane 1619'.

261

(*b*) is the earliest BM, but the dedn. by the elder Kufner to Antony Fugger, printed in both (*a*) and (*b*), is dated 'Apud Aras Flavas 1545'. The *De Aegr. Infantium* has half-title, sep. pagn. and dedn. dated 'Reginoburgi 1543'. The *Praefatio* cites Avicenna's four canons for the treatment of children's diseases.

There is a sketch (p. 12 *seq*) of the life of de Victoriis.

714. VICTORIUS, BENEDICTUS. *Benedicti Victorii . . . in Hippocratis Prognostica commentarii. His accessit Theoricae Latitudinum Medicinae liber, ad Galeni scopum in Arte Medicinali.*

Laurentinus Torrentinus Ducales, Florence 1551

Prov. GORDON.

715. VICTORIUS, B. . . . *Medicationis empeiricae libri tres. Itemque Camilli Tomai Ravennatis Curandorum morborum rationalis methodus. Hactenus diu desiderata, . . . recognita . . ., nunc primo in Germania prodeunt.*

Io. Collizius for Io. Schoenwetter, Frankfurt 1598

Not BM, but presumably based on *Empirica* 2nd edn. Venice 1554, which also contains Toma's work (BM).

Prov. LIDDEL.

Cont. pagn. but secn. heading to Toma's work p. 325.

716. VIDIUS, VIDUS. . . . *Universae artis medicinalis ab ipso nonaginta quinque libris comprehensae pars quae ad curationem morborum spectat. Continet haec pars quadraginta quinque libros quorum triginta quatuor ab ipso authore ante obitum editi sunt, reliqui undecim vix inchoati a V . . . V . . . nepote . . . conscripti.*

Heirs of Andr. Wechel, Frankfurt 1596

Not BM, except as part of *Opera omnia*, Frankfurt 1626.

The *Lectori* of Marnius and Aubrius, the printers, contains biographical details, with an interesting 'blurb' near the end in which they announce 'forthcoming titles'.

The *Tomus quintus* has sep. t.p. and pagn.

Since the younger Vidius refers to his 'patruus', 'nepos' presumably means nephew here. Guido Guidi became the first Professor of Medicine in the *Collège Royale*, enlarged by Francis I in 1545, now Collège de France.

717. VIGO, IOANNES DE [t.p. of First Part wanting]. *Ioannes de Vigo, secunda pars Practice in profectione chirurgica que compendiosa nuncupatur. Totius chirurgie documenta que in copiosa sive prima parte diffuse explicantur summatim complectens nuperrime compillata a Iohannetino de Vigo. . . .*

[Contents: (1) *De Vulneribus*, (2) *Apostematibus*, (3) *Ulceribus*, (4) *Auxiliis*. . . . Contents of First Part similar, but also *De Anatomia* and *De Morbo Gallico*.]

In quinto finem faciendo agit de variis et diversis auxiliis sub ordine Antidotarii et de parva chirurgicorum capsa navigantium pro eorum commoditate. . . .

Antonius de Ry, Lyon 1525

[*Ed. p.* Rome 1517, but completed only in Lyon 1522.]

Not BM. The dedn. is dated 'Tiburi sedente Leone X . . . 1517'. The col. to the two parts is on fo. LI*v*, but is undated. Then follows 'Marianus Barolita, *Compendium in chyrurgia* . . . *castigatum*', the work of a disciple of Vigo. Final col. fo. LXXXVI*v* listing the contents, with details of printing. Some sep. foln. but continuous signatures (not registered). The running titles in Part I are misleading.

Interesting for the history of 15th-16th cents., stressing the importance of anatomy.

718. VIGO, IO. *Practica in chirurgia . . . Practica in arte chirurgica copiosa Io. de Vigo . . . continens novem libros . . .* [Contents:] *De Anatbomia* [sic] *chirurgo necessaria, Apostematibus, Vulneribus, Ulceribus, Morbo Gallico, Fractura et dislocatione ossium, Natura simplicium, Natura compositorum, quibusdam additionibus totum complentibus.*

Jacobus Myt for Vincentius de Portonariis [Col. Lyon 1516].

Similar to No. 717.

Prov. Bound with No. 746 which bears marks of owners 1548, 1556, and Liddel 'empt . . . 1584'.

719. VIGO, IO. *The most excellent worckes of chirurgery, made and set forthe by Maister John Vigon, head chirurgien of oure tyme, in Italy, traunslated into Englishe. Whereunto is added an exposition of straunge termes and unknown symples, belongynge unto the arte . . .*

Edwarde Whytchurch, [London] 1550

[*Ed. p.* London 1543, STC 24720.]

Prov. BUTE.

No prelims. exc. detailed contents. No col. 'Finis' sig. AAa iii *v* without catchword.

720. VINET, ELIA. *Sphaera Ioannis de Sacro Bosco emendata . . . scholia in eandem sphaeram . . . Adiunximus huic libro compendium in Sphaeram per Pierium Valerianum. . . . Et Petri Nonii . . . demonstrationem eorum quae in extremo capite de Climatibus Sacroboscius scribit de inaequali climatum latitudine, eodem Vineto interprete.*

Hiero. de Marnef & widow of Gul. Cavellat, Paris 1577

[*Ed. p.* 1550 from dedn. BM cites three earlier than 1577.]
Prov. 'James Ogilvie', but the gloss looks like an earlier hand.
Dedn. by Vinetus contains the *locus classicus* concerning Io. de Sacrobosco's (alleged) native country, his works, sojourn in Paris, and burial in the church of St. Mathurin, the epitaph on his tomb therein being cited. Dated 1550.

Praefatio (presumably by Vinetus) contains the passage, 'neither Picus [No. 519] nor anyone else will make me believe that the stars have no effects on the elements and on the bodies of living creatures'.

The text is followed by: (1) The *scholia* of Vinetus. (2) The annotation on the Climes by Vinetus, 'written in Portuguese'. (3) *Expositio XXII ex libro Epitomae Ioannis de Monteregio in Almagestum Ptolemaei. Dies naturales duplici causa inaequales esse.* (4) *Ex Alfrag* [sic, see AHMAD IBN ABDULLAH] *De Ortu et Occasu Planetarum.* . . . (5) *De Ortu poetico . . . ex variis auctoribus collecta . . . incerto auctore.* (6) the *Compendium* of Valerianus. (7) *De Eclipsibus ex Proclo.*

MS sketch of Copernican system with satellites of Jupiter and Saturn in 17th cent. hand, p. 94.

721. VIRDUNG, IOANNES. *Nova medicinae methodus curandi morbos ex mathematica scientia deprompta, nunc denuo revisa & . . . emendata.* . . .

Valentin Kobian, Hague 1533

[*Ed. p.* Ettlingen 1532.]

No dedn. but *Ad lectorem* at end dated Heidelberg 1532.

An important source of the extremer form of 'astrological medicine'. See Allen, pp. 59, 250, 251, who emphasises the failure of the deluge of 1524 foretold in Virdung's *Practica Teütsch* . . . 1521. Especially useful for uroscopy, including 'urina non visa' sigs X ii–X iii *r*.

VITRUVIUS (=M.VITRUVIUS POLLIO). See Nos. 113, 516, 642.

722. WECKER, IOANNES IACOBUS. *Practica medicinae generalis . . . VII libris explicata.*

Hiero. Frobenius & relatives, Basel 1585

Ed. p.

Preceded by Aphorisms.

723. WECKER, IO. IAC. D. *Alexii Pedemontani de Secretis libri septem a Io . . . I . . . Vechero . . . ex Italico sermone in Latinum conversi & multis bonis secretis aucti. Accessit hac editione eiusdem weckeri* [sic] *opera, octavus de Artificiosis Vinis liber.*

P. Perna, Basel 1563

264

Ed. p. doubtful. Ferg I, 22 calls this the '3rd'. BM cites *Libri sex,* 1559, but not *Libri septem.* In his dedn. 'Weckher' writes '. . . ante annos plus minus tres, . . . cum selectorum sive secretorum ut vocant libros ex Italico in Latinum sermonem a me translatos in lucem ederem . . .'.

Alexis (see No. 21), in the *Ad lectorem,* claims a knowledge of Chaldee and Arabic as well as of Greek & Hebrew.

Incipit of text: 'Primo semper in animo et ante oculos positum esse oportet . . .' (Not ThK).

724. WECKER, IO. JAC. *Antidotarium generale . . . nunc primum . . .*
    digestum . . . .
    Eusebius Episcopius and heirs of Nich. Episcopius, Basel 1580
    [*Ed. p.* 1576 dedn.]
    Prov. LIDDEL.
Dedn. by Wecker to the Senate of Strassburg, referring to Paracelsus and his followers (*asseclas*).

Fine cuts of distillation apparatus. Text in four books with cont. pagn.

725. WECKER, IO. IAC. *Antidotarium speciale . . . Ex opt. authorum*
    tam veterum quam recentiorum scriptis . . . multis etiam simplicibus
    & compositionibus adauctum . . . [Imprint as in No. 724 1581.]
    [*Ed. p.* Colmar 1574 (from *Praefatio ad medicinae studiosos,* which contains much valuable historical matter)].
    [Earliest BM Basel 1588 'auctum'].
    Prov. LIDDEL.
This work commences within the same binding as No. 724 but *Liber II* is continued and concluded in a sep. vol. [Sigs. R8, R4 of Lib. II have been bound (out of order) in the first vol. *Liber* III follows *Liber* II; pagn. cont. throughout].

726. WECKER, IO. IAC. *Medicinae utriusque syntaxes ex Graecorum,*
    Latinorum, Arabumque thesauris. . . .
                              Stephanus Michaeles, Lyon 1583
Only edn. BM, but dedn. (to Wm. Landgrave of Hesse) dated Colmar 1576.
    Prov. GORDON.
The dedn. refers to the Paracelsian 'chymica schola'.

The text is arranged throughout in tabular form under the headings (in Greek) 'physiological', 'pathological', 'aetiological', 'therapeutic', 'prophylactic', and (in Latin) 'cure of external diseases'.

727. WERDMÜLLER, OTHO. *Similitudinum ab omni animalium genere desumptarum libri* VI. *Ex optimis quibusque authoribus sacris & profanis Graecis & Latinis per Othonem Vuerdmyllerum collecti & digesti.* [No imprint. *Epistola* from Zürich.]

Not BM.

Prov. 'Liber Coll. Regii Aberdonen. Ex dono Mgtri Jho Garden.' *Epistola* from Conrad Gesner 'Abel Vuerdmyllero, affini suo . . . Otho V . . . felicis memoriae pater tuus . . . '.

728. WIER, IOANNES [*syn.* Weyer]. *Artzney Buch von etlichen bisz anher unbekannten unnd unbeschrieben Kranckheyten deren Verzeichnung im folgenden Blatt zu finden. Durch . . . Herrn Johann Weyern . . . selbst verfertigt und in Teutsche Spraach verbracht. Jetzt aber auffs neuw gebessert und vermehret.*

Nich. Bassaeus, Frankfurt 1588

Only edn.

The dedn. to 'Frauwen Anna Grävin zu Teckhelnburgk Bentheim und Steinfurt' contains a history of the House of 'Teckhelnburgk'—from Clèves.

This is a work full of interesting allusions, e.g. *Löffelkraus* (clear description and fig.); 'English sweats', fo. 84*v* (see No. 119); the European pandemic of 1580 (fo. 79*v*); fig. with legend 'this is the appearance of the worm—upper and lower surface—which I regard as poisonous' (fo. 44*v*); *Solanum somniferum* (cut; fos. 46-7); 'abscheuwlichen Kranckheyt der Frantzosen' (= *Morbus Gallicus*) (fo. 51*r*); preparation of 'red precipitate' of mercury by means of nitric acid ('Scheidtwasser') (fo. 61*v*); fig. of still for 'Spiritus, Geist oder subtilste Krafft des Weins und aus Gewürtz abgezogen kan werden' (= distillation of *aqua vitae* from vegetable matter).

729. WIER, IO. [*syn.* Weyer]. (*a*) . . . *de Praestigiis Daemonum & incantationibus ac veneficiis libri sex, postrema editione quinta aucti & recogniti. Accessit liber apologeticus, et pseudomonarchia daemonum.* . . .

Off. Oporiniana, Basel 1577

[*Ed. p.* of original in V *lib.* Basel 15   , of enlarged version in VI *lib.* Basel 1564. Biog. Univ. Earliest BM (*Lib.* V) Basel 1566, (Lib. vi), Basel 1583.]

Prov. (i) 'Defens—praeceptoris monumentum habet. Thomas Cargill'.

(ii) 'Walterus Stewart'.

The *Liber Apologeticus* has half-title, sep. dedn. but cont. pagn.

266

(b) *Histoires disputes et discours, des illusions et impostures d s diables, des magiciens infames, sorcieres & empoissoneurs. Des ens rcelez & demoniaques, & de la guerison d'iceux. Item de la pu ition que meritent les magiciens, les empoisonneurs, & les sorciere . Le tout comprins en six livres (augmentez de moitié en ceste dernie re edition) par Jean Wier medecin de Duc de Cleves. Deux dialogues e Thomas Erastus* [No. 225] . . . *touchant le pouvoie des sorcieres & de la punition qu'elles meritent.*

Iaques Chouet (n.p.—believed Geneva, BM) 1579

The translator (unnamed, but alleged to be Simon C oulart in Biog. Univ.) says: 'Il ya neuf ans passez que cinq livres de l' mposture des diables prins du latin de Iean Wier & traduit en Fr nçois par Jaques Grevin furent imprimez à Paris . . .'. The trans. re erred to is dated 1567 in Biog. Univ, and BM cites an imperfect copy elieved to be of this date, also one of 1569 which fits the above acc unt more nearly.

The *Dialogues* of Erastus have sep. t.p. but cont. pagn.

The work has been described as the inspiration of human treatment of the mentally sick, though it was centuries before its deals were realised. It should however be pointed out that Wier's cor tribution is mainly in respect of *treatment*; for though he admits the 'nat ural causes' of many alleged sorceries, he is far from denying the exis ence of the devil and of a hierarchy of evil beings; see *Pseudomona chia* in (a). Wier also wrote an important work on the scurvy; the only copy of this in the Library is in *Opera Omnia*, Amsterdam 1660 (Greg ry).

730. WIGAND, IOANNES. *Vera historia de Succino Bor sico. De alce Borussica & de herbis in Borussia nascentibus. tem de Sale, creatura Dei saluberrima, consideratio methodica & heologica . . . Iam vero primum in studiosae iuventutis gratiam . . edita. Studio et opera Iohannis Rosini pastoris Wickerstadensis.*

Tobias Steinmann, Jer (Col. 1590)

Only edn. BM; Wigand's original dedn. is however da ed 1584, and that of Rosinus 1589.

Prov. READ.

Running titles *De Succino Prussiaco*. The study of amb r contains the statement: 'Whether the "ambra" found on the coa t of Ethiopia [probably = "Africa"] . . . is a certain species of "suc num" I leave to be discussed by those more experienced.' He refers to statement by de l'Ecluse that, when pricked with a pin, it exudes an oily liquor.

731. WILLICHIUS, IODOCUS. *Ars magirica coqui ria, de cibariis, ferculis, opsoniis, alimentis, & potibus diversis randis eorumque* *facultatibus. Liber medicis, philologis & sanitatis tuendae studiosis omnibus . . . utilis. . . . nunc primum editus. Hinc accedit Iacobi Bifrontis Rhaeti de Operibus Lactariis epistola.*

Iacobus Gesnerus, Zürich [Dedn. by C. Gesner 1563]
Only edn.

Prov. 'Dunaeus, D' (Patrick Dun?). Scored and glossed in Liddel's hand.

The dedn. by Conrad Gesner to Io. Pontisella ('apud Rhaetos alpinos civitatis Curiae Ludi moderatori') includes many refs. to his learned friends, e.g. Hiero. Brixinus. Preface by Willich. Letter by Bifrons, dated Valley of Engadine 1556. Bibliographical note by Gesner p. 227*v* (unnumbered).

732. WILLICHIUS, I. *Arithmeticae libri tres.*

Crato Mylius, Strassburg 1540
Not BM. See Smith, p. 197 *seq.* for valuable commentary.

The interest of this work is purely incidental, e.g. the tradition that 'Algebras' was an Arab mathematician is stated without comment.

733. [WIRSUNG, CHRISTOPH *syn.* Wirtzung.] *The General Practise of Physicke conteyning all inward and outward parts of the body . . . Compiled and written by . . . Christopher Wirtzung, in the Germane tongue, and now translated into English, in divers places corrected, and with many additions.*

Thos. Adams, London 1617
Ed. p. German 1582 (earliest BM); English 1598, STC 25862.

Prov. '. . . Ex dono Mgri Thomae Merseri 1635' (M.A. Marischal College 1613, Burgess of New Aberdeen 1631).

To the Readers: '. . . first written . . . in the high Dutch or Germaine language . . . Afterwards translated into the low Dutch or Flemish tongue by Carolus Battus. . . .'

Useful for English equivalents. Black letter text, with 'additions' in *Italic*.

734. WIRTZ, FELIX. *Practica der Wundartzney* [descriptive t.p. too long to quote] *Alles nach newer Form unnd Art . . . durch selbs eigene Erfahrung beschrieben durch F . . . W . . ., Wundartzet zu Basel.*

[Col. Seb. Henricpetrus, Basel 1596]
Only edn.

Vorrede by the author's son.

Prov. German MS of 1676 on ft. end papers. MSS used as binding papers.

Another copy.

Prov. 'P.D.D.' in pencil (Patrick Dun).

735. WITTESTEYN, CAROLUS. . . . *Disceptatio philosophica de quinta chymicorum essentia. Accessit Alexandri Carerii . . . Quaestio an Metalla artis beneficio permutari possint.*

S. Henricpetrus, Basel n.d.

Only edn. BM. Ferg. II, 554 dates Wittesteyn's work 1583 (on authority of Schenkius), and Carerius's *Ed. p.* Padua 1574.

Prov. LIDDEL (?).

The *Quaestio* follows with half-title but cont. pagn.

Dedn. to Cardinal Antonius Perenotus.

Index of propositions to be discussed gives a useful indication of method.

736. WITTESTEYN, C. . . . *Vera totius medicinae forma.*

Off. Chris. Plantin, Antwerp 1588

Only edn.

Prov. LIDDEL, and a second copy, probably Liddel's.

Dedn. to Alexander Farnese.

An interesting attempt at a 'natural history' of medicine, with refs. to its earliest history, but none, so far as I could discover, to Celsus. See p. 73 for movements of the heart.

In the *Ad lectorem* the author writes '. . . sciant inter arma Germaniae inferioris ac quasi in castris pleraque conscripta fuisse', i.e. mistakes must be excused since it was for the most part written 'in the field'.

737. [WITTICH, IOHANNES *ed.*] *Methodus tum simplicium tum compositorum medicamentorum quae apud recentiores sunt in usu . . . Adiungitur & brevis enarratio de Chirurgicis administrationibus a medico excellenti . . . Pataviano anno 1566 studiosis medicinae dictata. Accessit Paraenesis de praecipuis & maxime vulgaribus digestivis . . . decoctionibus minorativis & alterantibus remediis, cum certa & limitata rei mensura ac dosi incerti authoris. Nunc primum edita studio & labore Io . . . W . . . .*

Henningus Grosius, Leipzig [1596 from dedn.]

Ed. p.

Dedn. by Wittich to Io. Frid. Aurifaber.

738. [WOLFF, CASPAR *ed.*] *Gynaeciorum sive de mulierum affectibus commentarii Graecorum, Latinorum, Barbarorum, iam olim & nunc recens editorum. In tres tomos digesti et necessariis passim imaginibus illustrati.*

Conrad Waldkirch, Basel 1586

Enlarged version of *Ed. p.* Basel 1566 BM. Dedn. by Wolff is however dated Zürich 1564.

Prov. LIDDEL.

Vols. I and II are bound together. Vol. I, containing eight works, has dedn. by Wolff (Zürich, 1564) and a note (undated) by the printer on the circumstances attending the making of the collection. Vol. II, with sep. t.p. and pagn., has dedn. by Caspar Bauhin (Basel 1586). Vol. III (bound with Vol. IV) has title *Tomus III Gynaeciorum in quo Hippocratis Coi . . . Liber prior de Morbis Mulierum a Mauricio Cordaeo . . . commentariis . . . explicatur . . .* with note by printer about Cordaeus. Vol. IV has title: *Tomus IV Gynaeciorum libri iiii De Morbis Mulierum communibus, virginum, viduarum, sterilium, praegnantium, puerperarum, nutricum. Autore Ludovico Mercato . . .*

C. Waldkirch, Basel 1588

Prov. LIDDEL.

The work by 'Hippocrates' was regarded by Francis Adams as a pre-Aristotelian source of great importance but not by Hippocrates himself.

739. WOTTON, EDWARD. . . . *de Differentiis Animalium libri* x.

Mich. Vascosanus, Paris 1552

Only Edn.

Prov. Purchased 1953.

The dedn. to Edward VI refers to the author's friends, Georgius Agricola and [Sir] John Mason, Ambassador to Henry II of France. It includes indications of the method of classification. Dated, London 1551.

The first 'scientific' work on animals by an Englishman ('Oxoniensis'). It consists of a summary of the opinions of the leading ancient writers (including medical), prefixed by an account of the structure of animals on which a classification may be based. *Man* is dealt with in Book IV, thereafter the systematic survey commences with the Elephant. Final index of 'generic' and 'specific' names. For Gesner's opinion of the work, see No. 298.

740. YUHANNA IBN MASAWAIH [*syn.* Mesue]. *Mesue cum expositione Mondini super canones universales ac etiam cum expositione Christophori de Honestis in Antidotarium eiusdem. Additiones Petri Apponi. Additiones Francisci de Pedemontium.* Antidotarium Nicolai *cum expositione Platearii. Iohannes de Sancto Amando super* Antidotarium Nicolai. *Additiones nove super* Mesue *et* Nicolai Antidotariis. *Quotationes item marginales. Tractatus* Quid pro quo. *Tractatus* de Sinonimis. *Libellus Bulcasis sive Servitoris.* Compendium Aromatariorum Saladini. *Tabule due noviter addite . . . Emendata cum marginalibus adnotamentis nuperrime adiectis.*

(Col. Gilbert de Villiers for Vincent de Portonariis, Lyon 1519)
[*Ed. p.*—almost identical title, Venice 1495. *Ed. p.* of Mesue's works, Milan 1473, Hain 11105.]

    Prov. (i) 'Liber magistri roberti g[rey] medici.'
        (ii) 'Alexander Hay elder Burgess of Abredene 1585.'
Incipit (Of Mesue's text) 'In nomine dei misericordis cuius nutu . . .'
ThK 329, 14th cent.

    (Of *Ant. Nic.*) 'Ego Nicolaus rogatus a quibusdam in practica . . .'
ThK 231 13th-14th cents.

    (Of '*Bulcases*' = 'Albucasis') 'Dixit aggregator huius operis postquam ego collegi librum hunc . . .' ThK 208 13th-14th cents.

    (Of Saladinus—fo. CCCXVII)—'Quia solet aromatorium ignorantia . . . ', Th.K 572 and 515.

    (Of Jean de St. Amand): 'Operatio medicine secundum Ioannitium consistit in tribus . . .' ThK 471 Erfurt, A.D. 1341.

    For 'Mesue' see Vol. I, p. 240.

741. YUHANNA . . . [*syn.* Mesue]. *Mesue qui Graecorum ac Arabum postremus medicinam practicam illustravit . . . Ex duplici tralatione altera quidem antiqua, altera vero nova Iacobi Sylvii . . . Adiectae sunt etiam nunc primum annotationes in eundem Mesue Ioannis Manardi & Iacobi Sylvii. . . .*

                                 Juntae, Venice 1558
The table of contents is very similar to that of No. 740.
Prov. Glossed in red ink—probably Read. A second copy: 'Io. Gregorie 1769 2...6.'
For Manardi and Sylvius see Vol. I, p. 245 and p. 227 respectively.

742. ZACUTO, ABRAHAM BEN SAMUEL [*syn.* Abraham, etc. BM old catalogue]. *Almanach perpetuum exactissime nuper emendatum omnium celi motuum cum additionibus in eo factis tenens complementum.* (No imprint on t.p.) (Col. sig D [9] *v Opus Ephemeridium* [sic] *sive Almanach perpetuum Abraham Zacuti . . . noviter castigatum. Impressum est ac absolutum Venetiis . . . per Petrum Liechtenstein Coloniensem Anno . . .* 1502. . . . Followed by privilege, supplementary tables and register; last fo. blank.

[*Ed. p.* of *Tabulae astronomicae*, 1496—Hain 16267; of *Almanach perpetuum*—BM (Catalogue of exhibiton, 1960) cites three editions of 1496].

Bound with No. 556. See also Vol. I, pp. 136, 306.

    Prov. LIDDEL (?)
Sigs. a iiii—b viii *v* and c *r*—c vi *v* contain two sets of problems. The original work of Zacuto was circulated in Hebrew MSS from 1473 (3 extant). It was partially translated by Joseph Vicinho.
The BM copy lacks the fos. following col. on sig D [9] *v*.

743. ZALUZANSKY, ADAM. . . . *Methodi herbariae libri tres.*
Collegium Palthenianum, Frankfurt 1604
[*Ed. p.* Prague, 1592.]
Prov. Purchased 1954. 'Augustinus Lollius Adam Ph & Med D 1605.'
The dedn. to Io. Barwitz, of the Imperial Council, refers to the
unsatisfactory and confusing nature of the customary classification of
plants according to use, power, initial letter, etc.

Thoroughly Aristotelian, but in the best sense, insisting as it does on
the study of plants in their own right. On Sig. L *r* occurs an interesting
reference to the famous 'Silphium' in which a thriving export trade
developed from the ports of Cyrenaica, but which disappeared without
trace during the lifetime of Pliny (A. C. Andrews *Isis* **33,** 232).

After a reference to 'sex' the systematic review opens—the earliest
known instance—with 'fungi' and 'musci', the latter embracing lichens,
algae, etc.

744. ZECCHIUS, IOANNES. . . . *Liber primus Consultationum Medi-*
cinalium . . . Andromachi Zecchii ejus filii opera & studio . . . editus.
Gul. Facciottus, Rome 1599 (but the second dedn. by Io. Martin-
ellus is dated 15 Dec. 1600).
Only edn.
Prov. FORDYCE.

ZIEGLERUS, IACOBUS. See No. 752.

APPENDIX A

I. COLLECTIONS OF WORKS MADE BY PRINTERS

745. [ALDUS]. *Medici Antiqui omnes, qui Latinis literis diversorum mor-*
borum genera & remedia persecuti sunt, undique conquisiti & uno
volumine comprehensi, ut eorum, qui se medicinae studio dediderunt,
commodo consulatur.
Venice 1547 (Col. Apud Aldi filios)
Ed. p.
Prov. Three copies: (1) (i) 'Ioh. Franciscus D.' (ii) Liddel. (2) Forbes
(and an earlier). (3) Melvin.

Contents: Celsus, Serenus, 'Trotula', Marcellus, Scribonius,
Soranos, Pliny, L. Apuleius, A. Musa, A. Macer, Strabus Gallus,
Caelius Aurelianus, Theodorus Priscianus.

Of 'Trotula's' *Curandarum aegritudinum muliebrium, ante, in, et post, partum, liber unicus* it is said 'nusquam ante editus'. Of Aemilius Macer it is said 'De hoc Macro (candide Lector) nihil certi auctores tradidere, nisi quod coniectura est, plures Macros extitisse' (cf. No. 417). 'Strabus Gallus' was Walafrid Strabo, Abbot of Reichenau.

746. [DE CAPELLA, MICHAELES ed.] *Opus aureum et praeclarum . . . signa causas et curas febrium complectens. . . .*

(Col. Lyon, 1517. No printer's name, but the editorial work of M. de C. . . . is referred to.)

Ed. p.

Prov. LIDDEL.

Contents: *Prologus* from Gondisalvus of Toledo to Gabriel Mirus (first fo. mutilated). Marsilius de Sancta Sophia *De Febribus*; *De omnium modorum fluxu ventris; De omnium accidentium febrium cura; de febri pestilentiali.* Galeatius de Sancta Sophia (same contents as above except the last); Ricardus Parisiensis, *De Signis Febrium.* Anthonius d'Gradis, *De Febribus.* Christoferus Barisius, *Intentiones habende in febribus.*

747. [CAVELLAT, GULIELMUS]. *Annuli Astronomici . . . usus, ex variis authoribus.*

Gul. Cavellat, Paris 1558

Ed. p.

Contents: Petrus Beausardus, *A . . . A . . . usus.* Dedn. Louvain 1553. (Second Part, Franciscus Hymannus). Reinerus Gemma (Frisius), *Usus A . . . A . . .* Dedn. Louvain 1534.

Io. Dryander, *Annulorum trium diversi generis instrumentorum astronomicorum componendi ratio usus. . . .*

Dedn. Marburg 1536

Io. de Regiomonte, *Ad Bessarionem Cardinalem Nicenum ad Patriarcham Constantinopolitanum: De Compositione Meteoroscopi . . . Epistla* [sic]. Bonetus, *Delatis Hebraei medici provenzalis A . . . A . . . utilitatum liber ad Alexandrum Sextum Pontificem Maximum.* At end: Parce precor rudibus quae sunt errata Latino. Lex Hebraea mihi est lingua Latina minus' (n.d.).

M.T. *Compositio alterius annuli non universalis sed ad certam polarem elevationem instructi* (n.d.).

Burchardus Mithobius, *Annuli cum sphaerici tum mathematici usus et structura.*

Dedn. Marburg 1536

Orontius Finaeus, *. . . Compendiaria tractatio de fabrica et usu A . . . A . . .* Sep. half-titles, but cont. pagn. Numerous cuts. Table of latitudes & longitudes, which contains 'Eboracum' among 'Scotiae civitates';

moreover 'Efaguensis civitas' (which might be Glasgow) is given as east of Edinburgh (assuming, as seems to be the case, that all longitudes are measured *eastwards* from the 'Fortunate Isles').

748. [GARALDIS, BERNARDUS DE]. *Aureum opus et sublime ad medellam . . . Plinii . . . Nonnullaque opuscula, videlicet Ioannis Almenar; Nicolai Leoniceni interpretis fidelissimi ambobus, De Morbo Gallico, ut vulgo dicitur. Angeli Bolognini de Cura ulcerum exteriorum. Alexandri Benedicti de Peste. Et Dominici Arcignanei de Ponderibus et Mensuris nuper inventi. . . .*

(Col. B. de Garaldis, Pavia 1516)

Ed. p.

Prologue by Dominicus Massaria to the *de Ponderibus.*

Fine wood-cut heading to t.p. showing physicians and patient. Cont. pagn. Concerning this beautiful book see Arturo Castiglioni, *The School of Ferrara and the Controversy on Pliny* in [Singer], *Science, Medicine and History* **1,** 275, where he says that 'Kristeller knows of only three copies', two of which bear the imprint of Hyeronymus de Benedictis, Bologna 1516. The work of 'Pliny' is probably a compilation of the 3rd-4th cent. No. 748 has been rebound in contemporary (?) stamped covers with No. 1 'Abynzohar's' *Liber Servitoris* and *Colliget* (Averroes), G. de Gregoriis, Venice 1514. (See Plate VIII.)

749. [IUNTA, Thomas ed.]. [*De Balneis, omnia quae extant apud Graecos, Latinos, et Arabas tam medicos quam quoscunque caeterarum artium probatos scriptores qui vel integris libris, vel quoquo alio modo hanc materiam* tractaverunt: *nuper hinc inde accurate conquisita & excerpta . . . In quo aquarum ac thermarum omnium, quae in toto fere orbe terrarum sunt, metallorum item, & reliquorum mineralium naturae, vires atque usus . . . explicantur. Opus nostra hac aetate, in qua tam frequens est Thermarum usus, medicis quidem necessarium, ceteris vero omnibus . . . utile.*

Heirs of Lucantonius Iunta, Venice 1553.]

The t.p. and other prelims. wanting in No. 749: col. 'Venetiis apud haeredes Lucaeantonii Iuntae'. Title from BM copy.

Ed. p.

Prov. Bute.

The missing prelims. (seen at BM) include a dedn. by Thomas Iunta to Francesco Contareni, 'Patavinae urbis praefecto'.

This handsomely bound folio contains articles, of lengths varying from one to over a hundred (Avicenna) pages on all kinds of mineral and hot springs, mostly in Italy. Many of these (e.g. at Abano in the Euganean Hills near Padua) are still flourishing centres for hydrotherapy.

PLATE VIII

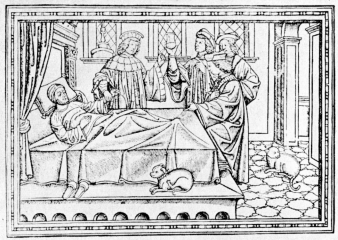

aureum opus ⁊ ſublime ad medellā non parum ytile
ꝓlinij philoſophi ⁊ medici itegerrimi.nōnullaq₃
opuſcula videlicet Joānis Almenar:Nicolai
Leoniceni iterpꝛetis fideliſſimi ambobus:
De moꝛbo gallico yt vulgo dicitur:An
geli Bolognini d cura ylcerū exterio
rum:Alexādri Benedicti de peſte:
⁊ Dominici Arcignanei de pō
deribus ⁊ mēſuris nuper
inuēti ad pꝛaxim quā⸗
maxie neceſſarij
feliciter in⸗
Cum gratia   cipiunt.   ⁊ ꝓiuilegio.

Title-page of a collection of medical works published in
1516 by the printer Bernardus de Garaldis of Pavia. A
rare work of great beauty (No. 748).

Though not indicated in the title, there are numerous articles by 'contemporary' (14th, 15th and 16th cent.) writers, including one by Conrad Gesner on the Swiss springs, the first fo. (289) of which, containing (BM) Gesner's dedication to T. Iunta, has been skilfully removed & the running titles all but obliterated. Fine cuts of mining galleries (fo. 287r, Agricola; article mutilated), pump and heating vessel (fo. 295r), mixed bathing of lame, etc., persons (fo. 299r), are included.

750. *Philosophiae Chemicae* IV *vetustissima scripta*
  I. Senioris Zadith F. Hamuelis, *Tabula Chymica.*
 II. Innominati philosophi, *Expositio Tabulae Chymicae.*
III. Hermetis Trismegisti, *Liber de Compositione.*
IV. Anonymi veteris philosophi, *Consilium coniugii seu de Massa solis et lunae libri tres.*
   Omnia ex Arabico sermone Latina facta & nunc primum . . . producta.
                        Io. Saur for Io. Bernerus, Frankfurt 1605
   Only edn.
   Prov. READ.
   Incipits: I. 'Intravi ego & Oboel charissima Barba in . . .' ThK, 365.
        Ferg. II, 563 quotes Schmieder as believing 'Zadith Senior' to be a 13th cent. Arab.
        II. 'Decem Aquilae, Sol & Luna, tres spherae & domus . . .'
        III. 'Hermes enim inquit': "In tam" ThK, 290. Preface by translator.
        IV. 'Non legitur a divina . . .' ThK 433.

751. [TORINUS, ALBANUS *ed.*]. *De Re Medica huic volumini insunt* Soranus—*In artem medendi isagoge hactenus non visa* Oribasius—*Fragmentum de Victus Ratione—antea nunquam aeditum* Pliny—*De Re Medica li. v accuratius recogniti. . . .*
   L. Apuleius—*De Herbarum Virtutibus . . . e tenebris eruta. . . . Accessit his vice coronidis libellus . . . de Betonica, quem quidam Antonio Musae, nonnulli L. Apuleio adscribendum autumant, nuper excusus.*
                        (Col. Andr. Cratander, Basel 1528.)
   Only edn. (BM lists under 'Thorer, Alban').
   Dedn. by editor.
   Fine ornamental borders of section-headings, also initials.
   Cont. pagn.

752. [VALDERUS, I. *ed.*]. *Sphaerae atque astrorum coelestium ratio, natura & motus ad totius mundi fabricationis cognitionem fundamenta.*
                        Valderus (n.p.) 1536 (No. col.)
   Only edn. BM (under 'Sphaera').

Contents: Iacobus Zieglerus, *De solidae sphaerae constructione.*
Proclus, *De Sphaera scholiis Ziegleri explicatus.* [*Ed. p.*]
De canonica per sphaeram operatione.
Berosus, *Hemicyclium* [*Ed. p.* 1525].

Aratus . . . *de Siderum natura et motu* (with comm. of Theon of
Alexandria).
Planisphaerium C. Ptolemaei & Iordani.
Text of Proclus and Aratus in Greek, commentaries in Latin.
There is a folding frontispiece by Guil. Morelius, Paris 1559, showing
South celestial pole, Zodiac signs and constellations near the equator:
it appears to have been pasted in. The *Planisphaerium* is a paraphrase of
an Arabic version of Ptolemy's work ('Ptolemy' and the 'Arabs' are
referred to *passim*) formerly attributed to Rodulphus Brughensis whose
Praefatio is dated Toulouse 1144. Since it is dedicated to 'Theodoricus
Platonicus' (Thierry of Chartres) 'praeceptor meus', and for other
reasons, it is now attributed to Rodolph's master, Hermann of Carin-
thia (Haskins, *Studies in Medieval Science*, Camb. Mass. 1924, p. 47 f).
The Greek text has never been found. Cantor (II, 53 *seq.*) accepts the
Planisphaerium Iordani as the work of Iordanus of Saxony (see Vol. I,
p. 154).

II. COLLECTIONS OF TRACTS, THESES, ETC.

753. [Collection of plague tracts in characteristic binding gilt stamped
with escutcheon of Bute. These are separate tracts, but catalogued
here on account of common subject-matter.]

(1) Ficino, Marsilio . . . *Il consiglio contro la pestilentia. Con altre cose
aggiunte appropriate alla medesima malattia.*
(Col. [Device of Juntae], Venice 1556)
[*Ed. p.* Siena 1523.]
At p. 77 starts (with secn. heading) *Consiglio di M. Thomaso del
Garbo. . . . Contro la pestilentia.*

(2) Tomitano, Bernardino, *Consiglio de . . . sopra la peste di Vinetia
l'anno* MDLVI. . . . Gratioso Perchacino, Padua 1556
Refs. to Thucydides and Boccaccio (1348).

(3) Landus, Basianus *De origine et causa pestis Patavinae anni* 1555.
Balthassares Constantinus, Venice 1555.
(Col. 'Ioan. Gryphius excudebat 1555')
Only edn.
Lectori by Blasius Sylva.

276

(4) Frigimelega, Francesco. *Consiglio sopra la pestilentia qui in Padoa dell'anno* MDLV. . . . *Fatto a richiesta di questi . . . signori e di questa alma cittá.*

G. Perchacino, Padua 1555
Only edn.

(5) Massa, Nicolò *Ragionamen'o . . . sopra le Infirmità che vengono dall'aere pestilentiale del presente anno* MDLV.
(Col. Giovanni Grissio for Giordan Ziletti, Venice 1556.)
Only edn.

754. FICINO, MARSILIO. *Contro alla Peste. Insieme con Tommasa del Garbo, Mengo da Faenza & altri autori, e ricette sopra la medesima materia. Aggiuntovi di nuovo una epistoia dell' eccellente Giovanni Manardi da Ferrara, & uno consiglio di Niccolò de Rainaldi da Sulmona non piu stampati.*

Junta, Florence 1576
[*Ed. p.* (without addition) Siena 1522]. This edn. not BM.
Prov. BUTE.
Opens: 'Charity towards my country moves me to write a *consilium* against the plague.' The writer continues that he will do so briefly in the Tuscan tongue and according to the authority of physicians ancient and modern and the experience of many, especially 'del nostro padre maestre Ficino'—which suggests that the *consiglio* by Ficino may be edited. It appears to be quite different from No. 753(1). With reference to the 'presente peste del 1478 & del 1479' the conjunction of Mars and Saturn is mentioned as significant, but in particular cases the evil may be derived from winds, evil vapours, and lakes. The *Lettori* by the translator (N. Lorenzini) from the Latin of Manardi's letter is dated 'Firenze Oct. 29, 1576'.

755. CAMERARIUS, IOACHIMUS (the younger). *Synopsis Commentariorum de Peste.*
Col. Off. Cath. Gerlach & heirs of Io. Montanus, Nürnberg 1583
Only Latin edn. BM., German (exc. Rincio) 1597.
Three copies.
Prov. of (1) Bute and 'du Poirier 1653'.
(2) Liddel.
(3) James Cargill and another name (early 16th cent. MS binding).
Contains plague tracts by Hiero. Donzellini; Io. Ph. Ingrassia (from Italian—plague in Palermo 1575-6); Cesare Rincio (on the

277

plague in Milan 1577); Camerarius (*De Bolo Armenia et Terra Lemnia;
De . . . ratione praeservandi a pestis contagio . . .*); Regulations in the
Venetian plague 1576-7—from Italian; *Ratio expurgandarum rerum
infectarum . . .*; List of works on the plague by Italian writers.

756. [Medical Faculty of Helmstadt—Plague Regulations.] *Kurtze und
notwendige Ordnung. Wie man sich in jetzt grassierender gesch-
winden Pestilentz mit Göttlicher Hulff praeservieren oder vor dem
Gifft verwaren müge. Auch so jemand damit befallen mit Gottes
Hülff durch ordentliche Mittel zu curieren oder wider auff zu helffen
sey. Gestellet zu unterricht der studierenden Jugend in der löbichen
Julius Universitet und der Bürgerschafft zu Helmstadt. Durch die
Professores Facultatis Medicae daselbst.*

Iac. Lucius, Helmstadt 1597

757. [Bound with:] [No author]. *Kurzer nützlicher und nötiger Bericht
wie man sich issiger Zeit der Pestilentz verhalten und die Artzeney
so Anno 82. Auff des Raths der Altenstadt Magdeburgs Apoteck
verordnet unnd jtzo dieses 97 Jahres aus Befehl eines erbarn Rahts
von den verordneten Medicis da selbst von newen Ubersehen und
verbessert zubekommen gebrauchen sol.*

(Col. Andr. Duncker, Magdeburg, 1597.)

Prov. LIDDEL (?)

758. [No author's name.] *Büchsenmeisterey. Geschoss, Büchsen, Pulver,
Kugeln, Salpeter, Feurwerk und Pfeil etc. Zum Schumpff und ernst
zu machen zuzurichten und nach jedes Gewicht Stein und Lot
zugebrauchen. Dabey gemeine Kriegsrecht, Rath, Regiment und
Ordnung. Sampt der lehre Keysers Maximiliani desz ersten so im in
seiner Jugent gemacht und durch einen erfarnen trefflichen Mann seiner
Kriegsräthe zugestellt ist. . . .*

(Col. Heirs of Christian Egenolph, Frankfurt 1582)

Not BM or Bibl. Nat.

Prov. Purchased 1956.

Vorred on the evils of war: 'Also dasz die Heyden den aller unbil-
lichsten Friden besser dann den aller billichsten Krieg nicht unrecht
gesagt haben.' No fortress is strong enough against the cruel instrument
of gunshot (*Büchsengeschoss*). Therefore the book is published and
contains matters previously kept secret.

Contains interesting questions on the quality, origin, etc., of saltpetre, and asks whether the 'fire' or the 'vapour' drives the projectile.
Followed by Rules & Customs of War.

278

The *Büchsenmeisterei* fo ws pretty closely the first printed version
(1529) of the 'Feuerwerɩ uch von 1420'. See Vol. I, p. 179, also
W. Hassenstein, *Das Feueɩ kbuch von* 1420, München 1941.

759. (A collection of theses ɩ ɩnd in rubricated MS parchment, many
glossed in Liddel's hand, soɩ inscribed by authors.)

(1) Erastus, T. *Disputatio de ɩ 'redine*, etc., see No. 226.

(2) Syblinus, Marcus (*Praes.* Planerus) *De . . . formatione humani
    foetus.* . . .
                            Nic. Wyriot, (Strassburg ?) 1576.

(3) Saelius, Henricus (*Praes.* H. ɩucaeus) . . . *De Apoplexia.*
                            ʿypis Myliandrinis, Rostock 1589.

(4) Randovius, David, and Sch eveldt, Stephanus à (*Praes.* H.
    Brucaeus) *De Scorbuto.* Im ɩnt and date as for (3).

(5) Rumbaum, G. *De Epilepsia* (*Prɩ Franciscus Parcovius*).
                            Iac. Lucius, Helmstadt 1593.

(6) Machildus H., Cunderdingus, A. ɩrnoldus, C. *De Pleuritide* . . .
    (*Praes.* Franciscus Parcovius)
                            Lucius, Helmstadt 1594.

(7) Turnetus, Iacobus, *Scotus . De Cathɩ ɩ. . . .*
        Andr. Eichorn, Frankfurt 15ɩ

(8) Pappus, Io. S. . . *De animae facultatibu ɩ. .* (*Prop.* Israel Spachius).
                            Ant. ɩ tramus, Strassburg 1591.

(9) Craig, Io. *Scotus.* . . *De hepatis disposit ɩbus* . . . (Prop. & de-
    fended by J . . . C. . . . *Mathem ɩm & Organi Aristot. in
    Academia Francofurdiana Professore puɩ ɩ).*
                            Leonhardɩ Ostenius, Basel 1580.
    Ded. to Ludovicus Ballendinus 'Iustitiae ɩr Scotiam Praesidi'
from Basel.

(10) Helvigius, Andr. *De partis affectae dignotiɩ* ('Praeside Divina
     Triade').
                            Leon. (ɩenius, Basel 1587.

(11) Cargillus, Iacobus, *Scotus. . . De intestinorum lɩ bricis. . . .* (*Praes.*
     Io. N. Stupanus).
                            C. Walɩ rch, Basel 1594.
    Ded. to Caspar Bauhin 'patrono et praeceptoɩ
    Inscribed: 'Viro magno D. Duncano Riddelio [*sic*] ɩ ɩthematum in
Helmstadiensi Academia Professori publico popul[ar meo, Basilea
mittebam—Iacobus Cargillus.'

279

(12) Solfleisch, Martinus *De ΓΗΡΟΚΟΜΙΚΗ et ΑΝΑΛΗΠΤΙΚΗ medicinae partibus.* (*Praes.* A. Planerus)

G. Gruppenbach, Tübingen 1585.

(13) Rentzius, G. *Disputatio Posteriorem doctrinae de differentiis symptomatum partem continens* . . . (*Praes.* A. Planerus).

Gruppenbach, Tübingen 1589.

(14) Agerius, Nich. . . *De elementis* . . . (*Praes.* Lud. Hawenreuter).

Ant. Bertramus, Strassburg n.d.

(15) Bokelius, Gul. . . *de peste* . . . (*Praes.* Io. Bokelius)

Iac. Lucius, Helmstadt 1591.

(16) Hastaeus, Christianus . . . *Complectens rationem ac methodum universalem dignoscendorum locorum male adfectorum & adfectuum praeter naturam* . . . (*Praes.* F. Parcovius).

I. Lucius, Helmstadt 1591.

(17) Lautenbach, Josephus . . . *De memoria laesa* . . . (*Praes.* F. Parcovius).

I. Lucius, Helmstadt 1591.

(18) Stiberus, Bernhardus. *De causis, differentiis et curatione febrium iuxta et pestis* (*Praes.* Dan. Moegling).

G. Gruppenbach, Tübingen 1591.

(19) Schiller, Wolfgang *De methodo medendi* . . . (*Praes.* Andr. Planerus).

G. Gruppenbach, Tübingen 1592.

(20) Laurenberg, Guilielmus, Dossius, Nic., Arsenius, Petrus, *De Hydrope* (*Praes.* H. Brucaeus).

Steph. Myliander, Rostock 1587.

(21) Pannonius, Valentinus . . . *De Epilepsia* . . .

Off. Geo. Osterberger. Königsberg 1588.

760. Bound in leather boards with vestiges of brass clamps—presumed early 17th cent. First thesis inscribed in usual hand & terms of Alex. Read's donation. Contents (some relating to logic are omitted):

1, 2, 3 [omitted.]

(4) Agerus, Nicolaus, *prop.*; Ludovicus Hawenreuterus *praes. Theses medicae physicae de elementis, horum mixtione, mixtorumque ratione seu temperamento.*

Antonius Bertramus, Strassburg, n.d.

(5) Saracenus, Enocus *prop.*; Henricus Smeatius *praes. Theses de humorum necessitate, et proximis humani corporis principiis.*

Heidelberg 1587.

(6) Spachius, Israel *prop.*; Kobenhaupt, Io. Sebaldus *resp.* D.N.A.
Disputatio medica de humoribus. . . .
Anton. Bertramus, Strassburg 1591.

(7) Planerus, Andreas *prop.*; Stiberus, Bernhardus *resp. De convulsione
seu spasmo, eiusque causis.*
Georgius Gruppenbachius, Tübingen 1590.

(8) Hettenbach, E. *resp.*; Albertus, Salomon *praes. De schorbuto.*
Zach. Lehman, Wittenberg n.d.

(9) Cunelius, Georgius *prop.*; D.O.M.F. *De haemorrhoidibus.* . . .
Mich. Lantzenberger, [Leipzig] 1591.

(10) Haeberlin, Oswaldus (Planerus, A. *praes.*). . . . *de influxu facultatum
animae; de causis & curatione methodica stuporis ac παραλυσεως.*
G. Gruppenbach, Tübingen 1589.

(11) Krentzheim, Leonhartus *resp.*; Gravius, Ludovicus, *praes.*; *De
dyspnoea.*
Heidelberg 1586.

(12) Zwingerus, Iac. *prop.*; Planerus A.; *praes.* . . *de arthritide*
G. Gruppenbach, Tübingen 1592.

(13) Hubnerus, Bartholomaeus. . . . *De Veris immotisque fundamentis Artis
Medicae, et Philosophiae, deque impietate, vanitate, portentosis et
pernitiosis erroribus Philippi Paracelsi, et sectatorum eius, quibus
Theologiam pariter & Philosophiam cum Medicina nefarie con-
spurcarunt.*
Io. Pistorius, Erfurt 1593.

' . . . . whereby he would contrive a method by the new Paracelsian
medication by means of metals & minerals, and pass off his "fumos
Alcumisticos" as more precious than balsam, alexipharmacas and the
most noble remedies . . .'.

(14) Grillus, Iro [?] . . . *de phantastismo eiusque speciebus* . . .
Eutrapelus Nugo, Nargenova 1594.

(15) Mögling, Io. Rudolphus, *resp.* Danielis Moegling *praes. Anatomicarum disputationum prima.*
G. Gruppenbach, Tübingen 1594.

(16) [Ditto] . . . *Altera* [same imprint].

[Appendix] Questions, mainly on value of Paracelsian remedies,
showing a more reasoned approach than No. (13) e.g. 'An Medicina
Paracelsica vel autoris illius libri retinendi sint in Schola nostra Medica
Dogmatica? Dubitamus'. But 'An ex Antimonio, Sandaraca, Arsenico,
& similibus venenis, omnibusque aliis metallis, arte Chymica, possit
extrahi venenum ut sine noxa gravi exhiberi queant? Affirmamus.'

The answer was of course wrong in detail, since nothing can be 'extracted' from antimony etc. but the metal itself, yet it reveals an open (if optimistic!) mind.

(17) Ringk, Io *resp.*; Mögling, D. *praes.* . . . *de sanguinis sputo ex pulmonibus.* . . .

G. Gruppenbach, Tübingen 1593.

(18) Langenmantelii fratres, Wolfgangus Henricus and Io. Sebastianus *prop. De sensibus externis et internis.*

David Sartorius, Ingolstadt 1594.

(19) Dyrrhius, Io. *def.;* Mögling, D. *praes. De apoplexia, seu, ut Latini vocant, morbo attonito.*

G. Gruppenbach, Tübingen, 1594.

Ded. to Frederic, Duke of Würtemberg and Founder of the University. 'Pars 1 Theorica' only.

APPENDIX B

This important book was received by the library too late to be included in its proper order:

TARTAGLIA, N. *La Prima parte del General Trattato di Numeri et Misure* . . . *nellaquale* . . . *si dichiera tutti gli alti operative, pratiche et regole necessarie, non solamente in tutta l'arte negotiaria & mercantile ma anchor in ogni altra arte, scientia, over disciplina dove intervenghi il calculo.*

La Seconda . . . *nellaquale* . . . *si notifica la piu elevata et speculativa parte della practica Arithmetica, laqual e tutte le regole & operationi praticali delle progressioni radici proportioni & quantita irrationali.*

Curtio Troiano de i Navo, Venice 1556

Ed. p.

Prov. Tarradale House 1960.

Two vols. ded. to Richard Wentworth (cf. No. 673) from Venice 1556. Fine engraved portrait of author. See Vol. I, p. 32.

Two small works (bound with others) were discovered too late for serial insertion:

CLUSIUS, CAROLUS. *Aliquot notae in Garciae Aromatum Historiam* . . .

Off. Chris. Plantin, Antwerp 1582

Only edn. BM.

Prov. LIDDEL.

Many references to plant specimens brought from America and especially those which Francis Drake had shown De L'Ecluse after his recent circumnavigation.

REYMERS, NICHOLAS. *Fundamentum Astronomicum, id est nova doctrina Sinuum et Triangulorum. Eaque* . . . *eiusque usus in astronomica calculatione & observatione.*

Bernhardus Jobin, Strassburg 1588

Only edn. BM (under URSUS).

Prov. LIDDEL.

An 'advanced text' for the solution of astronomical problems together with the author's 'new hypotheses' concerning the motion of the heavenly bodies.

INDEX OF NAMES

This index consists almost entirely of names of persons, and excludes the principal references already alphabetised in the text of the Bibliography. The main purpose is to assist readers interested in general history, bibliography, and scholarship. Most of the entries are names of translators, commentators, and of other people named incidentally, either in the citations of titles or in the notes appended to them. Names of printers as such, and of some minor figures have regretfully been omitted.

287

289

Leoniceno, N., 100, 108, 128, 143, 163, 204, 274
Leuschner, C., 172
Libavius, A., 186, 226
Liébaut, J., 139
Librarian, Universities of Aberdeen, see Garden, Gray, Paterson
Library, of King's College, xiv, xviii, Lumley, xvi n.
Luton Hoo, xviii
Marischal College, xiv
of Matthaeus Corvinus, 175
of H. Münzer, xvii
Phillips, xix
of Pico della Mirandola, xvii
of Nicholas Pol, xvii
Vatican, 87
Wellcome Historical Medical, 33
Liddel, D., xii f., xvi f., xix, 29, 69, 131, 260, 279
Chair of Mathematics, xviii
Linacre, T., 84, 108, 143
Lindsay, Lord of Balcarres, 110
L'Obel, L., 78
Lodge, T., 69
Lorenzini, N., 277
Louis XII (of France), 58
Louis XIII (of France), 91
Lublinus, V., 172
Ludwig, P., 225
Luis de Camöens, 184
Lull, R., 47, 256
Lumley, Lord, xvi n., 5, 113, 218, 260
Lupset, T., 108

Mabanus, Io., 77
Maccheli, N., 65
'Macer', 273
Maestlin, M., 122-3
Mackaile, M., xvi
MacNalty, Sir A., 84
Magdeburg Plague regulations, 278
Magellan, F., 239
Manardi, Io., 108, 271, 277
Manutius, P., 3
Marcellus, H., 161
Marinelli, Gio., 129
Marischal College, xiii f.

Marliani, G., 254
Marsilius of St. Sophia, 273
Martin IV, Pope, 40
Martius, H., 120
Mary I (Tudor), 211, 261
Mary, Queen of Scots, 61, 174
Mason, Sir J., 270
Massa, N., 132, 277
Mathesius, H., 2
Mattioli, P., 77, 83
Maurice, Landgrave of Hesse, 63
Maurice of Nassau, 126, 243
Maurice of Saxony, 3, 69
Maurolyco, F., 251
Maximilian I, Emperor, 206, 248, 254, 278
Maximilian II, Emperor, 52, 120, 163, 167
Medici, Catherine de', 97, 124, 208
Medici, Cosimo de', 15, 24, 36, 53
Melanchthon, P., 193, 194, 207, 212
Melvin, xviii
Mengo da Faenza, 277
Menher, V., 248
Mercado, L., 27, 61, 270
Mercator, G., 73, 199
Mercuriali, H., 166
'Mesue', 160
Michael, Archbishop of Salzburg, 235
Michael de Capella, 117
Mithobius, B., 273
Moir, J., xv
Molyneux globes, 92
Monavius, 235
Monardes, Io., 4
Mondino, 72, 102, 137
Mongius, Io., 132
Montagnana, B., 165
Montagnana, P. de, 136
Montanus, I. B., 241
Montmorency, N. and J. à, 60
Mordisius, H., 69
More, Sir T., 255
Morelius, G., 2, 276
Morley, H., 56
M(o)uf(f)et, T., 82, 235
Münzer, H., xvii